PARAMETRIC RANDOM VIBRATION

Raouf A. Ibrahim
Professor of Mechanical Engineering
Wayne State University
Detroit, Michigan

DOVER PUBLICATIONS, INC.
Mineola, New York

Bibliographical Note

This Dover edition, first published in 2007, is an unabridged republication of the work originally published in 1985 by Research Studies Press Ltd., Letchworth, England. The author has provided a new errata list for this edition.

Library of Congress Cataloging-in-Publication Data

Ibrahim, R. A. 1940–
Parametric random vibration / Raouf A. Ibrahim.
p. cm.
Reprint. Originally published: Letchworth, Hertfordshire, England : Research Studies Press, 1985.
Includes bibliographical references and index.
ISBN-13: 978-0-486-46262-2
ISBN-10: 0-486-46262-5
1. Random vibration. 2. Stochastic processes. I. Title.

TA355.I27 2007
620.3—dc22

2007021269

www.doverpublications.com

In memory of my beloved parents

Acknowledgments

It is my great pleasure to acknowledge with deep gratitude the help of many people preceding and during the writing of this research monograph. I was privileged to gain the basic foundation of the area of parametric vibration from Professor A. D. S. Barr for whom I have great respect. During my postdoctoral research fellowship at the University of Edinburgh I spent a profitable two and one-half years with Dr. J. W. Roberts. The kind invitation of Dr. J. B. Roberts to write this monograph was a great incentive. Sincere thanks are due to Professor J. H. Lawrence for his continual encouragement and support during the course of preparing the manuscript. I am also indebted to Professor W. Marcy for his expertise in transferring the manuscript from the Word Star disks to CPT word processor disks. I would like to thank my colleagues and students for proofreading the manuscript at various stages and for making valuable suggestions. In this regard many thanks are due to Dr. R. McLauchlan, Dr. A. Koh, and Mr. R. Blackstone. During the last stage of preparing the manuscript the staff of Mrs. V. A. Wallace of the Research Studies Press conducted a careful proofreading which resulted in correcting several mistakes. My students, A. Soundararajan and H. Heo, derived a number of relations presented in Chapter 2, and carried out the numerical calculations of a number of examples in Chapter 8.

I owe special thanks to Mrs. D. H. Wardrip for her excellent and careful typing. She typed several drafts and assisted me in editing the manuscript. Last, but not least, I am indebted to my wife, Sohair, for her initial encouragement and continual support.

R. A. Ibrahim
Lubbock, Texas
May, 1985

Preface

This book presents a coherent and systematic treatment of the linear and non-linear parametric random vibrations. The subject of the book is basically a combination of the theory of stochastic processes, stochastic differential equations and applied dynamics. Over the last few decades there has been a continuous interest in studying the behavior of dynamic systems subjected to parametric random excitations by mathematicians, automatic control engineers and dynamicists. Remarkable progress has been made in developing new stochastic stability theorems and analytical techniques for determining the random response of non-linear systems. However, controversies and paradoxes have arisen due to misinterpretation of certain stochastic processes or improper use of the methods of analysis. These problems are discussed and resolved in this book and the mathematical abstraction, frequently found in the stochastic processes literature, is avoided. The book is self-contained and presents the state-of-the-art which is scattered in over 500 published papers and technical reports.

A brief overview of the general topic of parametric vibration and its subclasses is presented in chapter 1. This chapter introduces the reader to the basic definitions of parametric and autoparametric instabilities, the chaotic motion, pseudo-random excitation, crypto-deterministic systems, and a brief review of parametric random vibration.

Chapters 2 through 5 constitute the fundamental concepts of random processes and their calculus rules with a sufficient emphasis on the main ingredients required for the analysis of parametric vibration problems. Chapter 2 outlines the statistical and probabilistic descriptions of random variables and random processes. It also dwells into the essential features and properties of Gaussian and non-Gaussian probability density functions. The construction of non-Gaussian probability density functions is demonstrated in terms of the Edgeworth or the Gram-Charlier asymptotic expansions. Chapter 3 provides basic concepts of stochastic calculus operations. It includes the common notations of advanced calculus, modes of stochastic convergence, the mean square derivative, and the Riemann-Stieltjes integrals. The properties of special classes of stochastic processes such as the Brownian motion, the white noise, and Markov processes are introduced in chapter 4. The stochastic calculus due to Itô and Stratonovich, in conjunction with the Wong-Zakai and Khas'miniskii limit theorems, is carefully clarified. A complete and clear understanding of this chapter is very essential in modeling and analyzing dynamic systems subjected to physical white noise excitations. Chapter 5 demonstrates the application of the Fokker-Planck-Kolmogorov equation and the Itô formula for stochastic differential to generate differential equations for the response statistics.

The next three chapters deal with the stability and response analyses of linear and non-linear systems with random coefficients. The stochastic averaging methods are outlined and supported by classical and real application problems in chapter 6. The methods include the first and second order stochastic averaging techniques and the energy envelope method. Chapter 7 is devoted to study various modes of stochastic stability with reference to stability of moments and almost sure stability. Comparisons between stability boundaries as obtained by various theorems are provided for particular types of problems. Chapter 8 presents a number of techniques frequently used in determining the random response of linear and non-linear dynamic systems. One main section is allocated to examine the response characteristics of helicopter rotor blades in various modes of their motions. Another section deals with the response of non-linear systems with parametric random coefficients and with autoparametric interaction. These problems result in a set of infinite hierarchy moment equations which are closed via non-Gaussian closure techniques.

The last chapter compiles the results and observations of experimental investigations reported in the literature. This chapter is followed by a bibliography in two parts. The first part lists the literature which is referenced in the main text of the book, the second part lists only the unreferenced literature.

Contents

Errata List

1. Page 6: Fig. (1.2) the label of the horizontal axis reads:

$$\left(\frac{2\omega}{\Omega}\right)^2$$

It should read:

$$\left(\frac{\Omega}{\omega}\right)^2$$

2. Page 37: Third equation of Equation (2.73) reads

$$\lambda_3[x^3] = E[x^3] - 3E[x]E[x^3] + 2(E[x])^3$$

It should read:

$$\lambda_3[x^3] = E[x^3] - 3E[x]E[x^2] + 2(E[x])^3$$

3. Page 37, Equation (2.78) reads

$$E\left[x_1^{k_1} x_2^{k_2} \ldots x_n^{k2}\right] = \left(\frac{1}{i}\right)^k \frac{\partial^K \ell n F_{\underline{x}}(\theta)}{\partial\theta_1^{k_1} \ldots \partial\theta_n^{k_n}} |_{\underline{\theta}=\underline{0}} \tag{2.78}$$

It should read:

$$E\left[x_1^{k_1} x_2^{k_2} \ldots x_n^{k2}\right] = \left(\frac{1}{i}\right)^k \frac{\partial^K F_{\underline{x}}(\theta)}{\partial\theta_1^{k_1} \ldots \partial\theta_n^{k_n}} |_{\underline{\theta}=\underline{0}} \tag{2.78}$$

4. Page 84, Equation (4.82) reads

$$dX_j(t) = \left\{ f_j(\underline{X}, t) + \frac{1}{2} \sum_{\ell}^{n} \sum_{k}^{n} G_{\ell k}(\underline{X}, t) \frac{\partial G_{j\ell}(\underline{X}, t)}{\partial X_k} \right\} dt + \sum_{\ell}^{n} G_{j\ell}(\underline{X}, t) dB_\ell(t) \tag{4.82}$$

It should read:

$$dX_j(t) = \left\{ f_j(\underline{X}, t) + \frac{1}{2} \sum_{\ell}^{n} \sum_{k}^{n} G_{\ell k}(\underline{X}, t) Q_{\ell k} \frac{\partial G_{j\ell}(\underline{X}, t)}{\partial X_k} \right\} dt + \sum_{\ell}^{n} G_{j\ell}(\underline{X}, t) dB_\ell(t) \tag{4.82}$$

$$\text{where } \underline{Q}dt = E\left[d\underline{B}(t) d\underline{B}^T(t)\right]$$

5. Page 84, the first equation of (4.86) reads:

$$\lim_{T\to\infty} \frac{1}{T} E\left[F_i^1(\underline{X}, \xi(t), t)\right] dt \qquad = F_i^1(\underline{X})$$

It should read:

$$\lim_{T\to\infty} \frac{1}{T} \int_{t_0}^{t_0+T} E\left[F_i^1(\underline{X}, \xi(t), t)\right] dt \qquad = F_i^1(\underline{X})$$

6. Page 87, the third line of the top equation (without number) reads

$$= \frac{1}{2}\left\{B^2(b) - B^2(a)\right\} - \frac{\sigma^2}{2}(b - a)$$

It should read:

$$=\frac{1}{2}\left\{B^2(b)-B^2(a)\right\}-\frac{\sigma^2}{2}(b-a)$$

7. Page 89, Equation (4.106) reads

$$\frac{\partial p(\underline{X},t)}{\partial t}=-\sum_{i=1}^{n}\frac{\partial}{\partial X_i}\{p(\underline{X},t)f_i(\underline{X},t)\}+\frac{1}{2}\sum_{i=1}^{n}\sum_{j=1}^{m}\frac{\partial^2}{\partial X_i\partial X_i}\left\{p(\underline{X},t)\left(\underline{G}\underline{Q}\underline{G}^T\right)_{ij}\right\} \qquad (4.106)$$

It should read:

$$\frac{\partial p(\underline{X},t)}{\partial t}=-\sum_{i=1}^{n}\frac{\partial}{\partial X_i}\{p(\underline{X},t)f_i(\underline{X},t)\}+\frac{1}{2}\sum_{i=1}^{n}\sum_{j=1}^{m}\frac{\partial^2}{\partial X_i\partial X_j}\left\{p(\underline{X},t)\left(\underline{G}\underline{Q}\underline{G}^T\right)_{ij}\right\} \qquad (4.106)$$

8. Page 93, Equation (5.8) reads:

$$N=n(n-1)(n-2)\ldots(n+K-1) \qquad (5.8)$$

It should read

$$N=n(n+1)(n+2)\ldots(n+K-1) \qquad (5.8)$$

9. Page 125, the first line of equation (vi) reads:

$$\left(\underline{G}(a)\underline{G}(a)^T\right)_{11}=\frac{1}{T}\int_0^T ds\int_{-\infty}^{\infty}E\left[\left(\omega\ddot{\xi}(s)a\cos\zeta\right)\left(\omega\ddot{\xi}(s+\tau)a\cos(\phi+\omega\tau)\right)\right]\cdot$$

It should read:

$$\left(\underline{G}(a)\underline{G}(a)^T\right)_{11}=\frac{1}{T}\int_0^T ds\int_{-\infty}^{\infty}E\left[\left(\omega\ddot{\xi}(s)a\cos\phi\right)\left(\omega\ddot{\xi}(s+\tau)a\cos(\phi+\omega\tau)\right)\right]\cdot$$

CHAPTER 1

Parametric Vibrations

1.1 INTRODUCTION

Parametric vibration refers to the oscillatory motion that occurs in a structure or a mechanical system as a result of time-dependent variation of such parameters as inertia, damping, or stiffness. This variation may be due to the influence of externally applied forces or acceleration fields which are referred to as "parametric excitations." Although parametric vibration is regarded to be of a secondary interest, it can have catastrophic effects on mechanical systems near the critical regions of parametric instability.

Parametric vibration may be classified into two main categories, deterministic and random. However, there are intermediate situations which do not belong to either one of these categories. For example, the response of a dynamic system to deterministic parametric excitation may result (under certain conditions) in a chaotic motion which is random. Other examples include the pseudo-random excitation and crypto-deterministic response. The purpose of this chapter is to provide an overview to the various classes of parametric vibration.

1.2 MECHANICS OF DETERMINISTIC PROBLEMS

The deterministic approach to the analysis of parametric vibration problems is based upon the mathematical theory of differential equations with periodic coefficients which is outlined by McLachlan (1947) and Yakubovich and Starzhinskii (1975). The mechanics of parametric vibration in various mechanical engineering problems is well documented in books by Bolotin (1964), Schmidt (1975), Evan-Iwanowski (1976), Tondl (1978), and Nayfeh and Mook (1979). In addition, an extensive literature review in four parts, covering various aspects and current problems of parametric vibration, has been given by Ibrahim and Barr (1978a,b) and Ibrahim (1978a,b).

In a sense, a single degree-of-freedom system with a non-linear restoring force could be said to have a time-dependent stiffness because the stiffness is displacement dependent and the displacement varies with time. This relationship is implicit, however, and the problem is not classed as parametric. Alternatively, linear systems with time-dependent coefficients are similar in some ways to non-linear systems and can be said to be classified between constant coefficient linear systems and non-linear systems. This is most clearly seen when a linear periodic parametric system displays a half-order subharmonic response, an effect not found in a time-invariant linear system. Furthermore, the resulting response is orthogonal to the direction of the excitation, a feature which is opposite to the response of constant coefficient linear systems under forced excitation.

Some time-dependent variations in parameters apparently result from geometric changes, and these are always forces or torques that can do work on the system. Energy can flow into a system from external sources during resonance, which is dependent upon the frequency of the parameter variation and the natural frequencies of the system; the state so created is called "parametric resonance." Thus the dynamic system will experience parametric instability when the excitation frequency Ω is twice (or any integer multiple of) the system natural frequency ω, i.e.,

$$\Omega = 2\omega \quad \text{or} \quad (\Omega = m\omega,\ m = 2,3,\ldots) \tag{1.1}$$

The relation $\Omega = 2\omega$ is known as "principal parametric resonance" and differs from normal resonance, in which the forcing frequency and a natural frequency coincide.

When some average quantity of energy flows into a system, the amplitude of the system response increases. This behavior is referred to as "parametric instability." The rate of increase in amplitude is generally exponential and thus potentially dangerous, more so in some ways than typical resonance in which the rate of increase is linear. Furthermore, damping effectively reduces the severity of typical resonance, but might reduce only the rate of increase during parametric resonance and thus have little or no effect on final amplitudes. Indeed, in some situations the introduction of damping can extend the instability region.

In parametric instability the amplitude of a linear system grows without limit. Because this growth cannot be accounted for in linear, time-invariant models, they must be made physically realistic by introducing various non-linear effects as the amplitude increases. These non-linearities limit the growth of the response and allow it to attain a limit cycle.

Parametric instability can occur in structural systems subjected to vertical ground motion, in aircraft structures and helicopter rotor blades subjected to turbulent flow, in a ship's roll motion due to the time-dependent restoring moment, in gun tubes during multiple firing, and in machine components and mechanisms subjected to parametric excitation. Other examples include liquid sloshing in rocket tanks subjected to longitudinal excitation generated from the rocket engine. Spinning satellites describing elliptic orbits can pass through a periodically varying gravitational field. Figure (1.1) shows some common systems which may exhibit parametric instability under parametric excitations. The general mathematical description appropriate to a structure or mechanical system is given by the n-differential equations

$$\underline{M}(t)\ddot{\underset{\sim}{q}} + \underline{B}(t)\dot{\underset{\sim}{q}} + \underline{C}(t)\dot{\underset{\sim}{q}} + \underline{K}(t)\underset{\sim}{q} = \underset{\sim}{F}(t) \tag{1.2}$$

where dots denote differentiation with respect to time t. $\underline{M}(t)$, $\underline{C}(t)$, and $\underline{K}(t)$ are the inertial, damping, and stiffness matrices, respectively. They are assumed symmetric. $\underline{B}(t)$ is a skew-symmetric gyroscopic matrix. In deterministic analysis, these matrices are assumed to be periodic with least period T>0; i.e., $\underline{M}(t+T)=M(t)$. They usually contain both constant and time-dependent elements. $\underset{\sim}{q}$ is the generalized coordinate vector, and $\underset{\sim}{F}(t)$ is the vector of the external forcing function. The homogeneous form of equations (1.2), that is $\underset{\sim}{F}(t)=\underset{\sim}{0}$, is usually transformed to 2n-first order differential equations written in the simple matrix form

$$\dot{\underset{\sim}{X}}(t) = \underline{A}(t)\underset{\sim}{X}(t), \qquad \text{with } \underline{A}(t+T) = \underline{A}(t) \tag{1.3}$$

Although the matrix $\underline{A}(t)$ is usually real, complex forms can arise if, for example, hysteretic damping or viscoelastic effects are considered. The solutions of (1.3) are contained in the solution of the related matrix equation

$$\dot{\underset{\sim}{Q}}(t) = \underline{A}(t)\underset{\sim}{Q}(t) \tag{1.4}$$

If the solution of equation (1.4) is $\underline{\Phi}(t)$, which is formed from independent solutions of (1.3), and if it can be assumed without losing generality that $\underline{\Phi}(0)=\underline{I}$, the unit matrix, the form of the solution matrix after one period $\underline{\Phi}(T)$ establishes whether or not the system is stable. The matrix $\underline{\Phi}(T)$ is referred to as the monodromy matrix (Meirovich, 1970) of equation (1.4). One of its several important properties is that it multiplies the solution at t, i.e., $\underline{\Phi}(t)$, to produce the solution at time (t+T). In other words,

$$\underline{\Phi}(t+T) = \underline{\Phi}(t)\underline{\Phi}(T) \tag{1.5}$$

This property holds for periods 2T, 3T,..., so that

$$\underline{\Phi}(t+nT) = \underline{\Phi}(t)\underline{\Phi}^n(T) \tag{1.6}$$

The powers of the monodromy matrix thus determine whether the solution will grow with successive periods. The solution will be stable provided the eigenvalues of $\underline{\Phi}(T)$ lie within a unit circle on the complex plane. The eigenvalues λ_i are commonly referred to as the characteristic multipliers of the system. Alternatively, stability can be expressed in terms of the characteristic exponents ρ_i of

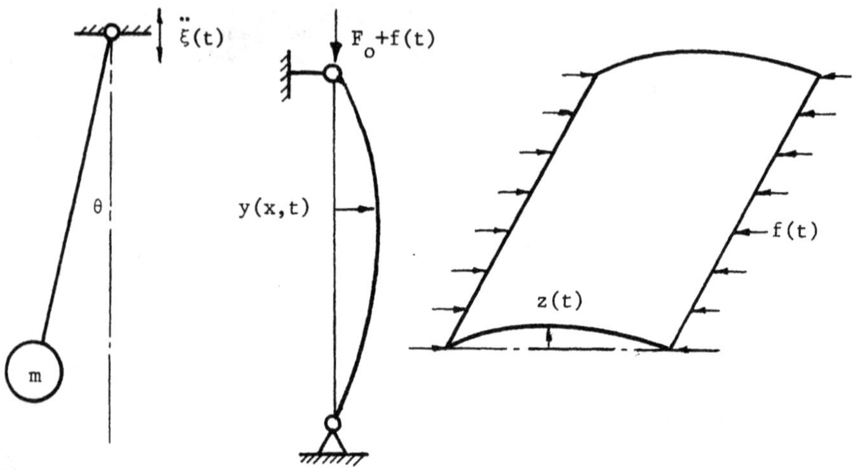

(a) Simple pendulum under vertical acceleration $\ddot{\xi}(t)$

(b) Beliaev column under longitudinal loading f(t)

(c) Flutter of flat plate under in-plane loading f(t)

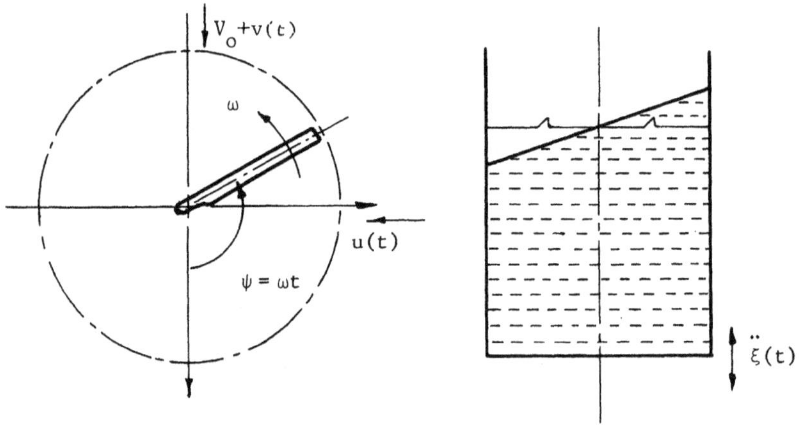

(d) Helicopter rotor blade in turbulent atmosphere

(e) Liquid free surface sloshing under vertical excitation

FIG. (1.1) Dynamic systems under parametric excitations

the system; these are defined by

$$\lambda_i = \exp(\rho_i) \tag{1.7}$$

In this case stability requires that the characteristic exponents have negative real parts. In fact, multipliers on the unit circle are also stable, except in certain situations involving repeated multipliers.

Another useful property of $\underline{\Phi}(t)$ is that it can be represented in the form

$$\underline{\Phi}(t) = \underline{P}(t)\exp(t\underline{R}) \qquad \text{with } \underline{P}(t+T) = \underline{P}(t) \tag{1.8}$$

where $\underline{P}(t)$ is a periodic non-singular matrix, and $\underline{R}$ is related to the monodromy matrix such that

$$\underline{\Phi}(T) = \exp(T\underline{R}) \tag{1.9}$$

That $\underline{\Phi}(t)$ is the product of a periodic matrix $\underline{P}(t)$ and an exponential matrix $\exp(t\underline{R})$, is a consequence of the basic work of Floquet (Cesari, 1963).

The mechanics of parametric instability falls into two classes of problems; single degree-of-freedom and multi-degree-of-freedom systems. The differential equation of a single degree-of-freedom system subjected to a parametric excitation $\xi(t)$ is usually written in the form

$$\ddot{Y} + 2\zeta\omega\dot{Y} + \left[\omega^2 + \varepsilon\xi(t)\right]Y = 0 \tag{1.10}$$

Equation (1.10) can be reduced into the Mathieu-Hill equation if $\xi(t)$ is purely harmonic, $\xi(t) = \cos\Omega t$, and the system experiences parametric instability if the parametric resonance condition (1.1) is satisfied. The instability regions bounded by periodic solutions with periods T or 2T have been well documented in several references; see, for example, Abramowitz and Stegen (1972). The stability-instability regions are usually presented in Ince-Strutt charts as shown in fig. (1.2). If the damping factor $\zeta=0$, the equilibrium solution $Y(t)=Y(t)=0$ of equation (1.10) is not stable and will, under the slightest initiation, pass into a diverging oscillation. In a stable region, on the other hand, the equilibrium solution is stable, and the general solution resulting from some set of initial conditions remains bounded with time. The influence of the system damping upon the stability-instability boundaries is to shrink the instability regions, and also results in a critical excitation amplitude below which the system remains stable.

Barr (1980) gave an account for more general cases in which the periodic stiffness of a single degree-of-freedom contains two or more periodic components whose frequencies are not simple multiples of one another. The combination of two sinusoidal inputs of equal size and

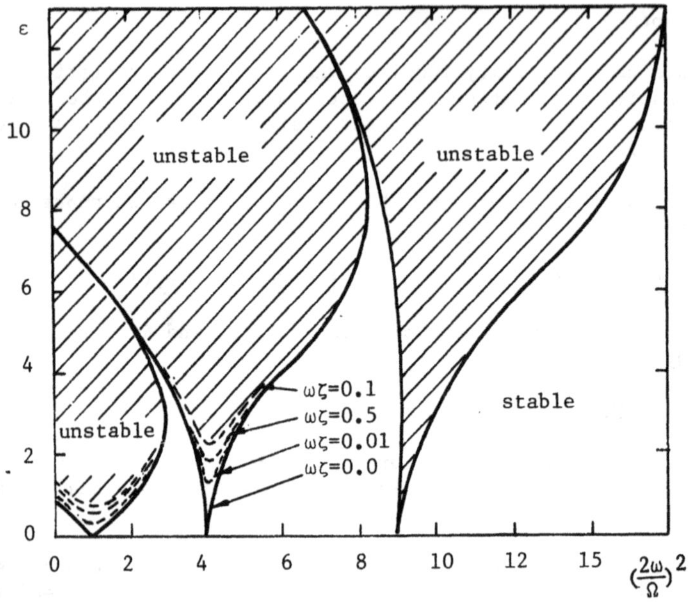

FIG. (1.2) Stability chart of systems described by Mathieu's equation $\ddot{X} + 2\zeta\omega\dot{X} + (\omega^2 - 2\varepsilon\cos\Omega t)X = 0$

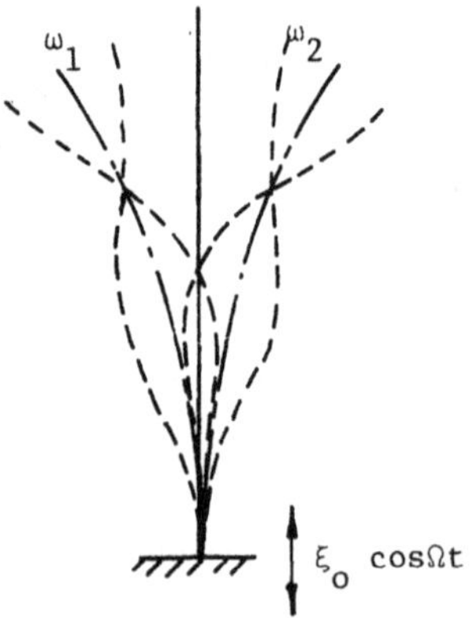

FIG. (1.3) Mode shapes of a cantilever under combination parametric resonance $\Omega=\omega_1+\omega_1$ (Jaeger and Barr, 1966)

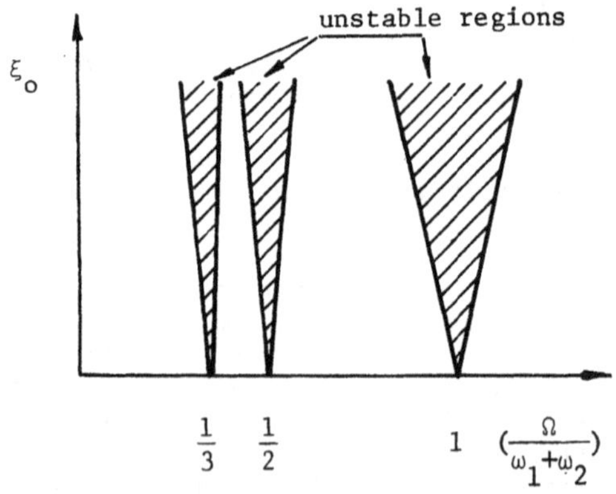

FIG. (1.4) Stability boundaries of a structure under combination resonance

nearly equal frequency has the appearance of a beating wave. The corresponding stability chart changes with additional new instability zone which is effectively due to the beating of two (close) excitation frequencies.

Systems with several degrees-of-freedom can exhibit parametric instability if the excitation frequency and two or more natural frequencies satisfy a linear relation with integer coefficients. Under this condition the system vibrates simultaneously in several modes. Figure (1.3) shows a typical mode shape for a simple vertical cantilever subjected to vertical excitation as observed by Jaeger and Barr (1966). Analytically the stability regions of these systems are obtained from the analysis of linear differential equations with periodic coefficients of the general form

$$\ddot{q}_i + \varepsilon \sum_{j=1}^{N} D_{ij}\dot{q}_i + \sum_{j=1}^{N} \{C_{ij} + \varepsilon \cos(\Omega t)G_{ij}\}q_j = 0 \tag{1.11}$$

where ε is a small parameter associated with the matrices $\underline{D}$ and $\underline{G}$; this implies that the parametric effects are small perturbations of the basic system. The second term is referred to as the velocity-dependent force. In general, $\underline{D}$ could be split into symmetric and skew-symmetric matrices. The symmetric part defines the dissipation forces and the skew-symmetric matrix indicates the presence of gyroscopic forces. The matrix $\underline{C}$ admits n real, positive and distinct eigenvalues. $\underline{G}$ depends on the characteristics of the system and the magnitude of the applied excitation. The analysis of equation (1.11) is simplified if the matrix $\underline{C}$ is diagonalized through a transformation matrix $\underline{R}$. This matrix in turn reduces $\underline{C}$ into the Jordan canonical form.

For a two degree-of-freedom system the solutions of (1.11) can be unbounded if the excitation frequency satisfies one of the relations

$$\Omega = \frac{1}{n}\left| \omega_i \pm \omega_j \right| \qquad (i \neq j) \tag{1.12}$$

This relation defines the condition of parametric combination resonance. It is said to be of the first order when n=1; otherwise, it is of n-th order. In most cases the system exhibits instability of the summed combination resonance if the matrix $\underline{G}$ is symmetric or non-symmetric provided that $G_{ij}G_{ji}>0$. For $G_{ij}G_{ji}<0$ parametric instability of the difference type occurs. The instability regions of conservative systems under the summed combination resonance are shown in fig. (1.4).

The linear modeling of parametric excited systems is quite adequate to predict the stability state of the equilibrium position. If the equilibrium is unstable, the linearized analysis shows that the amplitude grows exponentially without limit. However, most real systems possess a non-linearity which becomes predominant such that when the response reaches a certain level, the system ends up in bounded limit cycles. Bolotin (1964) and Barr (1980) classified the non-linearity into three categories: elastic, damping, and inertia. Elastic non-linearity exists as a result of non-linear strain displacement relations which are inevitable. This non-linear relationship can be expressed in a power series for which the first term represents the linear part, other terms such as quadratic and cubic, etc. account for the degree of stiffness non-linearity. Damping forces due to internal and external agencies are, in general, non-linear and the common representation of viscous linear damping is an idealization. Inertial non-linearity appears in the equations of motion of holonomic systems because the kinetic energy, in a Lagrangian formulation, has coefficients which are really functions of the generalized coordinates. In this case the equations of motion are written in the form

$$\sum_{j=1}^{n} m_{ij}\ddot{q}_j + \sum_{j=1}^{n}\sum_{\ell=1}^{n} [j\ell,i]\dot{q}_j\dot{q}_\ell + \frac{\partial V}{\partial q_i} = 0, \qquad i=1,2,\ldots,n \tag{1.13}$$

where $[j\ell,i]$ is the Christoffel symbol of the first kind and is given by the expression

$$[j\ell,i] = \frac{1}{2}\left(\frac{\partial m_{ij}}{\partial q_\ell} + \frac{\partial m_{i\ell}}{\partial q_j} - \frac{\partial m_{j\ell}}{\partial q_i}\right) \tag{1.14}$$

and the metric tensors m_{ij}, $m_{i\ell}$, and $m_{j\ell}$ are functions of the q's. V is the potential energy of the system.

If the linear coupling is removed, through transformation into principal coordinates, the resulting equations of motion will possess non-linear modal interaction and the effects of parametric excitations become of considerable significance under certain conditions. Barr (1980) pointed out that "the smallness of the non-linear and parametric terms does not prevent them from having an overwhelming effect on the response in the course of time. These added terms can be looked on as providing a coupling or an energy interchange mechanism between the linear modes." This type of coupling is referred to as "autoparametric interaction" when an externally excited mode can act as a parametric excitation for further modes. This situation may arise in many structural configurations, as those shown in fig. (1.5), and for such systems the deterministic case has been extensively studied. Both autonomous and non-autonomous systems have been investigated. In autonomous systems, a certain type of instability may occur when the following relationship exists between the frequencies of the normal modes

$$\sum_{i=1}^{n} k_i\omega_i = 0 \tag{1.15}$$

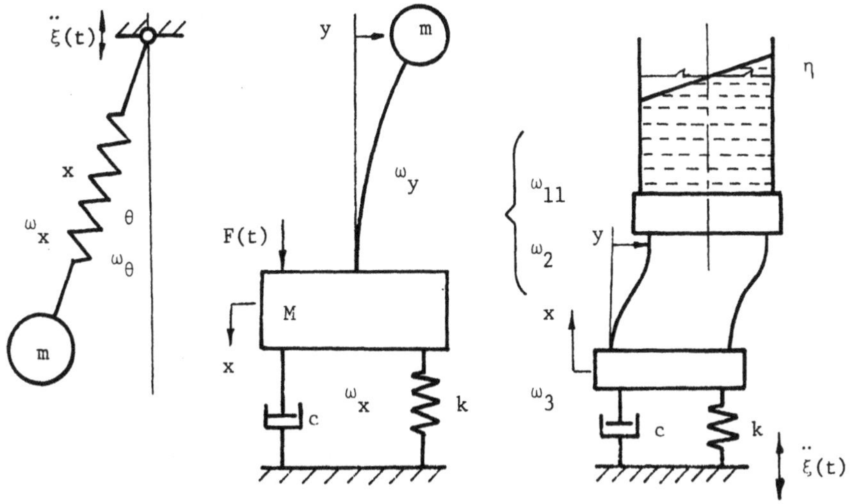

(a) Elastic pendulum

(b) Autoparametric vibration absorber

(c) Elastic structure carrying a liquid tank

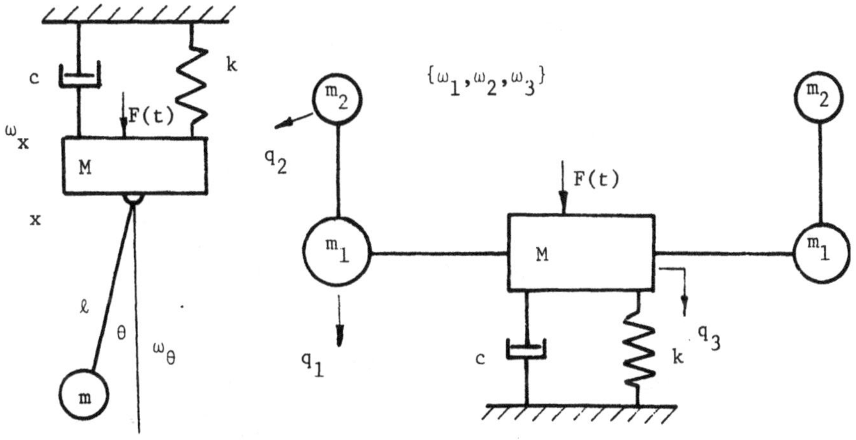

(d) Two degree-of-freedom autoparametric system (Hatwal, et al., 1983)

(e) Approximate airplane model

FIG. (1.5) Dynamic systems with autoparametric interaction and internal resonance $\sum_i^n K_i \omega_i = 0$

where the k_i are integers. This relation is referred to as "internal resonance condition." The vector $\underline{k}=(k_1,k_2,\ldots,k_n)$ is called the resonance vector, while the number $k = |k_1| + |k_2| + \ldots + |k_n|$ is known as the order of internal resonance. In the case of non-autonomous systems a state of "autoparametric resonance" may occur when the conditions of internal resonance (1.15) exist simultaneously with an external resonance as a result of external force. Another remarkable feature of autoparametric interaction is that the modes related by internal resonance may interact in such a way that, for example, ordinary forced vibration of one mode may result in exponential growth of a coupled mode, and this interacting mode may act as a vibration absorber to the excited mode.

As mentioned earlier the inclusion of the relevant non-linearity in the Mathieu equation will reduce the unstable response into a bounded limit cycle. Ibrahim and Barr (1978b) reviewed the influence of various types of non-linearities on the response of parametrically excited systems. If the stiffness non-linearity is included, the response of the one-half subharmonic increases to a certain maximum then decreases, resulting in a modulated one-half subharmonic. On the other hand, the inclusion of non-linear inertia may drag the state of an originally stable system into a catastrophic, unstable state.

1.3 CHAOTIC MOTION

The deterministic theory of parametric vibration predicts periodic responses when the excitation is periodic. However, several dynamic systems may respond in essentially random behavior to deterministic excitations. The term "chaos" is usually used to distinguish such behavior from a true random response produced from a random parametric excitation. Some researchers (see the review article by Holmes and Moon, 1983) have used the term "strange attractors" to distinguish them from classical limit cycles and forced (or parametric) periodic motions. Recently several investigators have examined the chaotic response motion of dynamic systems subjected to harmonic or periodic excitations. Chaotic motion is a form of response motion which is non-periodic in nature.

Hatwal, et al. (1983a,b) investigated the chaotic responses of dynamic systems involving autoparametric coupling shown in fig. (1.5d). They showed that for certain combinations of forcing amplitude and frequency the responses become random within the instability region III of fig. (1.6). The existence of this type of response was verified experimentally. The results of numerical integration were used to obtain mean square values and frequency content of the response. The statistical parameters were shown to be independent of the initial conditions but dependent on the values of the system parameters. The response of the primary mass revealed a strong periodic component at the forcing frequency when the pendulum reached a harmonic steady state. At an excitation frequency $\Omega/\omega_x \approx 0.97$ the pendulum response showed a wider spectrum characterized by irregular oscillations (random).

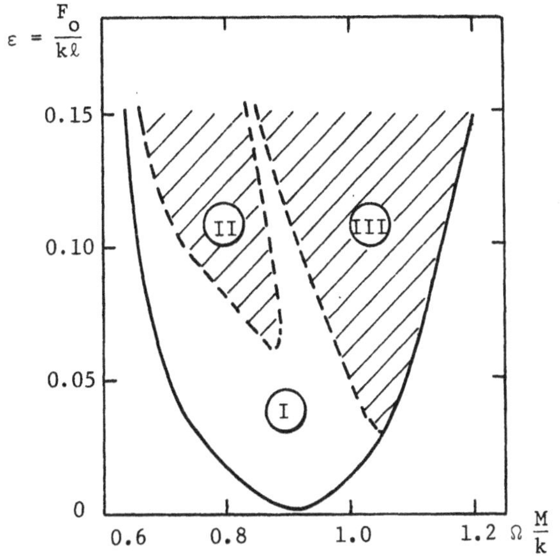

FIG. (1.6) Stability boundaries of system (1.5d) under harmonic excitation $F_0 \cos\Omega t$ (Hatwal, et al., 1983)

Region I stable harmonic solution
Region II unstable harmonic solution
Region III chaotic motion

The theory of chaotic responses of deterministic, non-linear systems is currently being developed. Moon and Holmes (1979) indicated that the classical techniques, such as averaging and perturbation methods, fail to predict these motions because these techniques assume periodic solutions. Various qualitative analyses, based on the Poincare map, have shown the existence and characteristics of chaotic motions in deterministic, non-linear systems. However, at present there is no theory to predict the range of parameters for which these motions can occur. The goal of current investigations is to develop threshold criteria for chaotic motion. Moon (1980) proposed a chaotic motion parameter (analogous to the Reynolds number in fluid flow) below which periodic motions can occur, and above which chaotic motions prevail in the system response. This "chaotic Reynolds number" is a function of driving frequency, forcing amplitude, and damping for most mechanical systems. Dowell (1984) discussed four descriptions to identify the evolution of chaos. These are: time histories, phase plane portraits, power spectral densities, and Poincare maps. The evolution of chaos is described in terms of three phases. The first begins with a limit cycle of a dominant single frequency. The second phase, known as the onset of chaos, may be characterized by the diffusion of the phase plane or the failure of the time history to repeat itself. The third is called the maturing process where the diffusion in the phase plane is more pronounced.

1.4 PSEUDO-RANDOM EXCITATION

The parametric excitation $\xi(t)$ in the Mathieu-Hill equation (1.10) was considered harmonic. An intermediate case between purely single harmonic input and random excitation is the artificial or pseudo-random process. One class of these processes is the binary signal (Newland, 1975) which is obtained by switching randomly a digital waveform generator between two output levels. Drexler and Kropác (1968,1969), Kropác (1971), and Kropác and Drexler (1967) conducted a series of analog simulation studies for a special class of parametrically excited vibratory electro-mechanical systems whose damping is controlled by a binary random signal that switches the system between damping and exciting states. These systems are described by differential equations of the form

$$\ddot{Y} + 2\omega\{\zeta + C\psi(f,p,t) + KA(y)\}\dot{Y} + \omega^2 Y = 0 \qquad (1.16)$$

where $\psi(f,p,t)$ is a symmetric pseudo-random binary signal having a binomial distribution of pulse length, f is the sampling frequency of the random generator, p is the +1 probability of the occurrence of the pulse, A(Y) is the envelope of Y(t), K and C are constants. The analog studies indicated that the mean value of the envelope process of the response changed significantly as a function of the damping coefficient, and the mean square remained approximately constant. The non-linear feedback in the damping term, which is dependent on the instantaneous value of the envelope, ensured the stability of the motion in the prescribed range and very effectively kept the envelope random process quasi-stationary with respect to its first probability density.

Another class of pseudo-random excitation is the envelope introduced by Rice (Wax, 1955) in the form

$$\xi(t) = \sum_{k=-n}^{n} C_k \cos(\omega_k t + \theta_k) \qquad (1.17)$$

where C_k and θ_k are random amplitudes and phase angles, respectively, and $\omega_k = \pm\left[\frac{1}{2}\Delta\omega(2k - 1)\right]$.

Watt and Barr (1983) considered unit amplitude sinusoids, C_k=1, and adopted the form

$$\xi(t) = \sum_{k=-n}^{n} \cos\{\Omega(1 + k\alpha)t + \theta_k\} \tag{1.18}$$

where Ω is the nominal frequency, α is the frequency spacing which is assumed to be small. The initial phase angles θ_k are fixed numbers which were chosen from a rectangular distribution on $[0,2\pi]$ generated by a digital computer. The stability boundaries of system (1.10), when $\xi(t)$ is expressed by the Rice noise (1.18), in the vicinity of twice the system natural frequency were determined by using deterministic mathematical tools such as the monodromy matrix method or asymptotic approximation techniques. The main feature of the results is that when the frequency spacing is relatively large $\alpha > \zeta$, the stability diagram is closely related to that for Mathieu's equation. However, the presence of multiple frequencies introduces new unstable zones. When $\alpha < \zeta$, on the other hand, the detailed shape of the stability boundaries becomes very complicated as the frequency spacing decreases.

1.5 CRYPTO-DETERMINISTIC SYSTEMS

The term crypto-deterministic is given to those time-variant deterministic systems with random initial states such that with increasing time the stochastic content in the response decreases. The joint probability density of the initial state and the Jacobian of the coordinate transformation should characterize the final state. Gaonkar (1971a) indicated that flight vehicle structures that pass through low terrain turbulence and enter into a quiet air region belong to this class.

Consider the state vector equations of n-degree-of-freedom systems described by equation (1.3) with zero mean value of the initial state. According to Gaonkar (1971a) the solution of the state transition matrix equation (see section 8.2.1)

$$\dot{\underline{\Phi}}(t,\tau) = \underline{A}(t)\underline{\Phi}(t,\tau), \qquad \underline{\Phi}(\tau,\tau) = \underline{I}_{2n} \tag{1.19}$$

for a given initial state $\underset{\sim}{X}(t_o)$ can be written in the form

$$\underset{\sim}{X}(t) = \underline{\Phi}(t,t_o)\underset{\sim}{X}(t_o) \tag{1.20}$$

The joint density function of the initial state is

$$p_o\big(\underset{\sim}{X}(t_o)\big) = p_o\big(X_1(t_o),\ X_2(t_o),\ldots,\ X_{2n}(t_o)\big) \tag{1.21}$$

The transformation from the initial state $\underset{\sim}{X}(t_o)$ to the final state $\underset{\sim}{X}(t)$ is uniquely governed by the state transition matrix $\underline{\Phi}$. Syski (1967) has shown that the joint density of $\underset{\sim}{X}(t)$ can be expressed by the relation

$$p(\underset{\sim}{X}(t)) = p(X_1(t),\ X_2(t),\ldots,\ X_{2n}(t))$$
$$= p_0(\underset{\sim}{X}(t))\ \frac{1}{|J(t)|} \qquad (1.22)$$

where the Jacobian J(t) is given by the expression

$$J(t) = \exp\{\int_{t_o}^{t} \text{Trace } \underline{A}(s)\ ds\} \qquad (1.23)$$

Relation (1.22) is expressed in terms of the components of the final state. Alternatively, the joint probability density function for random variables $X_1(t_1),\ X_1(t_2),\ldots,\ X_1(t_{2n})$ is

$$p(X_1(t_1),X_1(t_2),\ldots,X_1(t_{2n})) = p_0(\underset{\sim}{X}(t_o))\ \frac{1}{|J(t_1,t_2,\ldots,t_{2n})|} \qquad (1.24)$$

where the Jacobian J is

$$J(t_1,t_2,\ldots,t_{2n}) = \begin{vmatrix} \Phi_{11}(t_1) & \Phi_{12}(t_1)\cdots & \cdots\Phi_{1,2n}(t_1) \\ \vdots\cdots\cdots & \cdots\cdots\vdots & \cdots\vdots \\ \Phi_{11}(t_{2n}) & \Phi_{12}(t_{2n})\cdots & \cdots\Phi_{1,2n}(t_{2n}) \end{vmatrix} \qquad (1.25)$$

and $\Phi_{ij}(t_i)$ are the $2n \times 2n$ fundamental set of solutions which are normally the elements in the first row of the state transition matrix for which the response component, say $X_1(t)$, can be expressed at 2n different t values

$$X_1(t_i) = \sum_{j=1}^{2n} \Phi_{1j}(t_i)X_j(t_o) \qquad i = 1,2,\ldots,2n,\ t_o<t<T \qquad (1.26)$$

The Jacobian (1.25) gives a measure to the correlation of a certain pair of random variables. In addition, one can estimate the range after which the response process is essentially deterministic. This can be inferred from the fact that for stable systems which experience bounded responses under bounded inputs, the influence of the initial state decreases with the growth of the response. The degree of randomness of the response is governed by the order of the probability density function.

1.6 RANDOM PARAMETRIC VIBRATION

The deterministic approach to the analysis of parametric vibration is an idealization and simplification of the actual behavior of dynamic systems. In many real problems of engineering interest, however, either the excitation or the time variation of the system parameters is random. The random excitation may be classified into three categories: narrow band, wide band, and white noise processes. Furthermore, a random process may be stationary or non-stationary.

Recent developments in the mathematical theory of random processes and stochastic differential equations have promoted the study of response and stability of dynamical and control systems driven by random parametric excitation. These developments owe their origin to the pioneering works of Einstein (1905) and Kolmogorov (1933) on Brownian motion. Since 1933 there has been an extensive activity by engineers and mathematicians involved in the area of noise in communication devices and control systems. These activities resulted in what is known as the theory of stochastic processes which are reviewed by Lin (1969) and well documented by Doob (1953), Wax (1955), Parzen (1962), Yaglom (1962), Skorokhod (1965), and Sveshnikov (1966). The theory of stochastic processes is characterized mathematically by its own rules of stochastic calculus and differential equations. The mathematical theories of stochastic calculus and differential equations have been established by Itô (1944,1951a,1951b), Itô and Nisio (1964), Wong and Zakai (1965), and Stratonovich (1966). The theory of stochastic differential equations is now the subject of several books including Syski (1967), McKean (1969), Morozan (1969), McShane (1970), Bharucha-Reid (1972), Bunke (1972), Gikhman and Skorokhod (1972), Soong (1973), Arnold (1974), Friedman (1975), Kliatskin (1975,1980), and Khas'miniskii (1980).

Random vibration is essentially a combination of probability theory, stochastic calculus of random processes and applied dynamics. The probabilistic and statistical descriptions of the response of dynamic systems subject to random excitations are under continuous progress as reflected by a number of review articles by Crandall (1966), Caughey (1971), and Crandall and Zhu (1983). The basic concepts of the theory of random vibration of time-invariant systems are found in books by Crandall and Mark (1963), Robson (1963), Lin (1967), Newland (1975), and Nigam (1983). Unfortunately, none of these books has addressed the subject of random parametric vibration. However, there is an extensive body of published results on the stochastic stability and response of systems subjected to random parametric excitation, as indicated by review articles by Kozin (1969,1972), Ibrahim and J. W. Roberts (1978), and Ibrahim (1981). The mathematical theory of stochastic stability of differential equations has been reported in two monographs by Kushner (1967) and Khas'miniskii (1980). Bolotin (1984) and Dimentberg (1980b) have considered a number of problems in parametric random vibration.

Under random parametric excitations the system response and stability properties are characterized by probability measures, and problems of statistical estimation arise when attempts are made to relate the results of such procedures to experimental observations. Remarkable progress has been made in developing new stochastic stability theories and Gaussian and non-Gaussian closure techniques to truncate the infinite hierarchy moment equations to solve a number of problems that were regarded as difficult in the past. Controversies and different results are noticed in a number of papers due to misinterpretation of certain stochastic processes or improper use of the analytical methods. In an effort to resolve such problems, the International Union of Theoretical and Applied Mechanics (IUTAM) held a number of symposiums (see Curtain, 1972; Clarkson, 1977; and Hennig, 1983) to bring together mathematicians and engineers working on stochastic processes and random vibration problems. The purpose of these meetings was to present to the mathematicians a wide range of applications for their work from across the many engineering disciplines. Equally important was to bring to the engineers the latest developments which were being accomplished in the theory and methods of handling stochastic problems.

The main objective of this monograph is to present the state-of-the-art of the area of parametric random vibration with reference to wide band and white noise random excitations. This includes theory, methods of analysis, and important results. However, the rest of this section will provide the reader with a brief review of the methods and results pertaining to the behavior of dynamic systems subjected to narrow band parametric excitations.

For a very narrow band parametric excitation, the analysis of the system response is usually performed by employing one of the standard deterministic techniques of differential equations with periodic coefficients such as the averaging method of Bogoliubov and Mitropol'skii (1961), the Liapunov direct method described by Hahn (1963), asymptotic approximation techniques outlined by Nayfeh and Mook (1979), and the integro-differential equation method adopted by Schmidt (1975). The stability boundaries of linear systems whose random coefficients are bounded narrow band processes have been determined by Caughey and Dickerson (1967) who used the Liapunov direct method. Mirkina (1975,1977) employed the method of multiple scale expansion to determine the stability map of linear systems subjected to stationary parametric excitation represented by a function of a fast time variable and slow time variable. The stability conditions of the response root mean square were found to be identical to those obtained by Alekseyev and Valeev (1971) and Bolotin and Moskvin (1972). The stability boundaries exhibit the V-shape at the principal and secondary resonances. Baxter (1971) and Schmidt (1976) determined the probabilistic behavior of the response of non-linear systems with narrow band random coefficients.

Some experimental studies on a RLC circuit with randomly varying capacitance showed that any value of the spectral density of the capacitance variation would carry the circuit into unstable regions if the damping were zero as considered by Samuels and Eringen (1959). This instability was controlled by the circuit non-linearities when the response exceeded a critical level. These properties were verified by analog computer simulation by Samuels (1959) and Samuels and Eringen (1959). Bogdanoff and Citron (1965a,b) and Ness (1967) conducted a series of experiments to investigate the stability of an inverted pendulum subjected to support motion represented by a sum of several harmonics. The results supported the theoretical conclusion, obtained by Bogdanoff (1962), that only variation of the base velocity influences the stability of the inverted pendulum provided the amplitude of each harmonic is small.

An exploratory experimental investigation of the behavior of the free surface of a liquid free surface under narrow and wide band parametric excitations was conducted by Dalzell (1967). A very low level harmonic response was observed exactly as in sinusoidal excitation experiments, when the random excitation contained enough energy at twice the modal frequency of higher axisymmetric modes; such a mode was excited with a large amplitude. An important feature was that low level harmonic response to random Gaussian excitation was nearly Gaussian distributed. However, when a large amplitude subharmonic response was excited, the probability distribution changed abruptly into an extreme value distribution known as the double exponential distribution. Baxter and Evan-Iwanowski (1975) conducted a series of experiments to measure the response characteristics of a column excited by a narrow band random loading. Their observation supported theoretical results that the column vibrates near its natural frequency when the spectral density of the excitation is sufficiently high within the principal parametric zone of the column. They also found that the variance of the response amplitude decreases with increasing excitation bandwidth.

CHAPTER 2

Random Variables and Stochastic Processes

2.1 INTRODUCTION

The response and stability analyses of dynamic systems experiencing random vibration are based on the fundamental concepts of probability theory and stochastic calculus of random processes. This chapter deals with the probabilistic and statistical descriptions of random processes encountered in the broad area of random vibration. To serve the main subject of this monograph, the basic principles of random variables and random processes will be discussed briefly with an emphasis on the properties of Gaussian and non-Gaussian processes. For a rigorous treatment of the topics addressed in this chapter, the reader may refer to Parzen (1960), Loeve (1963), Papoulis (1965), and Gikhman and Skorokhod (1965).

2.2 RANDOM VARIABLES

The concept of random variables stems originally from the probability theory of events which is described by the ratio of the favorable outcomes to the total number of outcomes of equally likely trials or experiments. Thus a random variable $X(\omega)$ is regarded as the outcome of a random experiment ω from the total sample space Ω. The random variable is specified by an assigned real (or complex) number and its probability $P\{\omega:X(\omega)\leq x\}$, where x is a particular value. $X(\omega)$ can be considered as a real function that maps all elements of the sample space into points on the real line R extending from $-\infty$ to $+\infty$. With reference to the sample record shown in fig. (2.1) the values of the random response at times $t_1, t_2, \ldots, t_n$ represent a set of the random variables $X(t_1), X(t_1), \ldots, X(t_n)$. It is clear that the continuous random variable $X(\omega)$ may be considered as an element in an uncountable space having a continuous range of values and, as will be shown, its probability distribution is everywhere continuous. On the other hand, if the outcomes of the trials are countable distinct numbers on R, they are called discrete random variables. The probability description of random variables will be introduced in the next section.

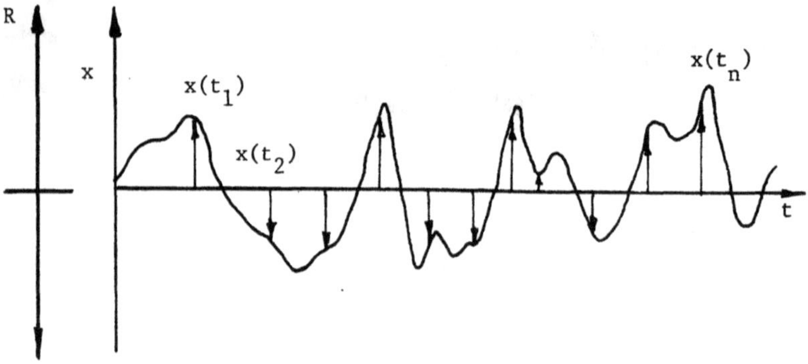

FIG. (2.1) Sample record of experiment ω

2.2.1 THE PROBABILITY DISTRIBUTION FUNCTION F(x)

The probability that the event $X(\omega)\leq x$ is expressed by the notation

$$F(x) = P\{\omega : X(\omega)\leq x\} \tag{2.1}$$

F(x) has the following properties:

$$\left.\begin{aligned} F(-\infty) &= 0 \\ F(+\infty) &= 1 \end{aligned}\right\} \tag{2.2}$$

$$\left.\begin{aligned} &F(x_1)<F(x_2) && \text{if } x_1<x_2 \\ &\text{and } P\{x_1<X\leq x_2\} && = F(x_2) - F(x_1) \end{aligned}\right\} \tag{2.3}$$

where the argument ω has been dropped from X.

2.2.2 THE PROBABILITY DENSITY FUNCTION p(x)

The probability density function is represented by the derivative of the probability distribution function

$$p(x) = \lim_{\Delta x\to 0} \left\{\frac{1}{\Delta x}\left[F(x+\Delta x) - F(x)\right]\right\}$$

$$= dF(x)/dx = dP(X\leq x)/dx \tag{2.4}$$

This definition implies

$$P(X \leq x_1) = \int_{-\infty}^{x_1} p(x)dx \tag{2.5}$$

Combining (2.2), (2.3), and (2.4) gives

$$p(x)\Delta x = P\{x < X \leq x+\Delta x\} \tag{2.6}$$

Figure (2.2) shows typical graphs for the probability distribution and probability density functions of a continuous random variable.

For discrete random variables the values of X will be denoted by $x_1, x_2, \ldots, x_n$. The probability that $X=x_i$ is shown in fig. (2.3a) and is known as the probability function $P(x_i)$, i.e.

$$P(x_i) = P(X = x_i) \tag{2.7}$$

The corresponding probability distribution function F(x) must have a stairstep form as shown in fig. (2.3b), and is related to the probability function by the formula

$$F(x) = \sum_{i=1}^{n} P\{X=x_i\}u(x-x_i) = \sum_{i=1}^{n} P(x_i)u(x-x_i) \tag{2.8}$$

where $u(x-x_i)$ is the Heaviside unit step function described by the two values

$$u(y) = \begin{cases} 1 & y>0 \\ 0 & y<0 \end{cases} \tag{2.9}$$

Because the probability distribution function is discontinuous, the derivative dF(x)/dx is not well defined. However, the concept of the unit-impulse function $\delta(x)$ or the Dirac delta function is introduced

$$p(x) = \sum_{i=1}^{n} P(x_i) \frac{du(x-x_i)}{dx} = \sum_{i=1}^{n} P(x_i)\ \delta(x-x_i) \tag{2.10}$$

where $\delta(x-x_i)$ has the following properties

$$\delta(x-x_i) = \begin{cases} 0 & \text{for } x \neq x_i \\ \infty & \text{for } x = x_i \end{cases} \tag{2.11}$$

2.2.3 THE ENTROPY OF RANDOM VARIABLES

In classical thermodynamics the variation of a system from one condition of thermodynamic equilibrium to another state of equilibrium is expressed in terms of the entropy. The concept of entropy has been adopted in other areas such as statistical mechanics and information theory. It represents a statistical mechanical

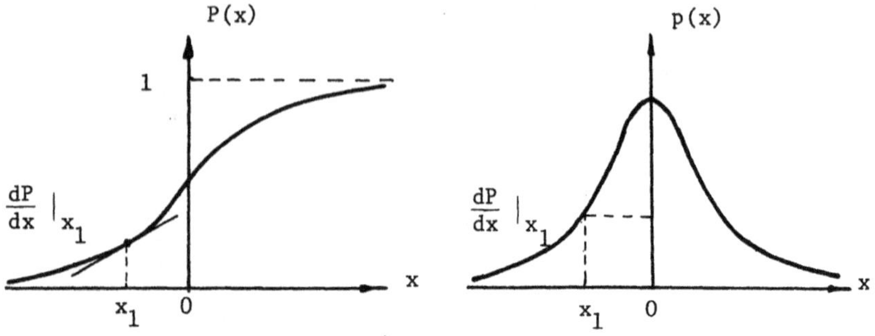

(a) Probability distribution function

(b) Probability density function

FIG. (2.2) Probability distribution and density functions of a continuous random variable

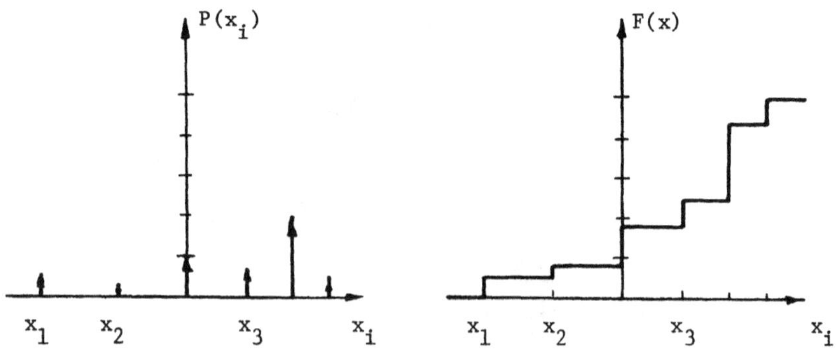

(a) Probability function

(b) Probability distribution function

FIG. (2.3) Probability functions of a discrete random variable

quantity $H(\underset{\sim}{x})$ of the state vector $\underset{\sim}{x}$ as well as the lack of information, both of which are not thermodynamic quantities. Shannon and Weaver (1949) defined the entropy as a measure of the degree of randomness or uncertainty in the system state. A highly organized state of a certain phenomenon possesses little entropy, whereas a disorganized one possesses more entropy. Shannon and Weaver defined the entropy as the average of the probability distribution function multiplied by the logarithm of the probability density function (Sveshnikov, 1978)

$$H(\underset{\sim}{x}) = - \int_{-\infty}^{\infty} \cdots \int p(\underset{\sim}{x}) \log\{p(\underset{\sim}{x})\} dx_1 \cdots dx_n \tag{2.12}$$

2.2.4 MATHEMATICAL EXPECTATION AND MOMENTS

Let the possible outcomes of a series of trials of a discrete random variable be $\underset{\sim}{X} = \{x_1, x_2, \ldots, x_n\}$. The arithmetic mean value of $\underset{\sim}{X}$ is

$$\bar{x} = \frac{1}{n} \sum_{i=1}^{n} x_i \tag{2.13}$$

Now let n_i denote the number of times that $X=x_i$. Thus the relative frequency of x is n_i/n and the average value of X becomes

$$\bar{x} = \sum_{i=1}^{m} \frac{n_i}{n} x_i \qquad m \leq n \tag{2.14}$$

The frequency n_i/n may be interpreted as the weighting factor, and in the limit, as $n \to \infty$, it represents the probability function $P(X=x_i)$, that is

$$\lim_{n \to \infty} (n_i/n) = P(X=x_i) \tag{2.15}$$

In terms of (2.15), the mean value $\bar{x}$ can be expressed as the mathematical expectation

$$E[x] = \sum_{i=1}^{\infty} x_i P(x_i) \tag{2.16}$$

where E[ ] denotes expectation.

The expected value of a continuous random variable is given in the form

$$E[x] = \int_{-\infty}^{\infty} x \, dF(x)$$

$$= \int_{-\infty}^{\infty} xp(x)dx \qquad (2.17)$$

The expected value of x is called the first moment. The second moment (or the mean square value) is

$$E[x^2] = \int_{-\infty}^{\infty} x^2 p(x)dx \qquad (2.18)$$

and the k-th moment is

$$E[x^k] = \int_{-\infty}^{\infty} x^k p(x)dx \qquad (2.19)$$

2.3 RANDOM PROCESSES

The random response of dynamic systems is usually given in the form of a time history record. Each record is called a realization or a sample function. The collection of all possible records is called the ensemble or random process. Figure (2.4) portrays a typical example of a random process. For a chosen value of time the values of the response across the ensemble constitute a set of continuous random variables. Thus a random process $X(\omega,t)$, (which is also called a time series, a stochastic process, or a random function), is defined as a parametered family of random variables with the parameters $\omega \subset \Omega$ and $t \subset T$ (where T is the entire time domain of the random process). In the study of random processes one must distinguish between two cases: (i) the time t can assume any value in a given interval and, (ii) the argument t can take only particular discrete values. The first case is referred to as a stochastic process while the second is called a random sequence.

In general, a single sample record may not be adequate to represent the actual behavior of the system (except for ergodic processes which will be described in section 2.3.5). For non-ergodic processes the statistical properties are determined across the ensemble at given values of time $t=t_1$, $t_2,\ldots,t_n$. If the random process is ergodic the statistical properties are determined along the sample provided the sample exhibits all statistical characteristics of the system. The statistical parameters of ergodic processes are called the temporal averages.

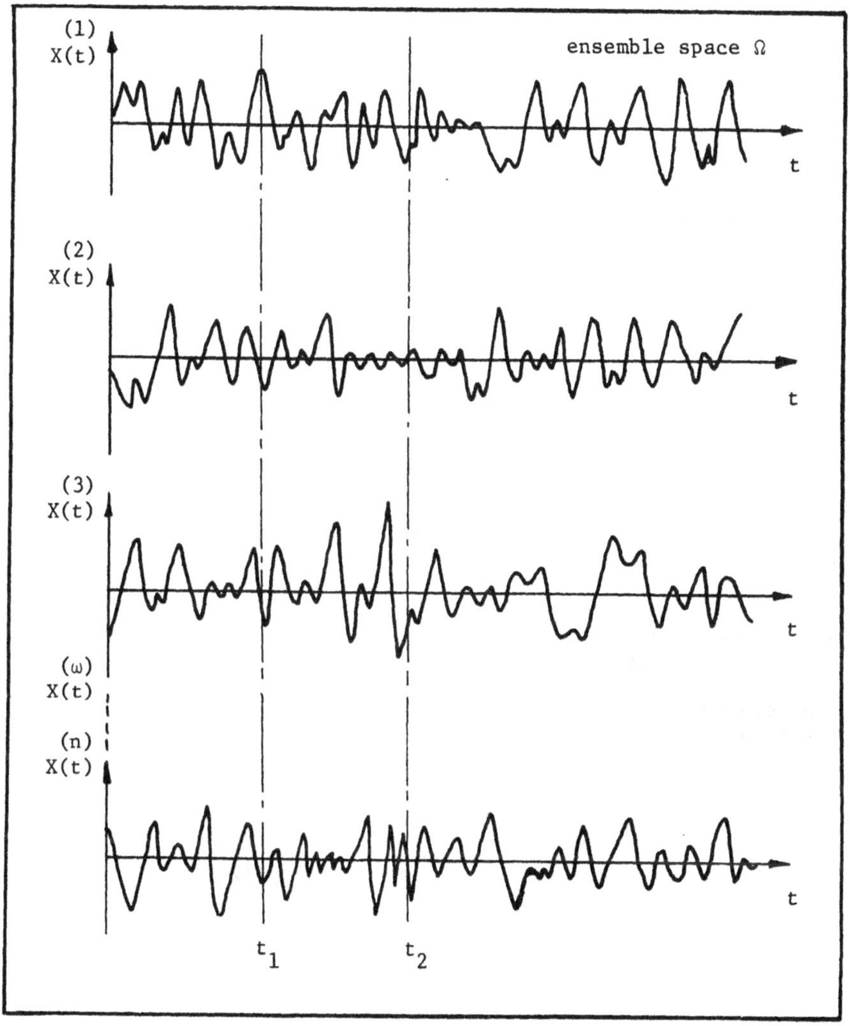

FIG. (2.4) Ensemble space of random process X(t)

Stochastic processes characterized by more than two indexing parameters are known as random fields $X(\omega,t,s)$, where s refers to the space coordinate. Thus a multiparametered random process is a random field. The statistical properties of random fields will be treated in chapter 8 in relation to the flapping motion of helicopter rotor blades.

2.3.1 PROBABILITY FUNCTIONS

The values of the random variable $X^{(i)}(t_1)$, $i=1,2,\ldots$, at $t=t_1$ across the ensemble constitute a continuous random variable vector $\underset{\sim}{X}(t_1)=\{x^{(1)}(t_1),x^{(2)}(t_1),\ldots,\ x^{(n)}(t_1),\ldots\}$ on R. The first probability distribution function of $X(t_1)$ is expressed in the form

$$F(x;t_1) = P\{X(t_1)\leq x\} \tag{2.20}$$

and the corresponding first order density function is

$$p(x;t_1) = \partial F(x;t_1)/\partial x \tag{2.21}$$

which satisfies the relation

$$p(x;t_1)\Delta x = P\{x<X(t_1)<x+\Delta x\} \tag{2.22}$$

Notice that the above relations depend on the time t_1. The physical importance of the probability density function in random vibration is that it provides information concerning the properties of the random response at $t=t_1$ in the amplitude domain.

The joint probability distribution function (or the second order) of two values of X across the ensemble at t_1 and t_2 is defined as the probability that $X(t_1)$ and $X(t_2)$ assume values between $x(t_1)$ and $x(t_1)+dx_1$, and $x(t_2)$ and $x(t_2)+dx_2$, respectively. Mathematically the second order probability distribution function is expressed in the form

$$\begin{aligned} F(x_1;t_1:x_2;t_2) &= P\{x_1<X(t_1)<x_1+dx_1:x_2<X(t_2)<x_2+dx_2\} \\ &= \int_{x_1}^{x_1+\Delta x_1}\int_{x_2}^{x_2+\Delta x_2} p\left(x_1(t_1),x_2(t_2)\right)dx_1\,dx_2 \end{aligned} \tag{2.23}$$

and the corresponding second order probability density function is

$$p(x_1;t_1:x_2;t_2) = \frac{\partial^2 F\left(x_1;t_1:x_2;t_2\right)}{\partial x_1\,\partial x_2} \tag{2.24}$$

Finally, the joint probability distribution of several random variables at $t_1, t_2, \ldots, t_n$ is given in the form

$$F\{x_1(t_1); x_2(t_2); \ldots; x_n(t_n)\} = P\{x_1 < X(t_1) < x_1 + dx_1; \ldots; x_n < X(t_n) < x_n + dx_n\}$$

$$= \int_{x_1}^{x_1+\Delta x_1} \cdots \int_{x_n}^{x_n+\Delta x_n} p(\underset{\sim}{x}; t)\, dx_1\, dx_2 \ldots dx_n \tag{2.25}$$

If the limits of integration in (2.25) are extended from $-\infty$ to $+\infty$, we obtain the normalized condition

$$\int_{-\infty}^{\infty} \cdots \int p(\underset{\sim}{x}; t)\, dx_1\, dx_2 \ldots dx_n = 1 \tag{2.26}$$

The n-th joint probability density function $p(\underset{\sim}{x}; t)$ is given by the mixed n-th derivative of the probability distribution function

$$p(\underset{\sim}{x}; t) = \frac{\partial^n F(\underset{\sim}{x}; t)}{\partial x_1 \partial x_2 \ldots \partial x_n} \tag{2.27}$$

It is evident that the joint probability distribution $F(\underset{\sim}{x}; t)$ is a positive non-decreasing function which is continuous everywhere to the right. It also possesses mixed derivatives, with respect to the state vector $\{\underset{\sim}{x}; t\}$, all of which are positive.

The probability density of a single random variable $x_i(t_i)$ can be obtained from the joint probability density function as

$$p(x_i; t_i) = \underbrace{\int_{-\infty}^{\infty} \cdots \int_{-\infty}^{\infty}}_{(n-1)} p(\underset{\sim}{x}; t)\, dx_1\, dx_2 \ldots dx_{i-1}\, dx_{i+1} \ldots dx_n \tag{2.28}$$

2.3.2 CONDITIONAL PROBABILITY

Consider a pair of two random processes $X(\omega; t)$ and $Y(\omega; t)$. The probability that $X(\omega; t_1)$ takes values in the range $x(t_1)$ to $x(t_1)+dx$ independently of the value of Y may be expressed in the form

$$\text{Prob}\{x < X(t_1) < x + dx;\; -\infty < Y(t_1) < +\infty\} = p(x)\, dx \tag{2.29}$$

Introducing (2.28) in terms of X and Y gives

$$p(x; t_1) = \int_{-\infty}^{\infty} p(x, y; t_1)\, dy \tag{2.30}$$

Instead of evaluating $p(x;t_1)$ over the whole range of Y, one may be interested in determining the probability of $X(\omega,t_1)$ corresponding to a limited range of Y, say $y(t_1)$ to $y(t_1)+dy$. This conditional probability is usually written in the form $p\big(x(t_1)\,\big|\,y(t_1)\big)$. It denotes the probability density of x given y at time $t=t_1$ across the two ensembles $X(\omega;t_1)$ and $Y(\omega;t_1)$. Mathematically we write

$$P\big(x(t_1)\,\big|\,y(t_1)\big) = \frac{P(x,y)}{P(y)} \tag{2.31}$$

Introducing (2.30) and (2.22) gives

$$p\big(x(t_1)\,\big|\,y(t_1)\big)dx = \frac{p(x,y)dxdy}{dy \int_{-\infty}^{\infty} p(x,y)dx}$$

or $$p\big(x(t_1)\,\big|\,y(t_1)\big) = p(x,y)/p(y) \tag{2.32}$$

The conditional probability density plays an important role in a number of stochastic processes such as Markov processes for which the conditional probability is termed as the transition probability density.

If $X(\omega;t)$ and $Y(\omega;t)$ are statistically independent, then

$$p\big(x(t_1)\,\big|\,y(t_1)\big) = p\big(x(t_1)\big) \tag{2.33}$$

which implies that

$$p(x,y) = p(x)p(y) \tag{2.34}$$

where the time t_1 has been dropped for abbreviation.

2.3.3 STATISTICAL EXPECTATIONS

Moments: According to the basic definition given in section (2.2.4) the mean value of a continuous random process $x(\omega,t)$ at time t_1 across the ensemble is given by the first moment of $p(x,t_1)$

$$E[x(t_1)] = \int_{-\infty}^{\infty} x(t_1)p(x,t_1)dx \tag{2.35}$$

In general, the mean value of x depends on the time t_1.

The mean square value (or the second moment) is

$$E[x^2(t_1)] = \int_{-\infty}^{\infty} x^2(t_1)p(x,t_1)dx \tag{2.36}$$

Similarly, the n-th moment is expressed in the form

$$E[x^n(t_1)] = \int_{-\infty}^{\infty} x^n(t_1)p(x,t_1)dx \tag{2.37}$$

For a multi-dimensional random process $\underset{\sim}{X} = \{x_1, x_2, \ldots, x_n\}$ the K-th joint moment is given by the relation

$$E[x_1^{k_1} x_2^{k_2} \ldots x_n^{k_n}] = \int \ldots \int_{-\infty}^{\infty} (x_1^{k_1} x_2^{k_2} \ldots x_n^{k_n}) p(\underset{\sim}{x};t)dx_1 \ldots dx_n \tag{2.38}$$

where $K = \sum_{i=1}^{n} k_i$

<u>The Variance σ_x^2:</u> is the mean square about the mean

$$\begin{aligned} \sigma_x^2 &= E[\{x(t_1) - E[x(t_1)]\}^2] \\ &= \int_{-\infty}^{\infty} \{x-E[x]\}^2 \, p(x;t)dx \\ &= E[x^2] - \{E[x]\}^2 \end{aligned} \tag{2.39}$$

Frequently the positive square root of the variance is used as a statistical function referred to as the <u>standard deviation</u>.

<u>The Autocorrelation Function</u>: The dependence of the values of a random process at time t_1 on its values at t_2 is given by the autocorrelation function

$$\begin{aligned} R_x(t_1,t_2) &= E[x(t_1)x(t_2)] \\ &= \int_{-\infty}^{\infty} x_1 x_2 p(x_1,x_2;t_1,t_2)dx_1 dx_2 \end{aligned} \tag{2.40}$$

In general, R depends on t_1 and t_2. In the limit when $t_1 = t_2$ the autocorrelation function becomes the mean square value

$$R_x(t_1,t_1) = E[x^2(t_1)] \tag{2.41}$$

If the autocorrelation function depends only on the time difference $t_2-t_1 = \tau$, not on t_1 and t_2, as in the case of stationary random processes, then $R_x(\tau)$ is a real-valued even function (symmetric about $\tau = 0$) with a positive maximum at $\tau = 0$.

As $(t_2 - t_1) \to \infty$, $X(t_1)$ and $X(t_2)$ become independent if the random process does not contain any periodic components. In this case the autocorrelation function is reduced to the square of the mean, if the process is stationary

$$R_x\big((t_2 - t_1) \to \infty\big) = (E[x])^2 \tag{2.42}$$

The dependence of a typical autocorrelation function on $t_2 - t_1 = \tau$ is shown in fig. (2.5). It is seen that for purely random processes (free of any harmonics) the autocorrelation exists up to a time interval τ_c known as the correlation time defined by the relation

$$\tau_c = \frac{1}{R_x(0)} \int_0^\infty \left| R_x(\tau) \right| d\tau$$

$$= \int_0^\infty \left| r(\tau) \right| d\tau \tag{2.43}$$

where $r(\tau) = R_x(\tau)/R_x(0)$ is the normalized correlation coefficient.

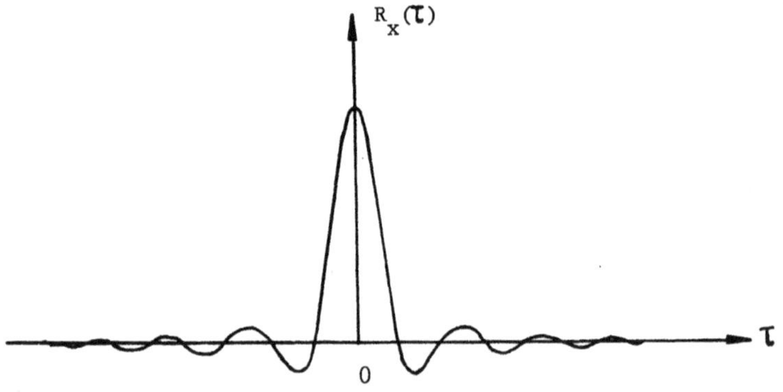

FIG. (2.5) Autocorrelation function of stationary random process

The Cross-Correlation Function $R_{xy}(\tau)$: The statistical function which measures the dependence of one random process X(t) at time t_1 on the value of another random process Y(t) at time $t_1+\tau$ is called the cross-correlation function and is written in the form

$$R_{xy}(t,\tau) = E[x(t_1)y(t_1+\tau)]$$

$$= \int_{-\infty}^{\infty}\!\!\cdots\!\int x(t_1)y(t_1+\tau)p(x,y)dxdy \quad (2.44)$$

The cross-correlation function is real-valued and exhibits symmetry if both X and Y processes are stationary and interchangeable, that is

$$R_{xy}(\tau) = E[x(t_1-\tau)y(t_1)]$$

$$= R_{yx}(-\tau) \quad (2.45)$$

The Covariance Function (or The Autocovariance Kernel): This function is defined as the expected value of the product of the deviation of x at time t_1 by its deviation at time $t_1+\tau$

$$C_x(t,\tau) = E\big[\{x(t) - E[x(t)]\}\{x(t+\tau) - E[x(t+\tau)]\}\big] \quad (2.46)$$

$$= R_x(\tau) - E[x(t)]E[x(t+\tau)] \quad (2.47)$$

If the processes are stationary, the covariance of two random processes X(t) and Y(t) becomes the joint central moment (or the cross-covariance)

$$C_x(\tau) = E\big[\{x - E[x]\}\{y - E[y]\}\big] \quad (2.48)$$

2.3.4 STATIONARITY OF RANDOM PROCESSES

The probability density functions and statistical parameters described in sections (2.3.1) through (2.3.3) are, in general, functions of the assigned time instants $t_1, t_2, \ldots, t_n$. This means that these functions are varying according to the time instant t_i across the ensemble, and therefore the random process is not stationary. On the other hand, if the random properties are independent of the origin of time measurements, the process is classified as stationary. The stationarity of the random process is an important property for which a large body of analytical methods can be applied in investigating various problems in random vibrations and other engineering areas.

The first step towards the concept of stationarity is that the probability density functions are to be time-invariant, that is

$$p(x;t_1) = p(x;t_2) = \cdots = p(x;t_n) \quad (2.49)$$

$$p(x_1,x_2,\ldots,x_n;t_1,t_2,\ldots,t_n) = p(x_1,x_2,\ldots,x_n;t_1',t_2',\ldots,t_n')$$

$$t_i \neq t_i' \qquad (2.50)$$

and for the two-dimensional case the joint probability density depends on the difference of the instants of time at which the amplitudes of the random processes were chosen, i.e.,

$$p(x_1,x_2;t_1,t_2) = p(x_1,x_2;t_2-t_1) \qquad (2.51)$$

Substituting (2.49) and (2.51) into the mean (2.35) and autocorrelation (2.40) gives

$$E[x(t)] = \int_{-\infty}^{\infty} xp(x)dx = \text{constant} \qquad (2.52)$$

$$R_{xy}(\tau) = E[x(t_1-\tau)y(t_1)]$$

$$R_x(t_1,t_2) = \int_{-\infty}^{\infty}\cdots\int x(t_1)x(t_2)p(x_1,x_2;t_2-t_1)dx_1dx_2$$

$$= R_x(t_2-t_1) \qquad (2.53)$$

If the mean and autocorrelation function as given by (2.52) and (2.53) do not depend on t_1 and t_2 (but on the difference $\tau = t_2-t_1$), the process is said to be weakly stationary (or stationary in the wide sense). This definition was originally proposed by Khinchin (Sveshnikov, 1966). The weakly stationary conditions serve one class of random processes which are governed by the Gaussian probability density.

A random process is said to be strongly stationary (or stationary in the strict sense) if all possible moments of various orders, joint moments, and variances are independent of time.

2.3.5 ERGODICITY AND TEMPORAL AVERAGES

Consider a sample record of a typical random process shown in fig. (2.6). The probability of $X(\omega,t)$ to take values within the range x and $x + \Delta x$ may be obtained by dividing the sum of all time intervals Δt_i of which Δx occupies by the total time period of the sample, that is

$$\text{Prob}\left(x \leq X(\omega,t) < x+\Delta x\right) = \lim_{T\to\infty} \frac{1}{T} \sum_{i=1}^{n} \Delta t_i \qquad (2.54)$$

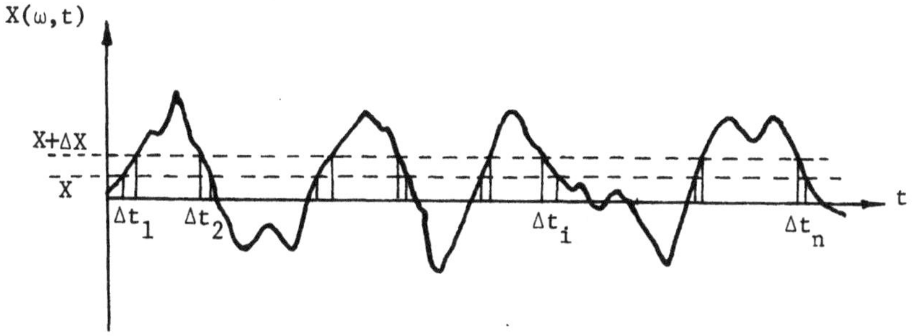

FIG. (2.6) Sample record of ergodic process

The period T is supposed to be very large, i.e., large enough to exhibit all dynamic characteristics of the system.

The average value of $X(\omega,t)$ may also be determined by taking the time average over the sample interval T

$$E[x] = \lim_{T\to\infty} \frac{1}{T} \int_o^\infty x(t)dt \tag{2.55a}$$

The mean square is also defined by the temporal averaging

$$E[x^2] = \lim_{T\to\infty} \frac{1}{T} \int_o^\infty x^2(t)dt \tag{2.55b}$$

The same concept* of time averaging can also be employed for all other statistical parameters.

Now if the random process is stationary, and the ensemble statistical expectations defined in section (2.3) are identical to the corresponding time averages as computed along one sample record, the

---

* According to the strong law of large number the ergodic process may be stated by the relation (Khas'miniskii, 1980)

$$P\left\{\lim_{T\to\infty} \frac{1}{T} \int_o^T f\big(x(t)\big)dt = \int_{\text{ensemble}} f(x)p(x)dx\right\} = 1 \tag{2.56}$$

where $f\big(x(t)\big)$ is an integrable function with respect to $p(x)$.

random process is said to be ergodic. In this case each realization is regarded as a typical representation of the ensemble, and constitutes a random process by itself.

2.3.6 SPECTRAL DENSITY FUNCTION

The response analysis of linear mechanical systems subjected to periodic excitations is usually determined by resolving the excitation into a series of harmonic components by using Fourier analysis. For non-periodic excitations of limited time duration it is possible to obtain the frequency composition by the Fourier transform

$$X(\omega) = \int_{-\infty}^{\infty} X(t)e^{-i\omega t}\,dt \tag{2.57}$$

where ω is the angular frequency. This transform results in a continuous spectrum and exists if

$$\int_{-\infty}^{\infty} \left| X(t) \right| dt < \infty \tag{2.58}$$

If X(t) is one realization of a stationary random process which has an infinite duration, it will not possess a Fourier transform since the integral of condition (2.58) will not be finite. In order to obtain some information about the frequency composition of a random stationary process, it is convenient to perform the Fourier transform on the autocorrelation function instead of X(t). This choice is based on the fact that the autocorrelation function diminishes as the correlation time increases and the condition

$$\int_{-\infty}^{\infty} \left| R_X(\tau) \right| d\tau < \infty \tag{2.59}$$

is satisfied for random processes which do not include periodic components.

The spectral density function $S_X(\omega)$ of a stationary random process X(t) is given by the Fourier transform of its correlation function

$$S_X(\omega) = \frac{1}{2\pi}\int_{-\infty}^{\infty} R_X(\tau)e^{-i\omega\tau}\,d\tau \tag{2.60}$$

If $S_X(\omega)$ is known, the autocorrelation function can be obtained by the inverse of Fourier transform

$$R_X(\tau) = \int_{-\infty}^{\infty} S_X(\omega)e^{i\omega\tau}\,d\omega \tag{2.61}$$

The pair of Fourier transform (2.60) and (2.61) is known as the Wiener-Khinchin relations, and they afford a transformation from the time domain into the frequency domain and vice versa. It is clear that when $\tau=0$, $R_x(0)$ becomes the mean square, and according to (2.61) it represents the area under the spectral density curve

$$R_x(0) = E[X^2] = \int_{-\infty}^{\infty} S_x(\omega)d\omega \tag{2.62}$$

Introducing the relation

$$e^{-i\omega\tau} = \cos\omega\tau - i\sin\omega\tau \tag{2.63}$$

into (2.60) and recalling the fact that $R_x(\tau)$ is an even function of τ, the product $R_x(\tau)\sin\omega\tau$ is an odd function, and their integral vanishes. Thus the spectral density function can be written in the form

$$S_x(\omega) = \frac{1}{2\pi}\int_{-\infty}^{\infty} R_x(\tau)\cos\omega\tau\, d\tau \tag{2.64}$$

which is a real even function of ω. Furthermore, it can be shown that $S_x(\omega)$ is always positive, a fact which is supported by relation (2.62).

Cross-Spectral Density: The cross-spectral density of a pair of random processes X(t) and Y(t) is given by the Fourier transform of their cross-correlation function $R_{xy}(\tau)$

$$S_{xy}(\omega) = \frac{1}{2\pi}\int_{-\infty}^{\infty} R_{xy}(\tau)e^{-i\omega\tau}\, d\tau \tag{2.65}$$

The inverse Fourier transform gives the cross-correlation function

$$R_{xy}(\tau) = \int_{-\infty}^{\infty} S_{xy}(\omega)e^{i\omega\tau}\, d\omega \tag{2.66}$$

provided that $\int_{-\infty}^{\infty} \left| R_{xy}(\tau) \right| d\tau < \infty$.

2.3.7 CHARACTERISTIC FUNCTIONS AND SEMI-INVARIANTS (CUMULANTS)

The characteristic function is defined as the expected value of $\exp(i\theta x)$

$$F_x(\theta) = E[\exp(i\theta x)]$$

$$= \int_{-\infty}^{\infty} \exp(i\theta x)p(x,t)dx \tag{2.67}$$

where θ is an arbitrary dummy real parameter. Since $p(x,t)$ is non-negative, and the magnitude of $\exp(i\theta x)$ does not exceed unity, it is not difficult to verify that the characteristic function always exists. Relation (2.67) states that the characteristic function is the Fourier transform of the probability density function of the random process $X(t)$. Conversely, the probability density can be defined as the inverse Fourier transform, that is

$$p(x,t) = \frac{1}{2\pi}\int_{-\infty}^{\infty} F_x(\theta)\exp(-i\theta x)d\theta \tag{2.68}$$

The characteristic function provides exhaustive information concerning the random process. For example, one can generate all the moments and semi-invariants of the random process. For this reason it is referred to as the 'moment generating function.' The n-th moment of $X(t)$ is obtained by taking the n-th derivative of $F_x(\theta)$ with respect to θ

$$E[x^n] = \frac{1}{i^n}\left.\frac{d^n F_x(\theta)}{d\theta^n}\right|_{\theta=0} \tag{2.69}$$

Expanding $\exp(i\theta x)$ in a Maclaurin series, the characteristic function may be written in the form

$$\begin{aligned} F_x(\theta) &= 1 + i\theta E[x] + \frac{(i\theta)^2}{2!}E[x^2] + \frac{(i\theta)^3}{3!}E[x^3] + \ldots \\ &= 1 + \sum_{n=1}^{\infty}\frac{(i\theta)^n}{n!}E[x^n] \end{aligned} \tag{2.70}$$

This expansion suggests that $F_x(\theta)$ can be written as an exponential function

$$F_x(\theta) = \exp\left\{\sum_{n=1}^{\infty}\frac{(i\theta)^n}{n!}\lambda_n\right\} \tag{2.71}$$

where λ_n are referred to as semi-invariants (or cumulants) of the random process $X(t)$. λ_n can be determined by taking the n-th derivative of the principal logarithm of $F_x(\theta)$ with respect to θ

$$\lambda_n[x^n] = \frac{1}{i^n}\left.\frac{d^n \ln F_x(\theta)}{d\theta^n}\right|_{\theta=0} \tag{2.72}$$

Definitions (2.69) and (2.72) provide a systematic technique to express the cumulant of order n in terms of moments of the same and less orders. For example, the first four order semi-invariants are:

$$\begin{aligned}
\lambda_1[x] &= E[x] \\
\lambda_2[x^2] &= E[x^2] - (E[x])^2 \\
\lambda_3[x^3] &= E[x^3] - 3E[x]E[x^3] + 2(E[x])^3 \\
\lambda_4[x^4] &= E[x^4] - 3E[x^2]^2 - 4E[x]E[x^3] + 12(E[x])^2\, E[x^2] - 6(E[x])^4
\end{aligned} \tag{2.73}$$

Introducing the definition of the central moment as the moment about the mean, the n-th order central moment is

$$\mu_n = E\big[(x-E[x])^n\big] \tag{2.74}$$

Alternatively, the semi-invariants can be written in terms of central moments:

$$\begin{aligned}
\lambda_1[x] &= 0 \\
\lambda_2[x^2] &= \mu_2 \\
\lambda_3[x^3] &= \mu_3 \\
\lambda_4[x^4] &= \mu_4 - 3\mu_2^2
\end{aligned} \tag{2.75}$$

For a multi-dimensional state vector $\underset{\sim}{X}(t)=\{X_1(t),X_2(t),\ldots,X_n(t)\}^T$, where T denotes transpose, the characteristic function takes the form

$$\begin{aligned}
F_{\underset{\sim}{x}}(\underset{\sim}{\theta}) &= E[\exp(i\theta_1 x_1+i\theta_2 x_2+\ldots+i\theta_n x_n)] \\
&= \int_{-\infty}^{\infty}\!\!\ldots\!\int \exp(i\sum_{j=1}^{n} \theta_j x_j)p(\underset{\sim}{x},t)dx_1 dx_2\ldots dx_n
\end{aligned} \tag{2.76}$$

and the joint probability density p(x,t) is given by the inverse Fourier transformation

$$p(\underset{\sim}{x},t) = \frac{1}{(2\pi)^n}\int_{-\infty}^{\infty}\!\!\ldots\!\int \exp(-i\sum_{j=1}^{n} \theta_j x_j)F_{\underset{\sim}{x}}(\underset{\sim}{\theta})d\theta_1 d\theta_2\ldots d\theta_n \tag{2.77}$$

The moments and joint moments of order $K = \sum_{j=1}^{n} k_j$ are generated by taking the K-th derivatives of $F_{\underset{\sim}{x}}(\underset{\sim}{\theta})$ with respect to the dummy parameters $\underset{\sim}{\theta}$

$$E[x_1^{k_1}\, x_2^{k_2}\ldots x_n^{k_n}] = \left(\frac{1}{i}\right)^K \left.\frac{\partial^K \ln F_{\underset{\sim}{x}}(\underset{\sim}{\theta})}{\partial\theta_1^{k_1}\ldots\partial\theta_n^{k_n}}\right|_{\underset{\sim}{\theta}=\underset{\sim}{0}} \tag{2.78}$$

Expanding the characteristic function in Maclaurin series

$$F_{\underset{\sim}{x}}(\theta) = 1 + \sum_{j=1}^{\infty} \frac{1}{k!} E[(i\theta_1 x_1 + i\theta_2 x_2 + \ldots + i\theta_n x_n)^j] \tag{2.79}$$

Again this expansion can be written in the exponential form

$$\begin{aligned} F_{\underset{\sim}{x}}(\underset{\sim}{\theta}) = \exp\Big\{ & \sum_{j=1}^{n} i\theta_j \lambda_1(x_j) + \frac{(i)^2}{2!} \sum_{\ell=1}^{n} \sum_{j=1}^{n} \theta_j \theta_\ell \lambda_2(x_j x_\ell) \\ & + \frac{(i)^3}{3!} \sum_{q=1}^{n} \sum_{\ell=1}^{n} \sum_{j=1}^{n} \theta_j \theta_\ell \theta_q \lambda_3(x_j x_\ell x_q) \\ & + \frac{(i)^4}{4!} \sum_{r=1}^{n} \sum_{q=1}^{n} \sum_{\ell=1}^{n} \sum_{j=1}^{n} \theta_j \theta_\ell \theta_q \theta_r \lambda_4(x_j x_\ell x_q x_r) + \ldots \Big\} \end{aligned} \tag{2.80}$$

where the semi-invariants of order K are written in the form

$$\lambda_K[x_1^{k_1} x_2^{k_2} \ldots x_n^{k_n}] = \left(\frac{1}{i}\right)^K \frac{\partial^K \ell n\, F_{\underset{\sim}{x}}(\underset{\sim}{\theta})}{\partial\theta_1^{k_1}\, \partial\theta_2^{k_2} \ldots \partial\theta_n^{k_n}} \Bigg|_{\theta = 0} \tag{2.81}$$

Relations (2.78), (2.79) and (2.81) provide relationships between the K-th joint semi-invariants and joint moments of order K and less. The joint semi-invariants give a measure of the correlations among the random variable $x_1, x_2, \ldots, x_k$. For example, the joint semi-invariants of orders up to eight are given by the relations:

$$\lambda_1[X_i] = E[X_i], \quad \lambda_2[X_i X_j] = E[X_i X_j] - E[X_i]E[X_j]$$

$$\lambda_3[X_i X_j X_k] = E[X_i X_j X_k] - \sum^{3} E[X_i]E[X_j X_k] + 2E[X_i]E[X_j]E[X_k]$$

$$\begin{aligned} \lambda_4[X_i X_j X_k X_\ell] = {} & E[X_i X_j X_k X_\ell] - \sum^{4} E[X_i]E[X_j X_k X_\ell] \\ & + 2\sum^{6} E[X_i]E[X_j]E[X_k X_\ell] - \sum^{3} E[X_i X_j]E[X_k X_\ell] \\ & - 6E[X_i]E[X_j]E[X_k]E[X_\ell] \end{aligned}$$

$$\lambda_5[X_iX_jX_kX_\ell X_m] = E[X_iX_jX_kX_\ell X_m] - \sum^{5} E[X_i]E[X_jX_kX_\ell X_m]$$

$$+ 2\sum^{10} E[X_i]E[X_j]E[X_kX_\ell X_m]$$

$$- 6\sum^{10} E[X_i]E[X_j]E[X_k]E[X_\ell X_m] + 2\sum^{15} E[X_i]E[X_jX_k]E[X_\ell X_m]$$

$$- \sum^{10} E[X_iX_j]E[X_kX_\ell X_m] + 24E[X_i]E[X_j]E[X_k]E[X_\ell]E[X_m]$$

$$\lambda_6[X_iX_jX_kX_\ell X_mX_n] = E[X_iX_jX_kX_\ell X_mX_n] - \sum^{6} E[X_i]E[X_jX_kX_\ell X_mX_n]$$

$$+ 2\sum^{15} E[X_i]E[X_j]E[X_kX_\ell X_mX_n] - 6\sum^{20} E[X_i]E[X_j]E[X_k]E[X_\ell X_mX_n]$$

$$+ 24\sum^{15} E[X_i]E[X_j]E[X_k]E[X_\ell]E[X_mX_n] - 6\sum^{45} E[X_i]E[X_j]E[X_kX_\ell]E[X_mX_n]$$

$$- \sum^{15} E[X_iX_j]E[X_kX_\ell X_mX_n] - \sum^{10} E[X_iX_jX_k]E[X_\ell X_mX_n]$$

$$+ 2\sum^{15} E[X_iX_j]E[X_kX_\ell]E[X_mX_n] + 2\sum^{60} E[X_i]E[X_jX_k]E[X_\ell X_mX_n]$$

$$- 120\, E[X_i]E[X_j]E[X_k]E[X_\ell]E[X_m]E[X_n]$$

$$\lambda_7[X_iX_jX_kX_\ell X_mX_nX_p] = E[X_iX_jX_kX_\ell X_mX_nX_p] - \sum^{7} E[X_i]E[X_jX_kX_\ell X_mX_nX_p]$$

$$+ 2\sum^{21} E[X_i]E[X_j]E[X_kX_\ell X_mX_nX_p] - 6\sum^{35} E[X_i]E[X_j]E[X_k]E[X_\ell X_mX_nX_p]$$

$$+ 24\sum^{35} E[X_i]E[X_j]E[X_k]E[X_\ell]E[X_mX_nX_p]$$

$$- 120\sum^{21} E[X_i]E[X_j]E[X_k]E[X_\ell]E[X_m]E[X_nX_p]$$

$$+ 2\sum^{70} E[X_i]E[X_jX_kX_\ell]E[X_mX_nX_p] + 2\sum^{105} E[X_i]E[X_jX_k]E[X_\ell X_mX_nX_p]$$

$$- 6\sum^{105} E[X_i]E[X_jX_k]E[X_\ell X_m]E[X_nX_p]$$

$$- 6 \sum^{210} E[X_i]E[X_j]E[X_k X_\ell]E[X_m X_n X_p]$$

$$+ 24 \sum^{105} E[X_i]E[X_j]E[X_k]E[X_\ell X_m]E[X_n X_p]$$

$$+ 2 \sum^{105} E[X_i X_j]E[X_k X_\ell]E[X_m X_n X_p]$$

$$- \sum^{21} E[X_i X_j]E[X_k X_\ell X_m X_n X_p] - \sum^{35} E[X_i X_j X_k]E[X_\ell X_m X_n X_p]$$

$$+ 720 E[X_i]E[X_j]E[X_k]E[X_\ell]E[X_m]E[X_n]E[X_p]$$

$$\lambda_8[X_i X_j X_k X_\ell X_m X_n X_p X_q] = E[X_i X_j X_k X_\ell X_m X_n X_p X_q]$$

$$- \sum^{8} E[X_i]E[X_j X_k X_\ell X_m X_n X_p X_q]$$

$$+ 2\sum^{28} E[X_i]E[X_j]E[X_k X_\ell X_m X_n X_p X_q]$$

$$- 6\sum^{56} E[X_i]E[X_j]E[X_k]E[X_\ell X_m X_n X_p X_q]$$

$$+ 24\sum^{70} E[X_i]E[X_j]E[X_k]E[X_\ell]E[X_m X_n X_p X_q]$$

$$- 120\sum^{56} E[X_i]E[X_j]E[X_k]E[X_\ell]E[X_m]E[X_n X_p X_q]$$

$$+ 720\sum^{28} E[X_i]E[X_j]E[X_k]E[X_\ell]E[X_m]E[X_n]E[X_p X_q]$$

$$- 120 \sum^{210} E[X_i]E[X_j]E[X_k]E[X_\ell]E[X_m X_n]E[X_p X_q]$$

$$+ 24 \sum^{420} E[X_i]E[X_j]E[X_k X_\ell]E[X_m X_n]E[X_p X_q]$$

$$- 6 \sum^{105} E[X_i X_j]E[X_k X_\ell]E[X_m X_n]E[X_p X_q]$$

$$+ 24 \sum^{560} E[X_i]E[X_j]E[X_k]E[X_\ell X_m]E[X_n X_p X_q]$$

$$- 6 \sum^{840} E[X_i]E[X_j X_k]E[X_\ell X_m]E[X_n X_p X_q]$$

$$- 6 \sum^{280} E[X_i]E[X_j]E[X_k X_\ell X_m]E[X_n X_p X_q]$$

$$+ 2 \sum^{280} E[X_i X_j]E[X_k X_\ell X_m]E[X_n X_p X_q]$$

$$- 6 \sum^{420} E[X_i]E[X_j]E[X_k X_\ell]E[X_m X_n X_p X_q]$$

$$+ 2 \sum^{210} E[X_i X_j]E[X_k X_\ell]E[X_m X_n X_p X_q]$$

$$+ 2 \sum^{280} E[X_i]E[X_j X_k X_\ell]E[X_m X_n X_p X_q]$$

$$- \sum^{35} E[X_i X_j X_k X_\ell]E[X_m X_n X_p X_q]$$

$$+ 2 \sum^{168} E[X_i]E[X_j X_k]E[X_\ell X_m X_n X_p X_q]$$

$$- \sum^{56} E[X_i X_j X_k]E[X_\ell X_m X_n X_p X_q]$$

$$- \sum^{18} E[X_i X_j]E[X_k X_\ell X_m X_n X_p X_q]$$

$$- 5040\, E[X_i]E[X_j]E[X_k]E[X_\ell]E[X_m]E[X_n]E[X_p]E[X_q] \tag{2.82}$$

where numbers over summation signs refer to the number of possible terms generated in the form of the indicated expressions without allowing permutation of indices. For example,

$$\sum^{3} E[X_i]E[X_j X_k] = E[X_i]E[X_j X_k] + E[X_j]E[X_i X_k] + E[X_k]E[X_i X_k]$$

As will be shown in section 2.3.10 semi-invariants of order greater than the second have an important role in measuring the deviation of the process from being Gaussian.

2.3.8 GAUSSIAN PROBABILITY DENSITY FUNCTION

A random process which possesses only the first and second order semi-invariants (cumulants) is known as a Gaussian process. For a single random process the Gaussian probability density is given by the expression

$$p(x) = \frac{1}{\sigma\sqrt{2\pi}} \exp\left\{-\frac{(x-m)^2}{2\sigma^2}\right\} \tag{2.83}$$

where m and σ^2 are the mean and variance of x, respectively.

When $m=0$ and $\sigma=1$ relation (2.83) is reduced into what is known as unit Gaussian probability density function.

It is evident that the Gaussian probability density function is an even function. The n-th moment of unit Gaussian process is

$$E[x^n] = \frac{1}{\sqrt{2\pi}} \int_{-\infty}^{\infty} x^n \exp(-x^2/2)dx$$

$$= \begin{cases} 0 & n \text{ odd} \\ 1.3.5\ldots(n-1) & n \text{ even} \geq 2 \end{cases} \tag{2.84}$$

Several random processes can be described by a normal distribution. For example, the random walk problem can be represented by a normal distribution law when the number of steps is very large and the length of each step is very small. A similar limiting approach is obtained in many practical applications when the number of observations is very large. This view of large numbers is supported by the central limit theorem which states that if $\{X_n\}$ is a sequence of independent random variables, the probability distribution of the sample approaches a Gaussian law as the number of independent random variables is increased without limit regardless of the probability distribution of the random variable. The mathematical justification of this theorem is given by Davenport and Root (1958) and Papoulis (1965).

The joint probability density of two random processes $X_1(t)$ and $X_2(t)$ is

$$P(X_1,X_2) = \frac{1}{2\pi\sqrt{C_{11}C_{22}-C_{12}^2}} \exp\{\frac{-1}{2(C_{11}C_{22}-C_{12}^2)} (C_{22}(X_1-m_1)^2$$

$$- 2C_{12}(X_1-m_1)(X_2-m_2) + C_{11}(X_2-m_2)^2)\} \tag{2.85}$$

where the covariance $C_{jk} = E[(X_j-m_j)(X_k-m_k)]$

and $m_j = E[X_j]$

For multi-dimensional random processes $\underset{\sim}{x}(t)=\{x_1(t),x_2(t),\ldots,x_n(t)\}^T$ the normal joint probability density is given by the relation

$$p(\underset{\sim}{X}(t)) = \frac{(2\pi)^{-n/2}}{|\Delta|^{1/2}} \exp\{-\frac{1}{2|\Delta|} \sum_{j=1}^{n} \sum_{k=1}^{n} |\Delta|_{jk} (X_j-m_j)(X_k-m_k)\} \tag{2.86}$$

where $|\Delta|_{jk}$ is the co-factor of the covariance element C_{jk} in the determinant $|\Delta|$ of the covariance matrix

$$\Delta = \begin{bmatrix} C_{11} & C_{12} \cdots\cdots\cdots\cdots C_{1n} \\ C_{21} & C_{22} \cdots\cdots\cdots\cdots C_{2n} \\ \vdots & \vdots \qquad\qquad\qquad \vdots \\ C_{n1} & C_{n2} \cdots\cdots\cdots\cdots C_{nn} \end{bmatrix} \tag{2.87}$$

The characteristic function of Gaussian processes can be established by taking the inverse Fourier transform of the Gaussian probability densities (2.83) and (2.86) which give, respectively

$$F_x(\theta) = \exp(im\theta - \frac{\sigma^2}{2}\theta^2) \tag{2.88}$$

$$F_{\underset{\sim}{x}}(\underset{\sim}{\theta}) = \exp\{i\sum_{j=1}^{n} m_j\,\theta_j - \frac{1}{2}\sum_{j=1}^{n}\sum_{k=1}^{n} C_{jk}\,\theta_j\,\theta_k\} \tag{2.89}$$

Comparing (2.88) and (2.89) with the characteristic functions of general random processes given by (2.71) and (2.80), respectively, reveals that all semi-invariants of order greater than two must vanish identically for Gaussian distributed processes. This property is very useful in establishing Gaussian and non-Gaussian closure schemes which will be discussed in section (5.4). Another important property is that any linear transformation of a normal process would yield another Gaussian process. For example, when dynamic systems are modeled by a set of linear differential equations with constant coefficients (time-invariant systems) and include non-homogeneous Gaussian random forcing terms, the response will be jointly Gaussian distributed as well. This latter property will not be valid in problems pertaining to random parametric vibration.

2.3.9 TRANSFORMATION OF THE PROBABILITY DENSITY FUNCTION (THE JACOBIAN)

In certain situations it is required to transform the probability density function from one set of state variables into another coordinate set. A direct relationship may be obtained when the two coordinate sets are related by one-to-one mapping. Let $\underset{\sim}{y}$ be the new coordinate vector which is related to the original random variable vector $\underset{\sim}{x}$ through single-valued continuous functions

$$\underset{\sim}{y} = \underset{\sim}{g}(\underset{\sim}{x}) \tag{2.90}$$

such that the inverse relationship is also expressed by single-valued continuous functions

$$\underset{\sim}{x} = \underset{\sim}{f}(\underset{\sim}{y}) \tag{2.91}$$

The transformation requires that the probability that a sample point takes values in the domain D_x of $\underset{\sim}{x}$ equals the probability that its mapping falls in the domain D_y of $\underset{\sim}{y}$. Furthermore, the joint probability density function of $\underset{\sim}{x}$ is related to the joint probability of $\underset{\sim}{y}$ by the relationship

$$\int_{D_x}\!\!\cdots\!\int p_1(x_1,\ldots,x_n)dx_1\ldots dx_n = \int_{D_y}\!\!\cdots\!\int p_2(y_1,\ldots,y_n)dy_1\ldots dy_n$$

$$= \int_{D_y}\!\!\cdots\!\int p_1(x_1{=}f_1,\ldots,x_n{=}f_n)\,\big|\,J\,\big|\,dy_1\ldots dy_n \tag{2.92}$$

where the Jacobian J is introduced as a consequence of the well known result of mapping.

From relation (2.92) one can write

$$p_2(\underset{\sim}{y}) = p_1(\underset{\sim}{x})\,\big|\big|\,J\,\big|\big|_{\underset{\sim}{x}=\underset{\sim}{f}(\underset{\sim}{y})} \tag{2.93}$$

where the Jacobian of the transformation is expressed by the absolute value of the determinant

$$\big|\,J\,\big| = \frac{\partial(x_1=f_1,\ldots,x_n=f_n)}{\partial(y_1,\ldots,\partial y_n)} = \begin{vmatrix} \dfrac{\partial f_1}{\partial y_1} & \cdots\cdots\cdots\cdots & \dfrac{\partial f_n}{\partial y_n} \\ \dfrac{\partial f_1}{\partial y_2} & \cdots\cdots\cdots\cdots & \dfrac{\partial f_n}{\partial y_2} \\ \vdots & & \vdots \\ \dfrac{\partial f_1}{\partial y_n} & \cdots\cdots\cdots\cdots & \dfrac{\partial f_n}{\partial y_n} \end{vmatrix} \tag{2.94}$$

In equation (2.93) the modulus $\big|\big|\,J\,\big|\big|$ is introduced to ensure that the joint probability density is positive.

This transformation is very useful in problems treated by the stochastic averaging and other techniques where the state vector $\underset{\sim}{x}$ is transformed into the polar coordinate state vector and vice versa.

2.3.10 NON-GAUSSIAN PROBABILITY DENSITY FUNCTION

Under non-linear transformation of Gaussian distributed processes the transformed processes are not Gaussian. This includes the response distribution of dynamic systems which involve random parametric excitations or non-linear terms. A proper estimation of the response probability density of such systems is not an easy task. The difficulties involved in dealing with non-Gaussian analysis are analogous, to a certain extent, to (or even more difficult than) those encountered in treating the deterministic response of non-linear vibrating systems. Differential equations of response moments (which will be treated in chapter 5) usually contain higher order moments. This infinite hierarchy must be truncated by an ad-hoc closure scheme. A proper treatment of the infinite hierarchy problem is to develop a non-Gaussian closure scheme to truncate the system differential equations of the response statistical functions to an acceptable degree of approximation.

Although the concept of developing an asymptotic expansion of a non-Gaussian probability density was established over forty years ago (Cramér, 1946) in the form of Edgeworth or Gram-Charlier expansions, it has not been employed until very recently for simple non-linear control and dynamic systems. Kuznetsov, et al. (1960), Stratonovich (1963), and Bover (1978a,b) developed Edgeworth expansions for single and multi-dimensional processes, in terms of quasi-moments which will be defined in this section. Sperling (1979) developed a set of linear differential equations for the response quasi-moments of discrete non-linear systems. Dashevskii (1967), Dashevskii and Liptser (1967), Nakamizo (1970) and Assaf and Zirkle (1976) developed truncated expansions in terms of the semi-invariants. It was indicated that the expansion is considered successful if it converges as the order of semi-invariant increases. In this section the Gram-Charlier and Edgeworth expansions for one- and multi-dimensional processes will be defined.

The Gram-Charlier Asymptotic Expansion: Introducing the standardized variable

$$z = \frac{x-m}{\sigma} \tag{2.95}$$

A non-Gaussian probability density function may be expressed by the Gram-Charlier expansion

$$p^*(z) = \sum_{n=0}^{\infty} \frac{C_n}{n!} \frac{d^n p(z)}{dz^n} \tag{2.96}$$

where p(z) is the Gaussian probability density, and C_n are constant coefficients to be determined.

Introducing the relationship between the unit Gaussian density function and the Chebychev-Hermite polynomials, expansion (2.96) takes the form

$$p^*(z) = \frac{1}{\sigma\sqrt{2\pi}} \exp(-z^2/2)\{1 + \sum_{n=1}^{\infty} \frac{C_n}{n!} H_n(z)\} \tag{2.97}$$

where the Chebychev-Hermite polynomials of order n are defined by

$$H_n(z) = (-1^n)\exp(z^2/2) \frac{d^n}{dz^n} \exp(-z^2/2) \tag{2.98}$$

with $H_0(z)=1$, the first six polynomials are:

$$\begin{aligned} H_1(z) &= z & H_4(z) &= z^4 - 6z^2 + 3 \\ H_2(z) &= z^2 - 1 & H_5(z) &= z^5 - 10z^3 + 15z \\ H_3(z) &= z^3 - 3z & H_6(z) &= z^6 - 15z^4 + 45z^2 - 15 \end{aligned} \tag{2.99}$$

These polynomials possess the following three main properties:

i. The differentiation law

$$\frac{d}{dz} H_n(z) = nH_{n-1}(z) \tag{2.100}$$

ii. The recurrence relation

$$H_{n+1}(z) = zH_n(z) - nH_{n-1}(z) \tag{2.101}$$

iii. The orthogonality relation with respect to the unit Gaussian density function

$$\frac{1}{2\pi} \int_{-\infty}^{\infty} H_m(z)H_n(z)\exp(-z^2/2)dz = \begin{cases} n! & m = n \\ 0 & m \neq n \end{cases} \tag{2.102}$$

The third property can be used to determine the constants C_n. Multiplying both sides of (2.97) by $H_j(z)$ and integrating over the whole space yields

$$C_j = \int_{-\infty}^{\infty} H_j(z)p^*(z)dz = E[H_j(z)] \tag{2.103}$$

Introducing (2.99) we obtain the corresponding values of C_n:

$$C_0 = 1 \qquad C_4 = \frac{\mu_4}{\sigma^4} - 3$$

$$C_1 = C_2 = 0 \qquad C_5 = \frac{\mu_5}{\sigma^5} - 10\frac{\mu_3}{\sigma^3} \tag{2.104}$$

$$C_3 = \frac{\mu_3}{\sigma^3} \qquad C_6 = \frac{\mu_6}{\sigma^6} - 15\frac{\mu_4}{\sigma^4} + 30$$

where $\mu_j = E[(x-m)^j]$ is the j-th central moment, and $m = E[x]$.

Quasi-Moment Functions: It is believed that these statistical functions were first introduced by Kuznetsov, et al., (1960). According to Stratonovich (1963), these functions occupy an intermediate role between the semi-invariant functions and moment functions. The quasi-moments $b_k(z)$ are related to the semi-invariants through the characteristic function

$$\exp\left\{\sum_{k=3}^{\infty} \frac{(i\theta)^k}{k!} \lambda_k(z)\right\} = 1 + \sum_{k=3}^{\infty} \frac{(i\theta)^k}{k!} b_k(z) \tag{2.105}$$

and the complete expression of the characteristic function in terms of the quasi-moments is

$$F_z(\theta) = \exp\left(im\theta - \frac{1}{2}(i\sigma\theta)^2\right)\left\{1 + \sum_{k=3}^{\infty} \frac{(i\theta)^k}{k!} b_k(z)\right\} \tag{2.106}$$

The probability density can be expressed by the Gram-Charlier expansion in terms of the quasi-moments

$$p^*(z) = \frac{1}{\sigma\sqrt{2\pi}} \exp(-z^2/2)\left\{1 + \sum_{n=1}^{\infty} \frac{1}{n!} \frac{b_n(z)}{\sigma^2} H_n(z)\right\} \tag{2.107}$$

Comparing (2.107) and (2.97) one can establish the relationship between the coefficients C_n and the quasi-moments

$$C_n = b_n/\sigma^n \tag{2.108}$$

Using equation (2.103), the quasi-moments are expressed by the formula

$$b_n(z) = \sigma^n E[H_n(z)] \tag{2.109}$$

Bover (1978a) derived the first ten quasi-moments of a single-dimension process. The first eight quasi-moments are listed below:

$$b_1 = b_2 = 0 \qquad b_3 = \sigma^3 E[z^3]$$

$$b_4 = \sigma^4\{E[z^4] - 3E[z^2]^2\} \qquad b_5 = \sigma^5\{E[z^5] - 10E[z^2]E[z^3]\}$$

$$b_6 = \sigma^6\{E[z^6] - 15E[z^2]E[z^4] + 30E[z^2]^3\} \tag{2.110}$$

$$b_7 = \sigma^7\{E[z^7] - 21E[z^2]E[z^5] + 105E[z^2]^2 E[z^3]\}$$

$$b_8 = \sigma^8\{E[z^8] - 28E[z^2]E[z^6] + 210E[z^2]^2 E[z^4] - 315E[z^2]^4\}$$

The constants C_n are related to the semi-invariants of one-dimensional process by the relation (Dashevskii and Liptser, 1967; Assaf and Zirkle, 1976)

$$C_n = \lambda_n/\sigma^n \tag{2.111}$$

The Gram-Charlier expansions given by (2.97) and (2.107) are usually truncated up to a limited number of terms which should guarantee asymptotic convergence. If convergence cannot be achieved, the Edgeworth expansion offers a good degree of accuracy.

The Edgeworth Expansion: This expansion is a rearrangement of the Gram-Charlier expansion so that the accuracy increases with the order of the terms. Assaf and Zirkle (1976) have indicated that the Edge-worth expansion yields sufficiently accurate representations of the probability density function if only the first four terms of the expansion are retained. The expansion is usually written in terms of the semi-invariants λ_j, and for the one-dimensional case it takes the form

$$p^*(z) = p(z)\{1 + \frac{1}{3!}\frac{\lambda_3}{\sigma_x^3} H_3(z) + \frac{1}{4!}\frac{\lambda_4}{\sigma_x^4} H_4(z) + \frac{10}{6!}\frac{\lambda_3^2}{\sigma_x^6} H_6(z) + \ldots\} \tag{2.112}$$

Multi-Dimensional Distributions: The multi-dimensional density function $p^*(\underset{\sim}{z})$ may be written in the Gram-Charlier expansion

$$p^*(\underset{\sim}{z}) = \sum_{k_1=0}^{\infty} \sum_{k_n=0}^{\infty} C_{k_1,k_2,\ldots,k_n} \frac{\partial^K p(\underset{\sim}{z})}{\partial z_1^{k_1} \ldots \partial z_n^{k_n}} \tag{2.113}$$

Introducing the change of variables

$$y_i = x_i - E[x_i] = \sigma_i z_i \tag{2.114}$$

the Chebychev-Hermite polynomials of multi-dimensional parameters may be written in the form

$$H_{k_1,k_2,\ldots,k_n}(\underset{\sim}{y}) = (-1)^K \exp\{\frac{1}{2} \sum_{j,\ell}^{n} a_{j\ell} y_j y_\ell\} \cdot \frac{\partial^K}{\partial y_1^{k_1} \ldots \partial y_n^{k_n}} \exp\{-\frac{1}{2} \sum_{j,\ell}^{n} a_{j\ell} y_j y_\ell\} \tag{2.115}$$

where $K = \sum_{i=1}^{n} k_i$ and the coefficients $a_{j\ell}$ are the elements of the matrix $\underline{A} = [E[\underset{\sim\sim}{yy^T}]]^{-1}$, which is the inverse of the covariance matrix of the central-moment variables y.

The multi-dimensional probability density function (2.113) may be expressed in terms of the quasi-moments $b_n(t)$ as (Kuznetsov, et al., 1967; Nakamizo, 1970; Ledwich, 1974; and Bover, 1978a)

$$p^*(\underset{\sim}{y}) = \frac{\Delta^{1/2}}{(2\pi)^{n/2}} \exp(-\frac{1}{2} \underset{\sim}{y}^T \underset{\sim}{Ay}) \sum_{L=0}^{\infty} \sum_{k_1+\ldots k_n=L} \{\frac{b_{k_1,\ldots,k_n}(t)}{k_1!\, k_2! \ldots k_n!} H_{k_1,\ldots k_n}(\underset{\sim}{y})\} \tag{2.116}$$

where $\Delta = \mathrm{Det} \left| a_{j\ell} \right|$.

In order to establish the orthogonality property, we introduce the adjoint Hermite polynomials $G_K(\underset{\sim}{y})$

$$G_{k_1,\ldots,k_n}(\underset{\sim}{y}) = (-1)^K \exp\{\frac{1}{2} \Psi(\underset{\sim}{u})\} \frac{\partial^K \exp\{-\frac{1}{2} \Psi(\underset{\sim}{u})\}}{\partial u_1^{k_1} \ldots \partial u_n^{k_n}} \tag{2.117}$$

where $u_j = \frac{1}{2} \frac{\partial \Phi(\underset{\sim}{y})}{\partial y_j} = \sum_{\ell=1}^{n} a_{j\ell} y_\ell$

$$\Phi(\underset{\sim}{y}) = \sum_{j,\ell}^{n} a_{j\ell} y_j y_\ell, \text{ and } \Psi(\underset{\sim}{u}) = \sum_{j,\ell}^{n} b_{j\ell} u_j u_\ell$$

$\Psi(\underset{\sim}{u})$ is the adjoint of $\Phi(\underset{\sim}{y})$ from which the relationship between $a_{j\ell}$ and $b_{j\ell}$ can be established.

Now the polynomials $H_{k_1,\ldots,k_n}(\underset{\sim}{y})$ and $G_{\ell_1,\ldots\ell_n}(\underset{\sim}{y})$ are orthogonal with respect to the Gaussian weighted function $\exp(-\frac{1}{2}\underset{\sim}{y}^T \underset{\sim}{A}\underset{\sim}{y})$

$$\int_{-\infty}^{\infty}\cdots\int \exp(-\frac{1}{2}\underset{\sim}{y}^T \underset{\sim}{A}\underset{\sim}{y})H_{k_1,k_2,\ldots,k_n}(\underset{\sim}{y})G_{\ell_1,\ell_2,\ldots,\ell_n}(\underset{\sim}{y})dy_1 dy_2\ldots dy_n$$

$$=\frac{(2\pi)^{n/2}}{\Delta^{1/2}}\prod_{i=1}^{n}\delta_{\ell_i}^{k_i}\,\ell_i! \qquad (2.118)$$

where $\delta_{\ell_i}^{k_i}$ is the Kronecker delta.

The joint quasi-moments $b_k(t)$ can be derived by multiplying both sides of (2.116) by $G_{k_1,k_2,\ldots k_n}(\underset{\sim}{y})$, integrating over the entire space $-\infty\leq \underset{\sim}{y}\leq\infty$, and using (2.118) yields

$$b_{k_1,k_2\ldots,k_n}(t) = E[G_{k_1,k_2\ldots,k_n}(\underset{\sim}{y})] \qquad (2.119)$$

Bover (1978a) derived the sixth order three-dimensional quasi-moments.

The Gram-Charlier expansion (2.116) can be rearranged to give the Edgeworth expansion in terms of the semi-invariants (Nakamizo, 1970)

$$p^*(\underset{\sim}{x}) = p(x) - \frac{1}{3!}\sum_{k,\ell,m}\lambda_{k,\ell,m}\frac{\partial^3 p(\underset{\sim}{x})}{\partial x_k\,\partial x_\ell\,\partial x_m} + \frac{1}{4!}\sum_{k,\ell,m,q}\lambda_{k,\ell,m,q}$$

$$\cdot\frac{\partial^4 p(\underset{\sim}{x})}{\partial x_k\,\partial x_\ell\,\partial x_m\,\partial x_q} + \ldots \qquad (2.120)$$

The two-dimensional probability density is of common use in control and dynamics problems. Assaf and Zirkle (1976) gave the Edgeworth expansion for two-dimensional case in the form

$$p^*(y_1,y_2) = p(y_1,y_2)\{\sum_{k=0}^{\infty}\frac{\rho^k}{k!}H_k(y_1/\sigma_1)H_k(y_2/\sigma_2)$$

$$+\sum_{j+\ell=3}\frac{1}{j!\ell!}\frac{\lambda_{j\ell}}{\sigma_1^j\sigma_2^\ell}\sum_{k=0}^{\infty}\frac{\rho^k}{k!}H_{k+j}(y_1/\sigma_1)H_{k+\ell}(y_2/\sigma_2)$$

$$+ \sum_{j+\ell=4} \frac{1}{j!\ell!} \frac{\lambda_{j\ell}}{\sigma_1^j \sigma_2^\ell} \sum_{k=0}^{\infty} \frac{\rho^k}{k!} H_{k+j}(y_1/\sigma_1) H_{k+\ell}(y_2/\sigma_2)$$

$$+ \frac{1}{2} \sum_{\substack{j+\ell=3 \\ r+s=4}} \frac{1}{j!\ell!r!s!} \frac{\lambda_{j\ell}}{\sigma_1^j \sigma_2^\ell} \frac{\lambda_{rs}}{\sigma_1^r \sigma_2^s} \sum_{k=0}^{\infty} \frac{\rho^k}{k!}$$

$$H_{k+j+r}(y_1/\sigma_1) H_{k+\ell+s}(y_2/\sigma_2)\} \tag{2.121}$$

where $\rho = \dfrac{\lambda_{11}}{\sigma_1\sigma_2} = \dfrac{E[y_1 y_2]}{\sigma_1\sigma_2}$ is the correlation coefficient of y_1 and y_2, and $p(y_1, y_2)$ is the two-dimensional normal density function for the processes y_1 and y_2 taken as independent

$$p(y_1, y_2) = \frac{1}{2\pi\sigma_1\sigma_2} \exp\{-\frac{1}{2} [(y_1/\sigma_1)^2 + (y_2/\sigma_2)^2]\} \tag{2.122}$$

CHAPTER 3

Stochastic Calculus Operations

3.1 INTRODUCTION

The theory of random vibration involves various types of mathematical operations on ensembles. These operations require a number of calculus rules which lead to information about the system response if the excitation is statistically well described. Operations such as differentiation and integration must exist in some stochastic sense. For example, differentiation is no longer simply defined as a limiting process for a single member, but as a limit which must hold for almost all samples with probability one. The concept of stochastic convergence can be used to examine the existence of operations such as continuity, differentiation, and integration.

This chapter begins with a brief review of the ordinary notations of limits and convergence encountered in advanced calculus. The discussion will be extended to cover the modes of stochastic convergence, mean square derivative, Riemann-Stieltjes and Lebesgue integrals.

3.2 BASIC NOTATIONS OF ADVANCED CALCULUS

A simple set S is said to be 'bounded above' if there is a number A such that

$$x \leq A \qquad\qquad x \subset S \tag{3.1}$$

On the other hand, S is 'bounded below' if there is a number B which satisfies the condition

$$x \geq B \qquad\qquad x \subset S \tag{3.2}$$

Suppose S is a set bounded above, and let A_0 be an upper bound of S which is smaller than or equal to any other bound, then A_0 is called the 'least upper bound' (LUB) or supremum (sup) of S and is expressed by the notation

$$A_o = \sup_S x \tag{3.3}$$

Similarly, if B_0 is the lower bound of S and is greater than or equal to any other lower bound, then B_0 is called the 'greatest lower bound' (GLB) or infimum (inf) of S and is given by the relation

$$B_0 = \inf_S x \tag{3.4}$$

Now let S be a set of cluster points of x_n, where x_n belongs to a sequence $\{x_n\}$ which is bounded above, then S is bounded above and has a supremum Λ given by the limit

$$\Lambda = \limsup_{n\to\infty} x_n \tag{3.5}$$

If x_n is bounded below, then its set S of cluster points is bounded below and has infimum which is called the limit of inferior of $\{x_n\}$ denoted by λ, i.e.,

$$\lambda = \liminf_{n\to\infty} x_n \tag{3.6}$$

Consider a sequence of real variables $\{x_n(t)\}$, $t \in [a,b]$, i.e., each x_i is a function of the independent variable t in the interval $a \leq t \leq b$. For a given $t=t_0$ the sequence $\{x_n(t_0)\}$ becomes a sequence of numbers and if

$$\lim_{n\to\infty} x_n(t_0) = x(t_0) \tag{3.7}$$

then the sequence is said to converge to $x(t_0)$ at $t=t_0$. This convergence is referred to as pointwise convergence.

The sequence $\{x_n(t)\}$ is said to have uniform convergence if for any $\varepsilon>0$, there exists a positive integer $N=N(\varepsilon)$ such that

$$\left| x_n(t) - x(t) \right| < \varepsilon \tag{3.8}$$

for $n \geq N(\varepsilon)$ and $t \in [a,b]$.

If the sequence is uniformly convergent, it is pointwise convergent. However, the converse is not true and a sequence may converge at every point in [a,b] without being uniformly convergent in [a,b].

3.3 MODES OF STOCHASTIC CONVERGENCE

The concept of stochastic convergence is completely different from the ordinary convergence described in section (3.2). There are four modes of stochastic convergence which serve to establish the stochastic stability of dynamic and control systems. These modes are:

i. Convergence in Distribution: Let $F_n(x)=P(X_n<x)$ be the distribution function of a random variable X_n. The sequence of random

variables $\{X_n\}$ is said to converge in distribution if $F_n(x)$ converges pointwise to F(x), i.e., if

$$\lim_{n\to\infty} F_n(x) = F(x) \tag{3.9}$$

where F(x) is a continuous distribution function.

ii. <u>Convergence in Probability</u>: For every $\varepsilon>0$, the sequence $\{X_n\}$ converges in probability to the random variable x if

$$\lim_{n\to\infty} P(|X_n - x| \geq \varepsilon) = 0 \tag{3.10}$$

Convergence in probability is known as stochastic convergence, and $\{X_n\}$ is said to converge stochastically to x.

iii. <u>Convergence with Probability One</u> (or <u>Almost Sure Convergence</u>):

The sequence $\{X_n\}$ of random variables is said to converge with probability one as $n\to\infty$ if

$$P(\lim_{n\to\infty} X_n = x) = 1 \tag{3.11}$$

where the limit x is unique. The terms almost sure, almost certain, or strong convergence are synonymous.

iv. <u>Convergence in Mean Square</u>: The sequence $\{X_n\}$ of random variables is said to converge in mean square to the random variable x if

$$\lim_{n\to\infty} E[|X_n - x|^2] = 0 \tag{3.12}$$

This definition is usually written in the form

$$\text{l.i.m. } X_n = x \tag{3.13}$$

which reads limit-in-the-mean.

According to (3.13) the random function X(t) is said to be continuous in mean square at time $t \in T$, and (3.13) can be written in the form

$$\underset{\Delta t\to\infty}{\text{l.i.m.}}\ X(t+\Delta t) = X(t) \tag{3.13a}$$

Convergence in mean square requires finite mean square, $E[X_n^2] < \infty$. The random variable is said to be of second order if its mean square exists.

Convergence in mean square implies convergence in probability. This can be verified directly by using the Bienayme-Chebychev inequality (Papoulis, 1965)

$$P\{\,|X_n - x| \geq \varepsilon\} < E[\,|X_n - x|^2]/\varepsilon^2 \tag{3.14}$$

Relations among the four modes are outlined in most references of probability theory and stochastic differential equations (Loeve, 1963; Lin, 1967; Thomas, 1971; and Soong, 1973). For example, almost sure convergence implies convergence in probability. However, almost sure convergence neither implies nor is implied by mean square convergence.

Convergence in mean square is very essential in establishing mean square calculus. Other modes are useful for developing stochastic stability modes.

3.4 STOCHASTIC LIMIT THEOREMS

i. The Weak Law of Large Numbers: The law of large numbers is a direct consequence of sums of independent random variables. For the random sequence $\{X_n\}$ we define the new sequence $\{Y_n\}$

$$Y_n = \sum_{i=1}^{n} X_i \qquad n = 1,2,\ldots \tag{3.15}$$

and $E[Y_n] = \sum_{i=1}^{n} E[X_n]$

The sequences $\{X_n - E[X_n]\}$ and $\{Y_n - E[Y_n]\}$ converge in probability to zero if

$$\lim_{n\to\infty} P\{\,|X_n - E[X_n]| \geq \varepsilon\} = 0 \tag{3.16}$$

$$\lim_{n\to\infty} P\{\tfrac{1}{n}|Y_n - E[Y_n]| \geq \varepsilon\} = 0 \tag{3.17}$$

respectively, where $\varepsilon > 0$.

ii. The Strong Law of Large Numbers: Relations (3.16) and (3.17) may be written in terms of almost sure convergence as

$$P\{\lim_{n\to\infty} (X_n = E[X_n])\} = 1 \tag{3.18}$$

$$P\{\lim_{n\to\infty} \frac{1}{n} (Y_n = E[Y_n])\} = 1 \tag{3.19}$$

It may be noticed that the ergodicity relation (2.56) is a special form of the strong law of large numbers.

iii. The Central Limit Theorem: Introducing the standardized (or normalized random variable

$$Z_n = (Y_n - E[Y_n])/\sigma_y \tag{3.20}$$

where $\sigma_y^2 = E[(Y_n - E(Y_n))^2]$

The random variable Z_n converges in distribution to a unit normal random variable. If X_i are continuous random variables, the density function of Z_n does converge to unit normal density.

iv. The Law of the Iterated Logarithm: Let the variance σ_x^2 of X_n exist, then with probability one the following two limits define the law of the iterated logarithm (Arnold, 1974)

$$\limsup_{n\to\infty} \frac{Y_n - nE[X_n]}{\sqrt{2n \log \log n}} = +\left|\sigma_x\right| \tag{3.21}$$

and

$$\liminf_{n\to\infty} \frac{Y_n - nE[X_n]}{\sqrt{2n \log \log n}} = -\left|\sigma_x\right| \tag{3.22}$$

where log denotes the natural logarithm.

3.5 MEAN SQUARE DERIVATIVE

The existence of a random process derivative at a point depends on whether the process is continuous with respect to its parameter. This will require the continuity of the random process $X(t)$, $t \subset T$, such that its mean is a continuous function of t and its covariance $C_x(t,s)$ is jointly continuous in t and s.

The random process $X(t)$ is said to be continuous in mean square if

$$\underset{\Delta t\to 0}{\text{l.i.m.}}\ X(t + \Delta t) = X(t) \tag{3.23}$$

This condition requires that the mean square of $X(t)$ be finite over the time period covering the sample record. A resulting consequence of (3.23) is that the correlation function

$$R_x(\tau) = E[X(t)X(t+\tau)] \tag{3.24}$$

approaches a finite limit as τ approaches zero. Obviously, the necessary and sufficient condition for the mean square continuity of $X(t)$ at time t is that its correlation function $R_x(\tau)$ be finite and continuous over the entire time domain τ.

The mean square derivative of the random process $X(t)$ is defined as

$$\underset{\Delta t \to 0}{\text{l.i.m.}} \frac{X(t + \Delta t) - X(t)}{\Delta t} = \frac{dX(t)}{dt} \tag{3.25}$$

According to the mean square convergence criterion, a sufficient condition for mean square differentiability of X(t) is to satisfy the condition

$$\lim_{\substack{\Delta t \to 0 \\ \Delta t' \to 0}} E\left[\left| \frac{X(t + \Delta t) - X(t)}{\Delta t} - \frac{X(t + \Delta t') - X(t)}{\Delta t'} \right|^2 \right] = 0 \tag{3.26}$$

This convergence will be guaranteed if, and only if, the following limit exists

$$\lim_{\substack{\Delta t \to 0 \\ \Delta t' \to 0}} E\left[\frac{X(t + \Delta t) - X(t)}{\Delta t} \cdot \frac{X(t + \Delta t') - X(t)}{\Delta t'}\right]$$

$$= \lim_{\substack{\Delta t \to 0 \\ \Delta t' \to 0}} \left\{\frac{1}{\Delta t} \left(\frac{R_x(\Delta t' - \Delta t) - R_x(\Delta t)}{\Delta t'} - \frac{R_x(\Delta t') - R_x(0)}{\Delta t'}\right)\right\}$$

$$= \frac{\partial^2 R_x(t-s)}{\partial t \partial s} < \infty \tag{3.27}$$

Thus the necessary and sufficient condition for X(t) to be differentiable in mean square is that the autocorrelation function has a finite continuous mixed second derivative. When $\Delta t = \Delta t'$ condition (3.27) takes the form

$$\lim_{\Delta t \to 0} E\left[\left| \frac{X(t + \Delta t) - X(t)}{\Delta t} \right|^2 \right] < \infty \tag{3.28}$$

If the limit of the mean square of the derivative X(t) is finite, the limit of the mean value will also be finite, i.e.,

$$\lim_{\Delta t \to 0} E\left[\frac{X(t + \Delta t) - X(t)}{\Delta t}\right] = E[\dot{X}(t)]$$

$$= \frac{dE[X(t)]}{dt} < \infty \tag{3.29}$$

where the commutative property between the expectation and the derivative is applied.

3.6 RIEMANN AND RIEMANN-STIELTJES INTEGRALS

In section (2.3.5) it has been indicated that the temporal average of ergodic processes is defined by the integral

$$E[X(t)] = \lim_{T\to\infty} \frac{1}{T}\int_0^T X(t)dt \qquad 0\leq t\leq T \tag{3.30}$$

This integral may be defined, over the interval $a<t<b$, as the area under the continuous curve $X(t)$ as shown in fig. (3.1). This can be interpreted from the elementary rules of calculus as

$$\int_a^b X(t)\, dt \cong \sum_{i=1}^{n} X(\bar{t}_i)\,(t_i - t_{i-1}) \tag{3.31}$$

where $\bar{t}_i$ is the mid-point in the i-th interval and

$$a = t_o < t_1 < t_2 <\ldots< t_n = b \tag{3.32}$$

The sum on the right side of (3.31) will approach the exact value of the integration on the left-hand side if the function $X(t)$ is well behaved and the time intervals $t_i - t_{i-1}$ are very small. For sufficiently well behaved $X(t)$ a unique limit exists which is referred to as the Riemann definite integral

$$\int_a^b X(t)\, dt = \lim_{\substack{\max \Delta t_i\to 0\\ \text{or } n\to\infty}} \sum_{i=1}^{n} X(\bar{t}_i)\,(t_i - t_{i-1}) \tag{3.33}$$

where $\Delta t_i = t_i - t_{i-1}$.

If $X(t)$ is a random process, the value of the integral (3.31) will be a random variable. Thus the convergence of (3.33) must be interpreted as a convergence of a sequence of random variables. The limit of (3.33) will be a mean square Riemann integral if

$$\underset{n\to\infty}{\text{l.i.m.}} \sum_{i=1}^{n} X(\bar{t}_i)\,(t_i - t_{i-1}) = \int_a^b X(t)\, dt \tag{3.34}$$

This convergence exists if the following limit is finite

$$\lim_{\substack{n\to\infty\\ m\to\infty}} E\Big[\Big\{\sum_{i=1}^{n} X(\bar{t}_i)\,(t_i - t_{i-1})\Big\}\Big\{\sum_{j=1}^{m} X(\bar{t}_j)\,(t_j - t_{j-1})\Big\}\Big] < \infty \tag{3.35}$$

Upon interchanging the order of summation and expectation, the limit (3.35) becomes the double Riemann integral of the autocorrelation function

$$\left| \int_a^b \int_a^b R_x(t-s)\, dt\, ds \right| < \infty \tag{3.36}$$

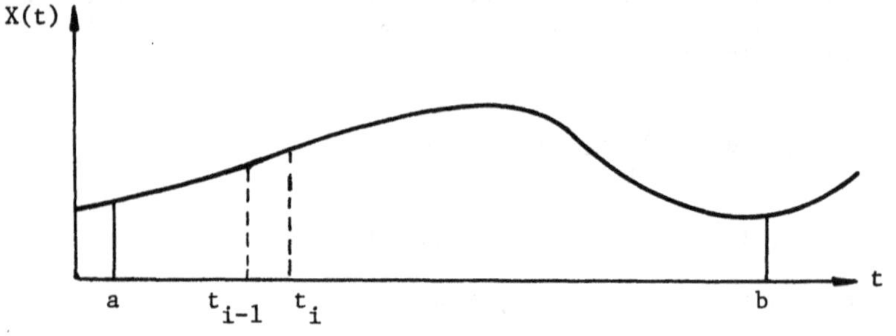

FIG. (3.1) Concept of Riemann integral

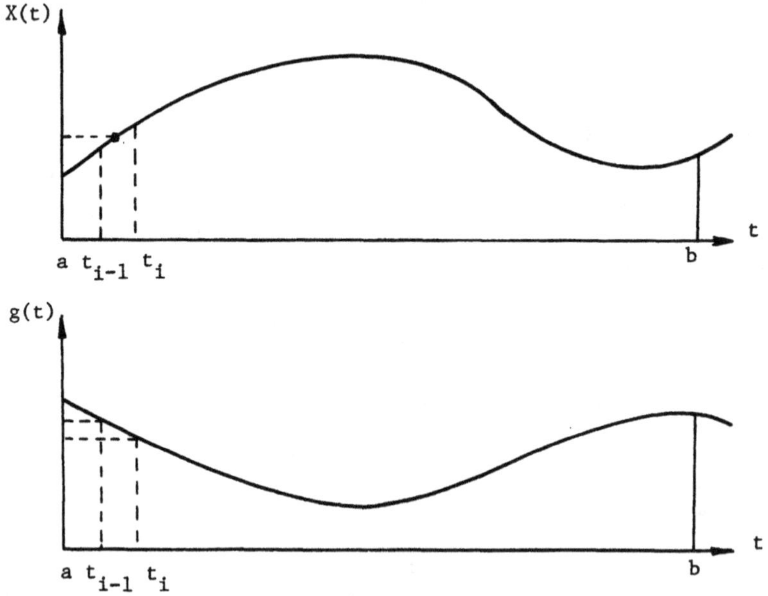

FIG. (3.2) Concept of Riemann-Stieltjes integral

The Riemann-Stieltjes integral may be defined as the integration of X(t) with respect to the function g(t), see fig. (3.2), over the interval [a,b] as

$$\int_a^b X(t)\, dg(t) = \lim_{\max|\Delta t_i| \to 0} \sum_{i=1}^{n} X(\bar{t}_i)\{g(t_i) - g(t_{i-1})\} \tag{3.37}$$

A sufficient condition for the existence of the limit is that g(t) possesses bounded variation, that is

$$\sum_{i=1}^{n} \left| g(t_i) - g(t_{i-1}) \right| < M \tag{3.38}$$

where M is a finite number. In addition, X(t) must be continuous over [a,b]. It is obvious that for the special case g(t)=ct, the integral (3.37) becomes the ordinary Riemann integral. The geometric meaning of the Riemann-Stieltjes integral may be gained by referring to fig. (3.2). Unlike the Riemann integral, the summation (3.37) does not represent the area under X(t), except when g(t)=ct. It is obvious that g(t) is used as a weighting function. The Riemann-Stieltjes integral becomes invalid when g(t) is a stochastic process with unbounded variation such as the Wiener process for which special rules of stochastic integration will be discussed in the next chapter.

3.7. LEBESGUE INTEGRAL

For an arbitrary measurable probability space* (Ω,e,P) the expected value of the random variable X may alternatively be written in terms of the Riemann integral

$$E[X] = \int_{-\infty}^{\infty} X\, p(X)\, dX \cong \sum_{i=-\infty}^{\infty} X_i p(X_i)\Delta X \tag{3.39}$$

Introducing the relation

$$p(X_i)\Delta X \cong P\{X_i < X < X_i + \Delta X\}$$

the mean value given by (3.39) takes the form

---

* the probability space is completely specified by the three entities (Ω,e,P) where Ω is the sample space, e is an appropriate event in the sample space, and P is a set probability function. Each event e involves a subset of the elements X_i which are also elements in the sample space. The field $\mathcal{F}$ which includes $(e_1, e_2, \ldots, e_n)$ is called a Borel field if the new sets $(e_1+e_2+\ldots+e_n)$ and $(e_1 e_2 \ldots e_n)$ belong also to $\mathcal{F}$.

$$E[X] \cong \sum_{i=-\infty}^{\infty} X_i P\{X_i < X < X_i + \Delta X\} \tag{3.40}$$

The increments $(X_i<X<X_i+\Delta X)$ can be viewed as "differential events", Papoulis (1965), and their sum is the certain event. For every element of the random process $(X_i<X<X_i+\Delta X)$ the random variable X may be replaced by X_i. If $\Delta X \to 0$ the resulting limit can be written in terms of the Lebesgue integral

$$E[X] = \int_{\Omega} X \, dP \tag{3.41}$$

It is not difficult to show that every bounded Riemann-integrable function defined on a finite interval is also Lebesgue-integrable and that the two integrals are identical.

CHAPTER 4

Calculus of Stochastic Processes

4.1 INTRODUCTION

This chapter deals with the description and mathematical treatment of special types of stochastic processes encountered in the area of parametric random vibration. In particular, the Markov processes will be the cornerstone of this chapter. These processes are defined on the basis of the conditional probability (or the transition probability), that is, the probabilistic structure of such processes in the future depends on its present state regardless of the past history of the process. The evolution of such conditional probability density is usually described by a partial differential equation known as the Fokker-Planck-Kolmogorov equation with time and state space vector as independent parameters. A sufficient condition for a random process to be Markovian is that it possesses independent increments. The Brownian motion of a free particle is an excellent example of a Markov process. The importance of Markov processes in random vibration is realized from the fact that when the excitation of a dynamic system is ideal white noise, the response constitutes a Markov process for which the Fokker-Planck-Kolmogorov equation or other techniques can be employed to determine the response statistics. This class of stochastic processes is handled analytically via special calculus rules. During the past two decades the literature contained a number of controversies and disputes which have been resolved by virtue of the stochastic rules developed by Itô (1951a), Wong and Zakai (1965), Gray and Caughey (1965), and Stratonovich (1966). The main results of these stochastic rules will also be outlined in this chapter.

4.2 THE BROWNIAN MOTION PROCESS

This process is named after the English botanist, Robert Brown, who conducted a series of experimental observations to study the motion of small particles suspended in a liquid through a high-powered microscope. The mathematical description of the Brownian motion was established by Wiener (1923,1958) and the process is synonymously called the Wiener process. Mathematically the velocity of the particle cannot be defined as a continuous analytic function which implies that the Brownian motion process is non-differentiable. Intuitively, if we assume that the interval between the collisions is too infinitesimal to the extent that the time derivative cannot be defined, one can reach the conclusion that the Wiener process does not have a derivative.

The successive displacements $B_i(t)$, with $B(0)=0$, are independent as a result of independent collision forces. This observation establishes the important property that the Brownian motion is a process with independent increments. In view of the small magnitudes of individual displacements, the central limit theorem (see section 3.4) can be applied to their sum with the conclusion that the Brownian motion is a Gaussian stationary process. According to Wiener the mathematical properties of the Brownian motion process $B(t)$ are:

i. The vector $\underset{\sim}{B}(t) = \{B(0), B(t_1), B(t_2), \ldots, B(t_n)\}^T$ constitutes a set of random variables with independent increments for arbitrary t_i (where $0<t_1<t_2<\ldots<t_n \leq T$). The random variables $B(0)$, $B(t_1)-B(0)$, $B(t_2)-B(t_1), \ldots, B(t_n)-B(t_{n-1})$ are mutually independent, i.e.,

$$E[\{B(t_i) - B(t_{i-1})\}\{B(t_{i+1}) - B(t_i)\}] = 0 \tag{4.1}$$

ii. $$P\{B(0) = 0\} = 1 \tag{4.2}$$

iii. The increments $\Delta B_i = B(t_{i+1}) - B(t_i)$ have normal distributions for every $t>0$.

iv. For every $t \geq 0$ the mean value of the increments is zero

$$E[\Delta B_i] = 0 \tag{4.3}$$

and their mean square is proportional to the time increment Δt_i, i.e.,

$$E[(\Delta B_i)^2] = \sigma^2 \Delta t_i, \qquad \Delta t_i = | t_{i+1} - t_i | \tag{4.4}$$

where σ^2 is the variance parameter which is a positive constant. It should be mentioned here that the fact that the mean square of ΔB_i is of order Δt_i is the main source of peculiarities involved in the rules of the Itô stochastic calculus which will be discussed in section (4.7).

Relation (4.4) can be generalized to the n-th order mean

$$E[(\Delta B_i)^n] = \begin{cases} 0 & n \text{ odd} \\ 1.3.5\ldots(n-1)\sigma^n(\Delta t)^{n/2} & n \text{ even} \end{cases} \tag{4.5}$$

v. The correlation function of the Brownian motion process is determined as follows

$$R_B(\Delta t_i) = E[B(t_{i+1})B(t_i)]$$
$$= E[\{(B(t_{i+1}) - B(t_i)) + B(t_i)\}B(t_i)]$$
$$= E[\Delta B_i(B(t_i) - B(0))] + E[B_i^2]$$
$$= 0 + \sigma^2\ t_i$$

or (Jazwinski, 1970)

$$R_B(t,s) = \sigma^2 \min(t,s) \tag{4.6}$$

where $\min(t,s) = \begin{cases} t & \text{if } t<s \\ s & \text{if } t>s \end{cases}$

which is continuous at every (t,s).

According to relation (3.13a) the Brownian motion process is mean square continuous on the time interval $[0,\infty)$.

vi. The Brownian motion process has continuous sample functions (Doob, 1953) with unbounded variation and is not differentiable in the mean square sense

$$E[(\frac{B(t + \Delta t) - B(t)}{\Delta t})^2] = \frac{\sigma^2}{|\Delta t|}$$

It follows that

$$\lim_{\Delta t \to 0} E[(\frac{\Delta B}{\Delta t})^2] = \infty \tag{4.7}$$

vii. According to the strong law of large numbers, the following limit vanishes with probability one

$$\lim_{t\to\infty} \frac{B(t)}{t} = 0 \tag{4.8}$$

The order of magnitude of the sample functions may be established from the law of the iterated logarithm. For each individual component of a single dimensional Brownian motion process, the following two limits are obtained with probability one

$$\limsup_{t\to\infty} \frac{B(t)}{\sqrt{2t \log \log t}} = 1$$

and

$$\liminf_{t\to\infty} \frac{B(t)}{\sqrt{2t \log \log t}} = -1 \tag{4.9}$$

Thus for every $0<\varepsilon<1$ and for every sample function B(t) we have (Arnold, 1974)

$$-(1+\varepsilon)\sqrt{2t \log \log t} \leq B(t) \leq (1+\varepsilon)\sqrt{2t \log \log t} \tag{4.10}$$

The local law of the iterated logarithm reveals enormous local fluctuations in B(t), and each portion of almost every sample function B(t) is of unbounded variation in a finite interval of time.

4.3 THE WHITE NOISE

The white noise is a purely mathematical process which is completely random, totally unpredictable, and not correlated with its earlier values. It must not be capable of extrapolation and, hence, it should not be differentiable. A continuous white noise W(t) is characterized by a constant spectral density $S(\omega)=S_0$ over the entire frequency range $-\infty<\omega<\infty$ as shown in fig. (4.1a). This is equivalent to say that the energy content of the white noise is uniformly distributed in analogy with the white light. It is evident that this process is unrealizable physically since its mean square is infinite, that is

$$E[W^2(t)] = \int_{-\infty}^{\infty} S_0 \, d\omega = \infty \tag{4.11}$$

The autocorrelation function of the white noise can be obtained from the inverse Fourier transform of the spectral density

$$\begin{aligned} R_w(\tau) &= \int_{-\infty}^{\infty} S_0 \exp(i\omega\tau) \, d\omega \\ &= \int_{-\infty}^{\infty} S_0 \cos\omega\tau \, d\omega \\ &= 2\pi \, S_0 \lim_{\omega\to\infty} \frac{\sin\omega\tau}{\pi\tau} = 2\pi \, S_0 \, \delta(\tau) \end{aligned} \tag{4.12}$$

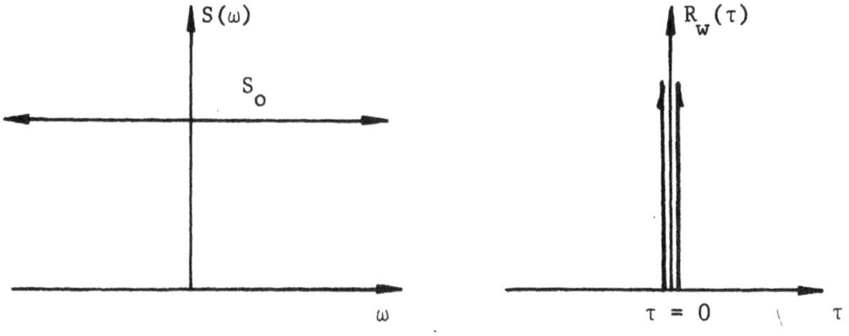

(a) Spectral density function (b) Autocorrelation function

FIG. (4.1) Properties of white noise process

This result simply states that the white noise is a delta-correlated process as shown in fig. (4.1b). It is a consequence of the white noise spectrum which yields infinite autocorrelation at $\tau=0$. The Gaussian white noise cannot be defined in the classical sense of the probability theory because of relation (4.12) which is regarded as a mathematical fiction. Although the white noise is a pathology, it is of great importance if one relates its properties with the characteristics of the Brownian motion process. Furthermore, the concept of this process is very useful in the theory of random parametric vibration and other engineering disciplines. For instance, many physical Gaussian random processes, having regular spectral density over a wide frequency range, can be approximated as a white noise process under certain rules of limit theorems. Such approximation requires that the correlation time must be smaller than the lowest characteristic time τ_n of the dynamic system. The mean square of a wide band random process $\xi(t)$ is

$$E[\xi^2(t)] = \int_0^{\omega_c} S_0 \, d\omega$$

$$= S_0 \, \omega_c = 2\pi \, S_0/\tau_c \tag{4.13}$$

where ω_c is the cut-off frequency, and τ_c is its period.

Replacing the actual process by a delta-correlated process implies that the frequency ω_c is not explicitly taken into account which is permissible if $\omega_c \gg \omega_n$ or $\tau_c \ll t_n$ (where ω_n and τ_c are the characteristic frequency and time period of the dynamic system). We may recall that the Brownian motion process is continuous and nowhere differentiable in the stochastic sense. However, it can be shown that the white noise is the "formal" or the ordinary derivative of the Brownian motion process

$$W(t) = dB(t)/dt \tag{4.14}$$

From definition (4.6) one can obtain the autocorrelation function of the process $\{dB(t)/dt,\ t \geq 0\}$ as

$$R_{\dot{B}}(t - s) = \frac{\partial^2 R_B(t - s)}{\partial t\, \partial s}$$

$$= \sigma^2 \frac{\partial^2 \min(t,s)}{\partial t\, \partial s} \tag{4.15}$$

$$\text{but } \frac{\partial}{\partial \tau} \min(t,s) = \begin{cases} 0 & s>t \\ 1 & s<t \end{cases} \tag{4.16}$$

which is the Heaviside unit step function and its derivative with respect to time is the Dirac-delta function so that (4.15) may be written in the form

$$R_{\dot{B}}(t - s) = \sigma^2\, \delta(t - s) \tag{4.17}$$

Accordingly, the white noise is the "formal" derivative of the Brownian motion process. Such a derivative is not regarded as a stochastic process in the usual sense of probability theory. However, $dB(t)=\dot{B}(t)dt$ is considered a functional that assigns values to the Riemann-Stieltjes integral.

Band-Limited White Noise: A random process having a non-zero and constant power spectrum over a finite frequency band and zero everywhere else, as shown in fig. (4.2a), is called a band-limited white noise

$$S(\omega) = \begin{cases} S_o & -\omega_c \leq \omega \leq \omega_c \\ 0 & \text{elsewhere} \end{cases}$$

The autocorrelation function of this process is given by the expression

$$R_\xi(\tau) = 2S_o \frac{\sin\omega_c\tau}{\tau}$$

The process is called bandpass if its power spectrum is specified as shown in fig. (4.2b). Generally, the term colored noise is given to any non-white noise process.

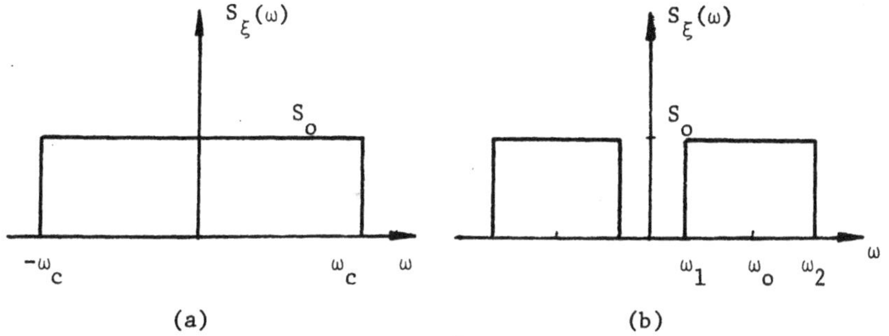

FIG. (4.2) Spectral densities of (a) band-limited white noise (b) band-pass white noise

It is important to realize that the correlation between the values of band-limited noise does not vanish for any correlation period. This property complicates the analysis and justifies the importance of adopting the white noise (which has uncorrelated values at distinct time intervals) in modeling random processes having essentially flat spectral density over a certain frequency range.

4.4 THE NARROW BAND PROCESS

The narrow band process plays an important role in the area of random vibration. It is encountered as an excitation to the system or as a response process. If the excitation is a wide band process the dynamic system acts as a mechanical filter which has a narrow band random response with a central frequency very close to the natural frequency of the system. Figure (4.3a) displays a sample record of a narrow band process and its spectral density. The process resembles a harmonic oscillation having a slow variation in the amplitude and phase. Mathematically it may be expressed by the formula

$$\begin{aligned} X(t) &= a(t)\cos\left(\omega_c t + \theta(t)\right) \\ &= a(t)\cos\theta(t)\cos\omega_c t - a(t)\sin\theta(t)\sin\omega_c t \\ &= x_c(t)\cos\omega_c t - x_s(t)\sin\omega_c t \end{aligned} \tag{4.18}$$

where $x_c(t) = a(t)\cos\theta(t)$, $x_s(t) = a(t)\sin\theta(t)$

$$a(t) = x_c^2 + x_s^2 \quad \text{and} \quad \theta(t) = \tan^{-1} x_s/x_c \tag{4.19}$$

$a(t)$ and $\theta(t)$ represent the slowly varying amplitude and phase shift, respectively. ω_c is the average (or central) frequency of the

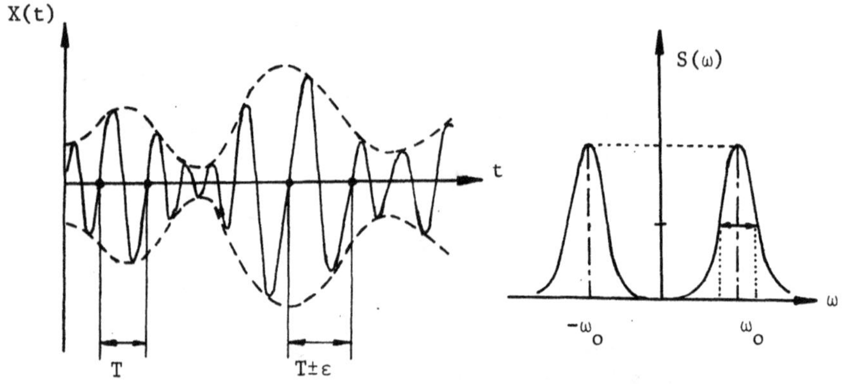

(a) Sample record (b) Spectral density

FIG. (4.3) Properties of narrow band process

spectral band. The power spectral density of this process is significant within a narrow frequency band, shown in fig. (4.3b), defined by the relation

$$\omega_c - \frac{1}{2}\Delta\omega \leq \omega \leq \omega_c + \frac{1}{2}\Delta\omega \tag{4.20}$$

where $\Delta\omega$ is the bandwidth corresponding to half the power and is much smaller than ω_c. The statistical properties of the envelope and phase of this process are treated rigorously by Davenport and Root (1958).

The random process X(t) can be represented by the Fourier series

$$X(t) = \sum_{n=1}^{\infty} (x_{cn} \cos n\omega t + x_{sn} \sin n\omega t) \tag{4.21}$$

$$\text{where } x_{cn} = \frac{2}{T}\int_0^T X(t) \cos n\omega t \, dt, \qquad T = 2\pi/\omega_n$$

$$x_{sn} = \frac{2}{T}\int_0^T X(t) \sin n\omega t \, dt \tag{4.22}$$

The coefficients x_{cn} and x_{sn} are Gaussian random variables which become uncorrelated as the duration of the expansion interval T increases without limit. The following relationships between x_{cn}, x_{sn} and x_c and x_s are introduced

$$x_c(t) = \sum_{n=1}^{\infty} \{x_{cn} \cos(n\omega - \omega_c)t + x_{sn} \sin(n\omega - \omega_c)t\}$$

$$x_s(t) = \sum_{n=1}^{\infty} \{x_{cn} \sin(n\omega - \omega_c)t - x_{sn} \cos(n\omega - \omega_c)t\} \tag{4.23}$$

The coefficients $x_c(t)$ and $x_s(t)$ are also Gaussian with the statistical properties

$$E[x_c(t)] = E[x_s(t)] = 0$$

$$E[x_c^2(t)] = E[x_s^2(t)] = \sigma_x^2 \tag{4.24}$$

$$p(x_c,x_s) = \frac{1}{2\pi\sigma_x^2} \exp\left\{-\frac{1}{2\sigma_x^2}\left(x_c^2(t) + x_s^2(t)\right)\right\}$$

The corresponding joint probability density of the amplitude and phase $p(a,\theta)$ is given by using relations (4.19) and the Jacobian of the transformation described in section (2.3.9) as

$$p(a,\theta) = p(x_c,x_s)\left|J\right| \tag{4.25}$$

where the Jacobian $\left|J\right|$ is given by the expression

$$\left|J\right| = \begin{vmatrix} \frac{\partial x_c}{\partial a} & \frac{\partial x_s}{\partial a} \\ \frac{\partial x_c}{\partial \theta} & \frac{\partial x_s}{\partial \theta} \end{vmatrix} = \begin{vmatrix} \cos\theta & \sin\theta \\ -a\sin\theta & a\cos\theta \end{vmatrix} \tag{4.26}$$

$$= a$$

Substituting in (4.25) gives

$$p(a,\theta) = \frac{a}{2\pi\sigma_x^2} \exp\left\{-\frac{a^2}{2\sigma_x^2}\right\} \qquad a \geq 0, \quad 0 \leq \theta \leq 2\pi \tag{4.27}$$

The probability density of the amplitude alone can be obtained by integrating (4.27) with respect to θ

$$p(a) = \int_0^{2\pi} p(a,\theta)d\theta = \frac{a}{\sigma_x^2} \exp\left\{-\frac{a^2}{2\sigma_x^2}\right\} \tag{4.28}$$

Formula (4.28) is known as the Rayleigh probability density which is shown in fig. (4.4a).

Similarly, the probability density of the phase θ is

$$p(\theta) = \int_0^\infty p(a,\theta)da = 1/2\pi \qquad 0 \leq \theta \leq 2\pi \tag{4.29}$$

which is represented by the uniform density given in fig. (4.4b).

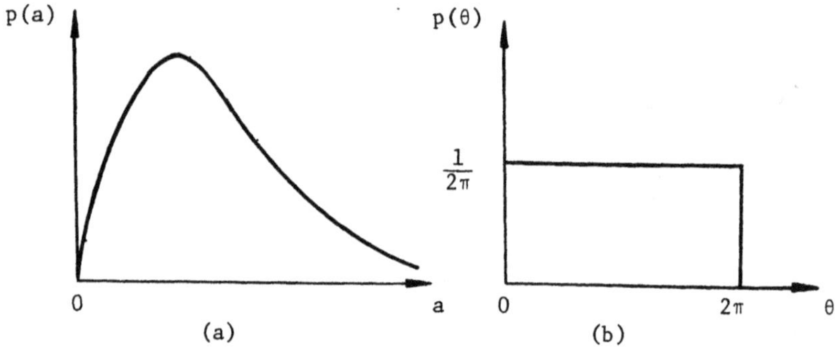

FIG. (4.4) Probability density functions of the (a) amplitude and (b) phase of a narrow band process

From (4.27) through (4.29) one can conclude that $p(a,\theta) = p(a)p(\theta)$ which implies that both a and θ are two independent random processes.

4.5 MARKOV PROCESSES

The degree of randomness of stochastic processes may be measured according to the way of their dependence on the behavior of their previous history. For example, a purely stochastic process $X(t)$ has no memory when its value at the present time t_n is completely independent of its value at any earlier time $t_{n-1}<t_n$ as in the case of the white noise. A continuous process with no memory is not physically realizable since the values $X(t_n)$ and $X(t_{n-1})$ would be independent regardless how short the time increment (t_n-t_{n-1}). However, the value of a physical stochastic process $X(t_n)$ is always influenced by the earlier value $X(t_{n-1})$. In the limit, as $(t_n-t_{n-1}) \to 0$ the process approaches a white noise.

When the future behavior at time t_{n+1} of a continuous random process $X(t)$, provided its present value $X(t_n)$ is given, does not change by additional information of its past, the process is known as a Markov process, or a diffusion process. This property led some people to name the Markov process as "one-step-memory stochastic process," or "process without after-effect." The Brownian motion is an ideal model of a Markov process.

Continuous Markov processes are described in terms of the conditional probability density

$$p(X_{n+1} \mid X_n, X_{n-1}, \ldots, X_o) = p(X_{n+1} \mid X_n) \tag{4.30}$$

The form of the conditional probability of Markov processes is called the transition probability.* The transition probability $p(X_{n+1} \mid X_n)$ can in turn be written in terms of the joint probability $p(x_{n+1}, x_n)$ as

$$p(X_{n+1} \mid X_n) = \frac{p(X_{n+1}, X_n)}{p(X_n)} \tag{4.31}$$

The first order probability density $p(X_{n+1})$ can be obtained from the corresponding joint probability $p(X_{n+1}, X_n)$ by employing relation (2.28), i.e.,

$$p(X_{n+1}) = \int_{-\infty}^{\infty} p(X_{n+1}, X_n) dX_n \tag{4.32}$$

The joint probability can also be obtained from a similar relation

$$p(X_{n+1}, X_{n-1}) = \int_{-\infty}^{\infty} p(X_{n+1}, X_n, X_{n-1}) dX_n \tag{4.33}$$

If X(t) is a Markov process, the third order joint probability becomes

$$p(X_{n+1}, X_n, X_{n-1}) = p(X_{n+1} \mid X_n)\, p(X_n \mid X_{n-1})\, p(X_{n-1}) \tag{4.34}$$

Substituting (4.34) into (4.33) and using (4.31) gives

$$p(X_{n+1} \mid X_{n-1}) = \int_{-\infty}^{\infty} p(X_{n+1} \mid X_n)\, p(X_n \mid X_{n-1}) dX_n \tag{4.35}$$

This relation is an integral equation governing the transition probability density of a Markov process. It is called the forward

---

*The term transition probability is usually used since $p(X_{n+1} \mid X_n)$ gives the density of probability of a transition from one point in the state space X_n at time t_n, to a point in the phase space X_{n+1} at time t_{n+1}.

Smoluchowski-Chapman-Kolmogorov equation. It has been used to derive the well known Fokker-Planck-Kolmogorov equations.

4.6 THE FOKKER-PLANCK-KOLMOGOROV (FPK) EQUATION

The solution of the Smoluchowski equation (4.35) is usually obtained by solving a partial differential equation established independently by Fokker, Planck, and Kolmogorov. The mathematical derivations of this equation have been given in several references (see, for example, Caughey, 1963a; Lin, 1967; and Melsa and Sage, 1973) and will not be presented here.

The FPK equation has two forms: the forward and backward equations. For a multi-dimensional process represented by the Markov vector $\underset{\sim}{X}(t)=\{X_1,X_2,\ldots,X_n\}^T$, the forward equation has the form

$$\frac{\partial p(\underset{\sim}{X},t)}{\partial t} = -\sum_{i=1}^{n} \frac{\partial}{\partial X_i} \{a_j(\underset{\sim}{X},t)\ p(\underset{\sim}{X},t)\}$$

$$+\frac{1}{2}\sum_{i=1}^{n}\sum_{j=1}^{n} \frac{\partial^2}{\partial X_i\,\partial X_j} \{b_{ij}(\underset{\sim}{X},t)\ p(\underset{\sim}{X},t)\} \tag{4.36}$$

where $p(\underset{\sim}{X},t)$ stands for the transition probability $p\{\underset{\sim}{X}(t_{n+1}) \mid \underset{\sim}{X}(t_n)\}$.

Equation (4.36) is subject to the initial condition

$$\lim_{t\to t_o} p(\underset{\sim}{X} \mid X_o) = \delta(\underset{\sim}{X} - \underset{\sim}{X}_o) \tag{4.37}$$

and the boundary conditions

$$p(\infty,t) = p(-\infty,t) = 0 \tag{4.38}$$

Equation (4.36) is referred to as the forward form because the time derivative of the transition probability $\partial p(X,t)/\partial t$ is taken with respect to the future time.

The backward FPK equation has the form

$$\frac{\partial p(X,t)}{\partial t} = -\sum_{i=1}^{n} a_i(\underset{\sim}{X},t)\frac{\partial p(\underset{\sim}{X},t)}{\partial X_i} - \frac{1}{2}\sum_{i=1}^{n}\sum_{j=1}^{n} b_{ij}(\underset{\sim}{X},t)\frac{\partial^2 p(\underset{\sim}{X},t)}{\partial X_i\,\partial X_j} \tag{4.39}$$

It is called backward because the derivative $\partial p(X,t)/\partial t$ is taken with respect to the earlier time.

The coefficients $a_i(\underset{\sim}{X},t)$ and $b_{ij}(\underset{\sim}{X},t)$ are the first and second incremental moments (or the drift and diffusion coefficients), respectively. These coefficients are evaluated as follows:

$$a_i(\underset{\sim}{X},t) = \lim_{\Delta t \to 0} \frac{1}{\Delta t} E[X_i(t + \Delta t) - X_i(t)] \tag{4.40}$$

$$b_{ij}(\underset{\sim}{X},t) = \lim_{\Delta t \to 0} \frac{1}{\Delta t} E[\{X_i(t + \Delta t) - X_i(t)\}\{X_j(t + \Delta t) - X_j(t)\}] \tag{4.41}$$

provided that all limits exist.

The two forms of the FPK equation can only be used if the random excitation is Gaussian white noise or filtered white noise. The solution of these equations gives the probabilistic behavior of the system response. In many cases, however, it is not possible to derive a closed form analytical solution to the FPK equation of the dynamical system. The existence and uniqueness of the solution have been discussed by Caughey (1971). Instead of seeking for stationary or non-stationary solutions of the FPK equation, one can generate a set of differential equations for the response moments. This alternative technique will be outlined in chapter 5.

In order to apply the FPK equation to dynamical systems the following conditions must be satisfied:

i. The damping forces must be proportional to the velocities.

ii. The correlation function matrix of the excitation should be proportional to the system damping matrix.

iii. The equations of motion should not include damping or inertia non-linear couplings. However, when such couplings exist they must be removed by successive elimination.

4.7 STOCHASTIC CALCULUS RULES

4.7.1 PRELIMINARIES

In view of the properties of the sample functions of the Brownian motion process, some mathematicians have established a number of theorems which define the rules of stochastic calculus. These rules differ from the ordinary calculus operations. Itô formulated the rules of stochastic differentiation and integration of the Brownian motion process. A rigorous mathematical treatment of the stochastic calculus has been given by Doob (1953), Skorokhod (1965), Gikhman and Skorokhod (1965,1972), and Arnold (1974). The treatment presented in this section is simple, oriented to engineering applications, and avoids mathematical abstraction.

Let us consider structural systems under wide band random parametric excitations. The equations of motion of such systems can be written as a set of first order differential equations in terms of the state vector $\underset{\sim}{X}(t)$

$$\dot{\underset{\sim}{X}}(t) = \underset{\sim}{f}(\underset{\sim}{X},t) + \underline{G}(\underset{\sim}{X},t)\,\underset{\sim}{\xi}(t) \tag{4.42}$$

where $\underset{\sim}{X}(t)$ is a 2n-state vector of the system response coordinates (n is the number of degrees of freedom), $\underset{\sim}{f}(\underset{\sim}{X},t)$ is a 2n-dimensional vector, $\underline{G}(\underset{\sim}{X},t)$ is a 2n×m matrix function, and $\underset{\sim}{\xi}(t)$ is an m-dimensional vector with elements which are physical wide band random processes. It is illegitimate to replace $\underset{\sim}{\xi}(t)$ by a white noise vector $\underset{\sim}{W}(t)$ since $\underset{\sim}{\xi}(t)$ is a physical random process. If $\underset{\sim}{\xi}(t)$ were a white noise vector, equation (4.42) would then constitute a Markov vector for which the FPK equation or the Itô stochastic calculus could be used. However, in order to apply these techniques, one must introduce some sort of correction terms to compensate for replacing the physical wide band excitation by a white noise. Kozin (1969) clarified this issue by the following simple example.

Example 4.1: Consider the first order differential equation

$$\dot{X} + \{a + \xi(t)\}X = 0, \quad \text{with } X(0) = X_0 \tag{i}$$

where a is a constant and $\xi(t)$ is a sample function from a wide band process with zero mean and satisfies the ergodic property

$$E[\xi] = \lim_{t\to\infty} \frac{1}{t}\int_0^t \xi(s)ds \tag{ii}$$

with probability one.

Applying the ordinary calculus rules, the solution of equation (i) may be written in the form

$$X(t) = X_0 \exp\{-at - \int_0^t \xi(s)ds\} \tag{iii}$$

Now in order to treat $\xi(t)$ as a white noise defined by the formal derivative of the Brownian motion $dB(t)/dt$, equation (i) must be written in the form

$$dX = -(a + \frac{1}{2}\sigma^2)Xdt - XdB(t) \tag{iv}$$

where σ^2 is the power spectral density of the white noise. Notice that the term $\frac{1}{2}\sigma^2\, Xdt$ has been introduced as a correction term and its origin will be demonstrated in this section. The solution of (iv) takes the form

$$X(t) = X_0 \exp\{-(a + \frac{1}{2}\sigma^2)t - B(t)\} \tag{v}$$

It is interesting to note that by extending the band width of $\xi(t)$ to a sufficiently large value, the process approaches the physical Gaussian white noise; however, the two solutions (iii) and (v) are still not identical. A fundamental difference between equations (i) and (iv) lies in the fact that dB(t) is independent of X(t), whereas $\xi(t)$ is correlated with the response. The analytical justifications of modeling physical systems by stochastic differential equations of the Itô type has been established by Wong and Zakai (1965).

4.7.2 THE ITÔ RULE OF STOCHASTIC DIFFERENTIAL

Stochastic Concept: Let $F\big(B(t)\big)$ be a continuous, real-valued, nonlinear function of the Brownian motion process B(t). The value of $F\big(B(t+\Delta t)\big)$ can be expressed in terms of $F\big(B(t)\big)$ by using the Taylor series

$$F\big(B(t+\Delta t)\big) = F\big(B(t) + \Delta B(t)\big)$$
$$= F\big(B(t)\big) + \frac{dF\big(B(t)\big)}{dB}\,\Delta B(t) + \frac{1}{2}\,\frac{d^2F\big(B(t)\big)}{dB^2}\,\big(\Delta B(t)\big)^2 + \ldots \tag{4.43}$$

The time derivative of $F\big(B(t)\big)$ can be defined by the limit

$$\frac{dF\big(B(t)\big)}{dt} = \lim_{\Delta t\to 0} \frac{1}{\Delta t}\left\{F\big(B(t+\Delta t)\big) - F\big(B(t)\big)\right\} \tag{4.44}$$

Taking expectation of both sides of (4.44) and introducing (4.43) yields

$$E\Big[\frac{dF\big(B(t)\big)}{dt}\Big] = \lim_{\Delta t\to 0} \frac{1}{\Delta t}\,E\Big[\frac{dF\big(B(t)\big)}{dB}\,\Delta B(t) + \frac{1}{2}\,\frac{d^2F\big(B(t)\big)}{dB^2}\,\big(\Delta B(t)\big)^2 + \ldots\Big]$$
$$= \frac{1}{2}\,\sigma^2\,E\Big[\frac{d^2F\big(B(t)\big)}{dB^2}\Big] \tag{4.45}$$

where the statistical properties of the Brownian motion process have been used. The fact that the mean square of $\Delta B(t)$ is of order Δt, not Δt^2, results in a non-zero expectation of (4.45).

Ordinary Concept: Applying the chain rule of ordinary calculus the differential $dF\big(B(t)\big)$ is written in the form

$$dF\big(B(t)\big) = \frac{dF\big(B(t)\big)}{dB}\,dB(t) \tag{4.46}$$

Dividing by dt and taking expectation of both sides gives

$$E\left[\frac{dF(B(t))}{dt}\right] = E\left[\frac{dF(B(t))}{dB} \cdot \frac{dB(t)}{dt}\right]$$

$$= E\left[\frac{dF(B(t))}{dB}\right] E\left[\frac{dB(t)}{dt}\right] = 0 \qquad (4.47)$$

which vanishes because the time derivative of the Brownian motion process has a zero mean.

In view of the two results given by (4.45) and (4.47), it is clear that the rules of ordinary differentiation must be modified to handle stochastic processes of independent increments. Because $dB(t)/dt$ does not exist anywhere in any stochastic sense, it is convenient to deal with the differential $dF(B(t))$ which involves $dB(t)$ rather than $dB(t)/dt$.

<u>Lemma</u>: (Gikhman and Skorokhod, 1972, p. 21). If the interval [a,b] is divided such that

$$a = t_0 < t_1 < \ldots < t_n = b, \qquad \lim_{i\to\infty} \max(t_{i+1} - t_i) = 0$$

then

$$\sum_{i=1}^{n-1} \{B(t_{i+1}) - B(t_i)\}^2 \to \sigma^2 (b-a) \text{ with probability one.} \qquad (4.48)$$

The Itô rule for the stochastic differential $dF(B(t))$ can be expressed in the form

$$dF(B(t)) = \frac{dF(B(t))}{dB} dB(t) + \frac{1}{2} \sigma^2 \frac{d^2F(B(t))}{dB^2} dt \qquad (4.49)$$

A consequent result of (4.49) is

$$dB^2(t) = 2B(t)\, dB(t) + \sigma^2\, dt \qquad (4.50)$$

$$dB^n(t) = n(B(t))^{n-1}\, dB(t) + \frac{1}{2} n(n-1)\sigma^2 (B(t))^{n-2}\, dt, \qquad n>2 \qquad (4.51)$$

4.7.3 THE ITÔ FORMULA OF STOCHASTIC DIFFERENTIAL

It is possible to extend the concept of stochastic differential to the response coordinates of dynamic systems described by the set of stochastic differential equations

$$d\underset{\sim}{X}(t) = \underset{\sim}{f}(\underset{\sim}{X}, t)\, dt + \underline{G}(\underset{\sim}{X}, t)\, d\underset{\sim}{B}(t) \qquad (4.52)$$

Consider first the case of one-dimensional systems for which a scalar function $F(X,t)$ will be assumed continuously differentiable in time and possessing continuous second partial derivative with respect to X. The differential $dF(X,t)$ was first formulated by Itô (1951a), and a complete proof is given by Gikhman and Skorokhod (1972) and Nevel'son and Khas'miniskii (1973).

Following the same partitioning of the interval [a,b] of the previous lemma, the differential $dF(X,t)$ bounded by this interval is

$$F(X,b) - F(X,a) = \sum_{i=0}^{n-1} \{F(X,t_{i+1}) - F(X,t_i)\} \tag{4.53}$$

Applying the Taylor expansion

$$F(X,t_{i+1}) - F(X,t_i) = \frac{\partial F(X,t_i)}{\partial t} \Delta t_i + \frac{\partial F(X,t_i)}{\partial X} \Delta X(t_i)$$

$$+ \frac{1}{2} \frac{\partial^2 F(X,t_i)}{\partial X^2} \left(\Delta X(t_i)\right)^2 + \ldots \tag{4.54}$$

The following sums will converge into the respective integrals:

$$\lim_{n\to\infty} \sum_{i=1}^{n-1} \frac{\partial F(X,t_i)}{\partial t} \Delta t_i = \int_a^b \frac{\partial F(X,s)}{\partial s} ds \tag{4.55}$$

$$\lim_{n\to\infty} \sum_{i=1}^{n-1} \frac{\partial F(X,t_i)}{\partial X} \Delta X(t_i) = \int_a^b \frac{\partial F(X,s)}{\partial X} dX(s)$$

$$= \int_a^b f(X,s) \frac{\partial F(X,s)}{\partial X} ds + \int_a^b G(X,s) \frac{\partial F(X,s)}{\partial X} dB(s) \tag{4.56}$$

where (4.52) has been used in (4.56).

The expression $\left(\Delta X(t_i)\right)^2$ in the third term in (4.54) is evaluated as follows

$$\left(\Delta X(t_i)\right)^2 = f^2(X,t_i)(\Delta t_i)^2 + 2f(X,t_i)G(X,t_i)\Delta t_i \Delta B(t_i)$$

$$+ G^2(X,t_i)\left(\Delta B(t_i)\right)^2$$

$$\cong G^2(X,t_i)\left(\Delta B(t_i)\right)^2 \tag{4.57}$$

where the first two terms have been dropped because they are of order higher than the increment Δt_i. The summation of the third term in (4.54) may be written, after using (4.48), as

$$\lim_{n\to\infty} \frac{1}{2} \sum_{i=0}^{n-1} \sigma^2 G^2 (X,t_i) \frac{\partial^2 F(X,t_i)}{\partial X^2} (t_{i+1} - t_i)$$

$$= \frac{\sigma^2}{2} \int_a^b G^2 (X,s) \frac{\partial^2 F(x,s)}{\partial X^2} ds \tag{4.58}$$

Substituting the three limits (4.55), (4.56), and (4.58) into (4.54) yields

$$F(X,b) - F(X,a) = \int_a^b \left\{\frac{\partial F}{\partial s} + f(X,s) \frac{\partial F}{\partial X} + \frac{\sigma^2}{2} G^2 (X,s) \frac{\partial^2 F}{\partial X^2}\right\} ds$$

$$+ \int_a^b G(X,t) \frac{\partial F}{\partial X} dB(s) \tag{4.59}$$

Relation (4.59) can be written in the differential form

$$dF(X,t) = \left\{\frac{\partial F}{\partial t} + f(X,t) \frac{\partial F}{\partial X} + \frac{\sigma^2}{2} G^2 (X,t) \frac{\partial^2 F}{\partial X^2}\right\} dt$$

$$+ G(X,t) \frac{\partial F}{\partial X} dB(t) \tag{4.60}$$

For multi-dimensional problems described by the state vector differential equations (4.52) the differential of the scalar function $F(\underset{\sim}{X},t)$ is given by the Itô formula (Jazwinski, 1970 or Arnold, 1974)

$$dF(\underset{\sim}{X},t) = \left\{\frac{\partial F}{\partial t} + \sum_{i=1}^{n} \frac{\partial F}{\partial X_i} f_i(\underset{\sim}{X},t) + \frac{\sigma^2}{2} \sum_{i,k}^{n} \sum_{j}^{n} \frac{\partial^2 F}{\partial X_i \partial X_j} G_{ij} G_{kj}\right\} dt$$

$$+ \sum_{i=1}^{n} \sum_{j=1}^{n} \frac{\partial F}{\partial X_i} G_{ij} dB_j(t) \tag{4.61}$$

which is equivalent to the compact form

$$dF(\underset{\sim}{X},t) = \left\{\frac{\partial F}{\partial t} + \frac{1}{2} \text{Trace } \underline{G}\ \underline{Q}\ \underline{G}^T F_{xx}\right\} dt + \underset{\sim}{F}_x^T d\underset{\sim}{X} \tag{4.62}$$

where $\underline{Q}\, dt = E[\{d\underset{\sim}{B}(t)\}\{d\underset{\sim}{B}(t)\}^T]$

$$\underset{\sim}{F}_x^T = \left\{\frac{\partial F}{\partial X_1}, \frac{\partial F}{\partial X_2}, \ldots, \frac{\partial F}{\partial X_n}\right\}$$

$$\text{and } \underline{F}_{xx} = \begin{vmatrix} \frac{\partial^2 F}{\partial X_1^2} & \frac{\partial^2 F}{\partial X_1 \partial X_2} & \cdots\cdots & \frac{\partial^2 F}{\partial X_1 \partial X_n} \\ \cdots & \cdots & \cdots & \cdots \\ \cdot & & & \cdot \\ \cdots & \cdots & \cdots & \cdots \\ \frac{\partial^2 F}{\partial X_n \partial X_1} & \frac{\partial^2 F}{\partial X_n \partial X_2} & \cdots\cdots & \frac{\partial^2 F}{\partial X_n^2} \end{vmatrix}$$

The expression involving the double summation in (4.61) is known as the Itô correction term.

4.7.4 THE WONG-ZAKAI CONVERGENCE THEOREM (WONG AND ZAKAI, 1965)

Wong and Zakai developed a theorem which establishes a relationship between the ordinary and stochastic differential equations. In particular, if $X^i(t)$ is the sequence of solutions of the differential equation

$$dX^i(t) = f(X^i,t)\,dt + G(X^i,t)\,db^i(t), \qquad X^i(t_0) = X^i_o \tag{4.63}$$

where $b^i(t)$ is a convergent sequence of continuous and piecewise linear approximation to the Brownian motion process $B(t)$, then $X^i(t)$ converges in the mean to the solution of the differential equation

$$dX(t) = \left\{f(X,t) + \frac{\sigma^2}{2}\,G(X,t)\,\frac{\partial G(X,t)}{\partial X}\right\}dt + G(X,t)\,dB(t) \tag{4.64}$$

The physical process $b^i(t)$ in (4.63) is a continuous function of bounded variation, while $B(t)$ is continuous with unbounded variation. A heuristic approach to verify this theorem involves a transformation which converts (4.63) into an equation including $db^i(t)$ rather than $G(X^i,t)\,db^i(t)$. Such transformation will avoid the subtleties of stochastic integrals. Introducing the transformation

$$Z^i(t) = g(X^i,t) \tag{4.65}$$

where $g(X,t)$ is defined such that

$$\frac{\partial g(X,t)}{\partial X} = \frac{1}{G(X,t)} \tag{4.66}$$

If the inverse $X^i(t) = H(Z^i,t)$ exists, we can write

$$dZ^i(t) = \left.\frac{\partial g(X,t)}{\partial t}\right|_{X=H} dt + \left.\frac{\partial g(X,t)}{\partial X}\right|_{X=H} dX^i(t) \tag{4.67}$$

Introducing (4.63) into (4.67) yields

$$dZ^i(t) = \frac{\partial g(X,t)}{\partial t}\Big|_{X=H} dt + \frac{\partial g(X,t)}{\partial X}\Big|_{X=H} \{f(X,t)\Big|_{X=H} dt$$

$$+ G(X,t)\Big|_{X=H} db^i(t)\}$$

$$= f_1(Z^i,t)dt + G_1(Z^i,t)db^i(t) \tag{4.68}$$

Introducing (4.66) the expressions $G_1(Z^i,t)$ and $f_1(Z^i,t)$ become

$$G_1(Z^i,t) = \frac{\partial g}{\partial X} G(X,t)\Big|_{X=H} = 1 \tag{4.69}$$

$$f_1(Z^i,t) = \{\frac{\partial g(X,t)}{\partial t} + \frac{\partial g(X,t)}{\partial X} f(X,t)\}\Big|_{X=H}$$

$$= \{\frac{\partial g(X,t)}{\partial t} + \frac{f(X,t)}{G(X,t)}\}\Big|_{X=H} \tag{4.70}$$

Substituting (4.69) into (4.68) gives

$$dZ^i(t) = f_1(Z_i,t)\, dt + db^i(t) \tag{4.71}$$

Integrating (4.71) over the interval (t_0,t) gives

$$Z^i(t) - Z^i(t_0) = \int_{t_0}^{t} f_1(Z^i,s)\, ds + \int_{t_0}^{t} db^i(s)$$

$$= \int_{t_0}^{t} f_1(Z^i,s)\, ds + b^i(t) - b^i(t_0) \tag{4.72}$$

Since $b^i(t)-b^i(t_0)$ converges with probability one to $B(t)-B(t_0)$ and there is no Itô's stochastic integral (see next section) in (4.72), the sequence of the response $Z^i(t)$ must converge to the solution of the differential equation

$$dZ(t) = f_1(Z,t)\, dt + dB(t) \tag{4.73}$$

provided $Z^i(t)$ always converges.

If in addition $X^i(t)$ converges, the limit $X(t)$ must be governed by the relation

$$X = H(Z,t) \tag{4.74}$$

Applying the Itô rule of differentiation to determine $dX(t)$ and introducing the convergence of $(dZ(t))^2 = 0 + (dB(t))^2$ from (4.73) we get

$$dX(t) = \left.\frac{\partial H(Z,t)}{\partial t}\right|_{Z=g} dt + \left.\frac{\partial H(Z,t)}{\partial z}\right|_{Z=g} dZ(t) + \frac{\sigma^2}{2} \left.\frac{\partial^2 H(Z,t)}{\partial Z^2}\right|_{Z=g} dt$$

$$= \left\{\frac{\partial H(Z,t)}{\partial t} + \frac{\partial^2}{2} \frac{\partial^2 H(Z,t)}{\partial Z^2}\right\}\Bigg|_{Z=g} dt$$

$$+ \left.\frac{\partial H(Z,t)}{\partial Z}\right|_{Z=g} \left\{\frac{\partial g(X,t)}{\partial t} + \frac{f(X,t)}{G(X,t)}\right\} dt + \left.\frac{\partial H(Z,t)}{\partial Z}\right|_{Z=g} dB(t) \tag{4.75}$$

where (4.70) has been used for $f_1(Z,t)$.

The limiting value of X can be written formally as

$$X = H(Z,t) = H\{g(X,t),t\} \tag{4.76}$$

Accordingly, we can write

$$\frac{\partial X}{\partial X} = 1 = \left.\frac{\partial H(Z,t)}{\partial Z}\right|_{Z=g} \frac{\partial Z}{\partial X} = \left.\frac{\partial H}{\partial Z}\right|_{Z=g} \frac{\partial g(X,t)}{\partial X}$$

$$= \frac{1}{G(X,t)} \cdot \left.\frac{\partial H(Z,t)}{\partial Z}\right|_{Z=g} \tag{4.77}$$

$$\frac{\partial X}{\partial t} = 0 = \left.\frac{\partial H(Z,t)}{\partial t}\right|_{Z=g} + \left.\frac{\partial H(Z,t)}{\partial Z}\right|_{Z=g} \frac{\partial g(x,t)}{\partial t} \tag{4.78}$$

$$\frac{\partial^2 X}{\partial X^2} = 0 = \frac{\partial^2 G(X,t)}{\partial X^2} \cdot \left.\frac{\partial H(Z,t)}{\partial Z}\right|_{Z=g} + \left(\frac{\partial g(X,t)}{\partial X}\right)^2 \left.\frac{\partial^2 H(Z,t)}{\partial Z^2}\right|_{Z=g} \tag{4.79}$$

Substituting (4.77) through (4.79) in (4.75) gives

$$dX(t) = f(X,t)\, dt + \frac{\sigma^2}{2} G(X,t) \frac{\partial G(X,t)}{\partial X} dt + G(X,t)\, dB(t) \tag{4.80}$$

which is the required result.

The theorem can be extended to a multi-dimensional stochastic process described by the vector differential equations

$$d\underset{\sim}{X}^i(t) = \underset{\sim}{f}(\underset{\sim}{X}^i,t)\, dt + \underline{G}(\underset{\sim}{X}^i,t)\, d\underset{\sim}{b}^i(t) \tag{4.81}$$

System (4.81) converges as $i \to \infty$ to the stochastic set of differential equations

$$dX_j(t) = \left\{f_j(\underset{\sim}{X},t) + \frac{1}{2}\sum_{\ell}^{n}\sum_{k}^{n} G_{\ell k}(\underset{\sim}{X},t)\frac{\partial G_{j\ell}(X,t)}{\partial X_k}\right\} dt$$

$$+ \sum_{\ell}^{n} G_{j\ell}(\underset{\sim}{X},t)\, dB_{\ell}(t) \tag{4.82}$$

This result has been verified rigorously by Khas'miniskii (1966).

4.7.5 THE KHAS'MINISKII LIMIT THEOREM (KHAS'MINISKII, 1966)

Khas'miniskii considered the following form of the state vector differential equations

$$\dot{X}_i = \varepsilon F_i(\underset{\sim}{X},\xi(t),t,\varepsilon) \qquad i=1,2,\ldots,2n \tag{4.83}$$

where ε is a small parameter and $\xi(t)$ is a stationary, measurable random process with zero mean and satisfies the condition

$$\left| P\{\xi(t) \,|\, Z_0\} - P\{\xi(t)\} \right| < \beta(T) \tag{4.84}$$

with probability one. It is assumed that a family Z_s^t, $0<s<t<\infty$ exists such that $Z_{s_1}^{t_1}$ belongs to Z_s^t if $s\leq s_1$, $t_1\leq t$. The function $\beta(T)$ is finite and decreases monotonically to zero as the total time duration $T \to \infty$.

Let the function $F_i(\underset{\sim}{X},\xi(t),t,\varepsilon)$ be measurable with respect to its arguments, bounded and satisfying the law of large numbers for fixed $\underset{\sim}{X}$. Suppose that as $\varepsilon \to 0$, the function F_i can be written in the form

$$F_i(\underset{\sim}{X},\xi(t),t,\varepsilon) = F_i^o(\underset{\sim}{X},\xi(t),t) + \varepsilon F_i^1(\underset{\sim}{X},\xi(t),t) + O(\varepsilon) \tag{4.85}$$

where the absolute values of the functions F_i^j and their first and second mixed partial derivatives with respect to $\underset{\sim}{X}$ are bounded. In problems involving parametric excitations of small intensities, the functions F_i^1 do not include $\xi(t)$ and F_i^o involve $\xi(t)$.

If, in addition, the following integrals exist:

$$\left.\begin{aligned}
&\lim_{T\to\infty} \frac{1}{T} E[F_i^1(\underset{\sim}{X},\xi(t),t)]dt && = \bar{F}_i^1(\underset{\sim}{X}) \\
&\lim_{T\to\infty} \frac{1}{T}\int_{t_0}^{t_0+T}\int_{t_0}^{t_0+T} E[F_i^o(\underset{\sim}{X},\xi(s),s)F_k^o(\underset{\sim}{X},\xi(t),t)]ds\, dt && = \bar{R}_{ik}(\underset{\sim}{X}) \\
&\lim_{T\to\infty} \frac{1}{T}\int_{t_0}^{t_0+T} ds \int_{t_0-T}^{s} \sum_{k=1}^{2n} E[\frac{\partial F_i^o(\underset{\sim}{X},\xi(s),s)}{\partial X_k} F_k^o(\underset{\sim}{X},\xi(t),t)]\, dt && \\
& && = \bar{K}_i(\underset{\sim}{X})
\end{aligned}\right\} \tag{4.86}$$

where t>s, then over the time interval $0<\varepsilon^2 t<t_0$ the solution $\underset{\sim}{X}(\varepsilon^2 t)$ converges weakly, as $\varepsilon \to 0$, to a Markov process X* which is continuous with probability one ($\tau = \varepsilon^2 t$) and governed by the solution of the Itô type equation

$$dX_i^*(\tau) = f_i(\underset{\sim}{X}^*)\, d\tau + \sum_{R=1}^{2n} G_{ik}(\underset{\sim}{X}^*)\, dB_k(\tau)$$

or

$$dX_i^*(\tau) = \{\bar{K}_i(\underset{\sim}{X}^*) + \bar{F}_i^1(\underset{\sim}{X}^*)\}\, d\tau + \sum_{k=1}^{2n} \bar{R}_{ik}(\underset{\sim}{X}^*)\, dB_k(\tau) \tag{4.87}$$

where $B_k(\tau)$ are independent Brownian motion processes with unit variance.

4.7.6 BASIC CONCEPT OF THE ITÔ INTEGRAL

The solution of the differential equation

$$dX(t) = f(X,t)\, dt + G(X,t)\, dB(t)$$

may be written in the form

$$X(t_1) = X(t_0) + \int_{t_0}^{t_1} f(X,t)\, dt + \int_{t_0}^{t_1} G(X,t)\, dB(t) \tag{4.88}$$

The second integral cannot be defined as an ordinary Riemann-Stieltjes integral since the Brownian motion process B(t) is continuous with unbounded variation. This integral is known as the "forward" Itô integral. Doob (1953) has defined the Itô integral such that

$$E[\int_{t_0}^{t_1} G(X,t)\, dB(t)] = 0 \tag{4.89}$$

and

$$E[\{\int_{t_0}^{t_1} G(X,t)\, dB(t)\}^2] = \sigma^2 \int_{t_0}^{t_1} E[G^2(X,t)]\, dt < \infty \tag{4.90}$$

The rules of the Itô integral can be conceived by considering the integration of the Itô stochastic differential (4.49), that is

$$\int_{t_0}^{t_1} dF\big(B(t)\big) = F\big(B(t_1)\big) - F\big(B(t_0)\big)$$
$$= \int_{t_0}^{t_1} \frac{dF\big(B(t)\big)}{dB}\, dB(t) + \frac{\sigma^2}{2} \int_{t_0}^{t_1} \frac{d^2F\big(B(t)\big)}{dB^2}\, dt \tag{4.91}$$

Rearranging (4.91) in the form

$$\int_{t_o}^{t_1} \frac{dF\big(B(t)\big)}{dB}\, dB(t) = F\big(B(t_1)\big) - F\big(B(t_o)\big) - \frac{\sigma^2}{2} \int_{t_o}^{t_1} \frac{d^2F\big(B(t)\big)}{dB^2}\, dt \tag{4.92}$$

where the integral on the left-hand side is identical to the Itô integral if the following substitution is introduced

$$\psi\big(B(t)\big) = dF\big(B(t)\big)/dB \tag{4.93}$$

thus equation (4.92) becomes

$$\int_{t_o}^{t_1} \Psi\big(B(t)\big)\, dB(t) = \int_{t_o}^{t_1} \psi\big(B(s)\big)\, dB(s) - \frac{\sigma^2}{2} \int_{t_o}^{t_1} \frac{d\psi\big(B(t)\big)}{dB}\, dt \tag{4.94}$$

The first integral on the right-hand side is an ordinary one where s is treated as a deterministic dummy variable. The Itô integral on the left-hand side can be defined as a stochastic limit of the Riemann integral

$$\int_{t_o}^{t_1} \psi\big(B(t)\big)\, dB(t) = \lim_{|\Delta t_i| \to 0} \sum_{i=0}^{n-1} \psi\big(B(t_i)\big)\{B(t_{i+1}) - B(t_i)\} \tag{4.95}$$

where $|\Delta t_i| = |t_{i+1} - t_i|$ and the $\psi\big(B(t_i)\big)$ is evaluated at the beginning of the interval $(t_{i+1}-t_i)$ over which the increment $\Delta B_i = B(t_{i+1}) - B(t_i)$ is taken. This implies that $\psi\big(B(t_i)\big)$ and ΔB_i are always independent, i.e.,

$$\begin{aligned} E\Big[\lim_{|\Delta t_i| \to 0} \sum_{i=0}^{n-1} \psi\big(B(t_i)\big)\{B(t_{i+1}) - B(t_i)\}\Big] &= \\ \lim_{|\Delta t_i| \to 0} \sum_{i=0}^{n-1} E[\psi\big(B(t_i)\big)]\, E[B(t_{i+1}) - B(t_i)] &= 0 \end{aligned} \tag{4.96}$$

which is identical to definition (4.89).

From relation (4.94) it may be concluded that the Itô integral gives entirely different results from those derived by the ordinary calculus. For example, (Doob, 1953) if $\psi\big(B(t)\big) = B(t)$, the Itô integration gives

$$\int_a^b B(t)\, dB(t) = \frac{1}{2}\{B^2(b) - B^2(a)\} - \frac{\sigma^2}{2}(b - a)$$

This result can be obtained by the equivalent limit of the sum

$$\int_a^b B(t)\, dB(t) = \lim_{i\to\infty} \sum_{i=1}^{n-1} B(t_i)\{B(t_{i+1}) - B(t_i)\}$$

$$= \lim_{i\to\infty} \frac{1}{2} \sum_{i=1}^{n-1} \{(B(t_{i+1}))^2 - (B(t_i))^2 - (B(t_{i+1}) - B(t_i))^2\}$$

$$= \frac{1}{2}\{B^2(b) - B^2(a)\} - \frac{\partial^2}{2}(b - a)$$

where $a = t_0 < t_1 < \ldots < t_n = b$

4.7.7 THE STRATONOVICH INTEGRAL (STRATONOVICH, 1966)

Stratonovich introduced a new symmetric (with respect to time) stochastic integral under rather restrictive conditions. The integral is distinct from the Itô integral in that it possesses the same rules of ordinary calculus. It is defined as the limit-in-the-mean

$$\int_{t_o}^{t_1} \psi(B(t))\, dB(t) = \underset{|\Delta t_i| \to 0}{\text{l.i.m.}} \sum_{i=0}^{n-1} \psi\left(\frac{1}{2}\left(B(t_i) + B(t_{i+1})\right)\right)\left(B(t_{i+1}) - B(t_i)\right) \tag{4.97}$$

Contrary to the Itô integral, definition (4.97) implies that $dB(t)$ and $\psi(B(t))$ are not independent, i.e.,

$$E\left[\int_{t_o}^{t_1} \psi(B(t))\, dB(t)\right] \neq 0 \tag{4.98}$$

and integral (4.97) has the same concept of the "formal" integral. Stratonovich showed that if $\psi(B(t))$ is continuous in time and has a continuous partial derivative $\partial\psi/\partial B$ with finite integral of the mean square,

$$\int_{-\infty}^{\infty} E[\psi^2(B(t))]\, dt < \infty \tag{4.99}$$

then the mean square limit (4.97) exists and is related to the Itô integral (4.94) through the relationship

$$(I)\int_{t_o}^{t_1} \psi(B(t))\, dB(t) = (S)\int_{t_o}^{t_1} \psi(B(t))\, dB(t) - \frac{\sigma^2}{2}\int_{t_o}^{t_1} \frac{d\psi(B(t))}{dB}\, dt \tag{4.100}$$

where (I) and (S) denote integrals in the Itô and Stratonovich senses, respectively.

The two definitions given by Itô (4.95) and Stratonovich (4.97) reveal that the Itô integral is established on the basis that $\Delta B(t)$ is independent of $B(t)$ while the Stratonovich integral is essentially based on their dependence.

Relation (4.100) is very useful to solve systems described by equation (4.52). Mortensen (1969) provides an excellent physical insight into the various aspects and interpretations of relation (4.100). He indicated that the solution given by the integral equation (4.88), in which the second integral can be viewed as Itô or Stratonovich integral, depends on the way of its interpretation. If one introduces the Wong-Zakai transformation, the Itô and Stratonovich solutions of (4.52) are related by one of the two equations

$$X_I(t_1) = X_I(t_0) + \int_{t_0}^{t_1} \left\{ f(X_I,t) - \frac{\sigma^2}{2} G(X_I,t) \frac{\partial G(X_I,t)}{\partial X} \right\} dt + (S) \int_{t_0}^{t_1} G(X_I,t)\, dB(t) \tag{4.101}$$

or

$$X_S(t_1) = X_S(t_0) + \int_{t_0}^{t_1} \left\{ f(X_S,t) + \frac{\sigma^2}{2} G(X_S,t) \frac{\partial G(X_S,t)}{\partial X} \right\} dt + (I) \int_{t_0}^{t_1} G(X_S,t)\, dB(t) \tag{4.102}$$

4.7.8 THE FOKKER-PLANCK-KOLMOGOROV EQUATION IN TERMS OF THE ITÔ EQUATION

The Itô stochastic differential equation

$$dX(t) = f(X,t)\, dt + G(X,t)\, dB(t) \tag{4.103}$$

constitutes a Markov process and its solution is characterized by the transition probability density function which is governed by the FPK equation

$$\frac{\partial p(X,t)}{\partial t} = - \frac{\partial}{\partial X} \{p(X,t)\, f(X,t)\} + \frac{\sigma^2}{2} \frac{\partial^2}{\partial X^2} \{p(X,t)\, G^2(X,t)\} \tag{4.104}$$

The derivation of this equation is given by Jazwinski (1970) and Gikhman and Skorokhod (1972).

For n-vector Itô differential equations

$$dX_i(t) = f_i(\underset{\sim}{X},t)\, dt + \sum_{j=1}^{m} G_{ij}(\underset{\sim}{X},t)\, dB_j(t), \quad i=1,2,\ldots,n \tag{4.105}$$

the corresponding FPK equation is given in the form

$$\frac{\partial p(\underset{\sim}{X},t)}{\partial t} = -\sum_{i=1}^{n} \frac{\partial}{\partial X_i} \{p(\underset{\sim}{X},t)\, f_i(\underset{\sim}{X},t)\}$$

$$+ \frac{1}{2} \sum_{i=1}^{n} \sum_{j=1}^{m} \frac{\partial^2}{\partial X_i \partial X_i} \{p(\underset{\sim}{X},t)(\underline{G}\ \underline{Q}\ \underline{G}^T)_{ij}\} \tag{4.106}$$

where $\underline{Q}\, dt = E[d\underset{\sim}{B}(t)\ d\underset{\sim}{B}^T(t)]$.

If (4.105) is interpreted in the Stratonovich sense, then the equivalent Itô equation becomes

$$dX_i(t) = \{f_i(\underset{\sim}{X},t) + \frac{1}{2} \sum_{k=1}^{n} \sum_{j=1}^{m} \sum_{\ell=1}^{n} G_{kj} Q_{j\ell} \frac{\partial G_{i\ell}}{\partial X_k}\} dt + \sum_{j=1}^{m} G_{ij} dB_j(t) \tag{4.107}$$

and the corresponding FPK equation is

$$\frac{\partial p(\underset{\sim}{X},t)}{\partial t} = -\sum_{i=1}^{n} \frac{\partial}{\partial X_i} (p(\underset{\sim}{X},t) f_i(\underset{\sim}{X},t))$$

$$+ \frac{1}{2} \sum_{i,k=1}^{n} \sum_{j=1}^{m} \frac{\partial}{\partial X_i} \{(GQ)_{ij} \frac{\partial}{\partial X_k} (p(\underset{\sim}{X},t)\ G(X,t)_{kj})\} \tag{4.108}$$

CHAPTER 5

Generation of Response Statistics

5.1 INTRODUCTION

It has been indicated in chapter 4 that the complete statistical description of the response of dynamical systems under random excitations can be obtained by solving the relevant FPK equation. In most cases it is not possible to obtain an analytical solution for the stationary or non-stationary probability density of the response. However, it is possible to generate differential equations for the response statistical functions. The response moments and correlation functions are of prime interest in the area of applied dynamics. In addition, statistical functions such as the semi-invariants (cumulants) and quasi-moments may be required to construct the probability density function of non-Gaussian response distributions.

In this chapter the generation of differential equations of the response statistical functions will be outlined. Two main approaches will be considered. These are the FPK equation method and the Itô stochastic calculus. These methods will be demonstrated by a number of examples of common use in the area of parametric random vibration. In general, the differential equations of the response moments may fall under one of the two categories: (i) consistent (or closed) equations or (ii) infinite coupled equations which must be closed via one of the closure schemes. The problem of the infinite hierarchy will be treated in section 5.4.

5.2 THE FOKKER-PLANCK-KOLMOGOROV EQUATION APPROACH

Let $\Phi(\underset{\sim}{X})$ be a general function of the response coordinate vector $\underset{\sim}{X}$

$$\Phi(\underset{\sim}{X}) = (X_1^{k_1} X_2^{k_2} \dots X_n^{k_n}) = \prod_{i=1}^{n} X_i^{k_i} \tag{5.1}$$

such that the moments of order $K=\sum_{i=1}^{n} k_i$ are expressed by the following notation

$$m_{k1,k2,\ldots,kn} = E[\Phi(\underset{\sim}{X})]$$

$$= \int_{-\infty}^{\infty}\!\!\cdots\!\int \Phi(\underset{\sim}{X})\; p(\underset{\sim}{X},t)\; dX_1 dX_2 \ldots dX_n \tag{5.2}$$

The differential equations of the response dynamic moments can be derived by multiplying both sides of the system FPK equation by $\Phi(\underset{\sim}{X})$ and integrating by parts over the entire state space $-\infty < \underset{\sim}{X} < \infty$

$$\dot{m}_{k1,k2,\ldots,kn} = \int_{-\infty}^{\infty}\!\!\cdots\!\int \Phi(\underset{\sim}{X}) \frac{\partial p(\underset{\sim}{X},t)}{\partial t}\, dX_1 dX_2 \ldots dX_n$$

$$= -\int_{-\infty}^{\infty}\!\!\cdots\!\int \Phi(\underset{\sim}{X}) \sum_{i=1}^{n} \frac{\partial}{\partial X_1} \{p(\underset{\sim}{X},t) f_i(\underset{\sim}{X},t)\}\; dX_1 \ldots dX_n$$

$$+ \frac{1}{2}\int_{-\infty}^{\infty}\!\!\cdots\!\int \Phi(\underset{\sim}{X}) \sum_{i-1}^{n} \sum_{j=1}^{n} \frac{\partial^2}{\partial X_i \partial X_j}$$

$$\{p(\underset{\sim}{X},t)(\underline{G}\; \underline{Q}\; \underline{G}^T)\}\; dX_1 \ldots dX_n \tag{5.3}$$

This approach is believed to have been first employed by Caughey and Dienes (1962), who examined the random behavior of dynamic systems subject to white noise parametric excitation. The method has become in common use to determine the response moments and their stability.

For the sake of illustration, consider the one-dimensional case of the Itô stochastic equation (4.52). The differential equation of the first order moment is obtained according to the following steps

$$\int_{-\infty}^{\infty} X \frac{\partial}{\partial t} p(X,t) dX = -\int_{-\infty}^{\infty} X \frac{\partial}{\partial X} \{p(X,t)\; f(X,t)\}\; dX$$

$$+ \frac{\sigma^2}{2} \int_{-\infty}^{\infty} X \frac{\partial^2}{\partial X^2} [p(X,t)\; G^2(X,t)]\; dX \tag{5.4}$$

It is legitimate to interchange the partial derivative and the integral on the left-hand side of equation (5.4)

$$\int_{-\infty}^{\infty} X \frac{\partial}{\partial t} p(X,t) dX = \frac{\partial}{\partial t} \int_{-\infty}^{\infty} Xp(X,t)\; dX$$

$$= \frac{\partial}{\partial t} m_1 \qquad = \dot{m}_1 \tag{5.5}$$

Applying integration by parts on the right-hand side of equation (5.4) gives

$$- \int_{-\infty}^{\infty} X \frac{\partial}{\partial X} \{p(X,t)\ f(X,t)\}\ dX + \frac{\sigma^2}{2} \int_{-\infty}^{\infty} X \frac{\partial^2}{\partial X^2} \{p(X,t)\ G(X,t)\}\ dX$$

$$= - Xp(X,t)\ f(X,t) \Big|_{-\infty}^{\infty} + \int_{-\infty}^{\infty} p(X,t)\ f(X,t)\ dX$$

$$+ \frac{\sigma^2}{2} X \frac{\partial}{\partial X} \{p(X,t)\ G^2(X,t)\} \Big|_{-\infty}^{\infty} - \frac{\sigma^2}{2} p(X,t)\ G(X,t) \Big|_{-\infty}^{\infty} \tag{5.6}$$

Introducing the boundary conditions $p(\infty,t) = p(-\infty,t) = 0$, every term on the right-hand side of (5.6) will vanish except the second expression which represents the expectation of $f(X,t)$. Substituting (5.5) and (5.6) in (5.4) gives

$$\dot{m}_1 = E[f(X,t)] \tag{5.7}$$

It is seen that if $f(X,t)$ is a linear function of the reponse coordinate X, equation (5.7) is consistent and can be solved for m_1. On the other hand, if $f(X,t)$ contains non-linear terms in X, equation (5.7) will be coupled with higher order moments.

If the number of the state coordinates $\underset{\sim}{X}$ is n, then the number of differential equations for the moments of order K is N, where N is given by the relation

$$N = n(n - 1)(n - 2)\ldots(n + K - 1)/K! \tag{5.8}$$

5.2.1 APPARENT PARADOX

The FPK equation has been extensively used to study the dynamic behavior of vibrating systems driven by Gaussian "physical" white noise or by the "formal" derivative of the Brownian motion process. The implications of these two cases have been discussed in chapter 4. In this section the distinction between the two processes will be further examined. It will be demonstrated that the resulting FPK equations of the same dynamic system driven by the two processes will not be the same. Gray and Caughey (1965) and Ariaratnam and Graefe (1965a,b) independently treated these two "apparent" identical problems and clarified the main differences. The coefficients of the FPK equations of the two problems are different and depend on whether the random coefficient to the next-to-the-highest derivative is a "physical" white noise or an increment Brownian motion process. The two cases have two different interpretations. When the random coefficient is represented by a Brownian process, the resulting model is referred to as strictly mathematical. On the other hand, if the excitation is approximated in the limit as a Gaussian white noise, the problem is referred to as a physical model. The ramifications of these modelings will be discussed.

The Mathematical Model: Consider the first order stochastic differential equation

$$dX = -aX\,dt - X\,dB(t), \qquad X(t=0) = X_o \tag{5.9}$$

where $B(t)$ is a Brownian motion process which possesses statistics given in section 4.2. The coefficients of the FPK equation for the one-dimensional Markov process (5.9) are

$$\begin{aligned} a_1 &= -aX \\ b_{11} &= \sigma^2 X^2 \end{aligned} \tag{5.10}$$

The FPK equation of (5.9) becomes

$$\frac{\partial p(X,t)}{\partial t} = a\frac{\partial}{\partial X}\{Xp(X,t)\} + \frac{\sigma^2}{2}\frac{\partial^2}{\partial X^2}\{X^2 p(X,t)\} \tag{5.11}$$

The Physical Model: Consider the first order system

$$\frac{\partial X}{\partial t} = -aX - X\xi(t) \tag{5.12}$$

where $\xi(t)$ is a physical Gaussian process with a very small correlation time τ_c. As $\tau_c \to 0$, $\xi(t)$ can be allowed to approach a Gaussian white noise and the response can be approximated by a Markov process. If the statistical properties of $\xi(t)$ are expressed in the form

$$E[\xi(t)] = 0, \qquad E\{\xi(t_1)\xi(t_2)\} = \sigma^2\delta(t_2 - t_1) \tag{5.13}$$

equation (5.12) may be transformed into the Itô type

$$dX = \{-aX + \frac{\sigma^2}{2}X\}\,dt - X\,dB(t) \tag{5.14}$$

where the "formal" derivative of the Brownian motion has been replaced for $\xi(t)$.

The coefficients of the FPK equation become

$$\begin{aligned} a_1 &= -aX + \frac{\sigma^2}{2}X \\ b_{11} &= \sigma^2 X \end{aligned} \tag{5.15}$$

and the resulting FPK equation is

$$\frac{\partial p(X,t)}{\partial t} = (a - \frac{\sigma^2}{2}) \frac{\partial}{\partial X} \{Xp(X,t)\} + \frac{\sigma^2}{2} \frac{\partial^2}{\partial X^2} \left(X^2 p(X,t)\right) \tag{5.16}$$

It is evident that the two systems, (5.9) and (5.12), are different and result in two different FPK equations. This difference is attributed mainly to the Wong-Zakai correction term, introduced in (5.14), in order to replace the physical wide band excitation by a white noise.

Example 5.1: Derive the general differential equation of the response moments of an elastic column subjected to the wide band axial load $P(t) = P_0 + \tilde{P}(t)$, where P_0 is the load mean value and $\tilde{P}(t)$ is the random fluctuation about P_0.

The transverse flexural displacement $Y(X,t)$ of the column is described by the partial differential equation (Bolotin, 1964)

$$EI \frac{\partial^4 Y(X,t)}{\partial X^4} + P(t) \frac{\partial^2 Y(X,t)}{\partial X^2} + m \frac{\partial^2 Y(X,t)}{\partial t^2} + C \frac{\partial Y(X,t)}{\partial t} = 0 \tag{i}$$

where EI is the flexural stiffness of the column, m is the mass per unit length, and C is the coefficient of viscous damping per unit length.

Let the response be represented by the first mode

$$Y(X,t) = y(t) \sin \pi X/L \tag{ii}$$

with the boundary conditions

$$Y(0,t) = Y(L,t) = \frac{\partial^2 Y(X,t)}{\partial X^2} \bigg|_{X=0,L} = 0 \tag{iii}$$

Substituting (ii) in (i) gives

$$\ddot{y}(t) + 2\zeta\omega_n \dot{y}(t) + \{\omega_n^2 - W(t)\} y(t) = 0 \tag{iv}$$

where

$$\omega_n^2 = \frac{\pi^2}{mL^2}(P_E - P_0), \quad \zeta = \frac{C}{2m\omega_n}$$

$$P_E = \pi^2 EI/L^2 \quad \text{the Euler Load,} \quad W(t) = \pi^2 \tilde{P}(t)/mL^2$$

The random component of the load $\tilde{P}(t)$ is assumed to have a flat spectral density S_0 over a wide band frequency such that $\omega \gg \omega_n$. In this case $\tilde{P}(t)$ may be assumed to approach the white noise process with the autocorrelation function

$$E[\tilde{P}(t_1)\tilde{P}(t_2)] = 2\pi S_0 \delta(t_2 - t_1) \tag{v}$$

and the autocorrelation function of $W(t)$ is

$$E[W(t_1)W(t_2)] = \sigma^2 \delta(t_2 - t_1) \tag{vi}$$

where $\sigma^2 = \pi^4(2\pi S_0)/(m^2 L^4)$

Equation (iv) can be written in terms of the state vector coordinates if we introduce the transformation $y = X_1$ and $\dot{y} = X_2$:

$$\begin{aligned} \dot{X}_1 &= X_2 \\ \dot{X}_2 &= -2\zeta\omega_n X_2 - \{\omega_n^2 - W(t)\}X_1 \end{aligned} \tag{vii}$$

It is not difficult to show that the coefficients of the FPK equation are

$$\begin{aligned} a_1 &= X_2, & a_2 &= -2\zeta\omega_n X_2 - \omega_n^2 X_1 \\ b_{11} &= b_{12} = b_{21} = 0, & b_{22} &= \sigma^2 X_1^2 \end{aligned} \tag{viii}$$

Substituting these coefficients in the forward FPK equation yields

$$\frac{\partial p}{\partial t} = -\frac{\partial}{\partial X_1}(X_2 p) + \frac{\partial}{\partial X_2}\{(2\zeta\omega_n X_2 + \omega_n^2 X_1)p\} + \frac{\sigma^2}{2}\frac{\partial^2}{\partial X_2^2}(X_1^2 p) \tag{ix}$$

Multiplying both sides of (ix) by $\Phi(\underset{\sim}{X}) = (X_1^{k_1} X_2^{k_2})$ and integrating by parts over the entire space $-\infty < \underset{\sim}{X} < \infty$ gives

$$\begin{aligned} \dot{m}_{k1,k2} = {} & k_1 m_{k1-1,k2+1} - 2\zeta\omega_n k_2 m_{k1,k2} - \omega_n^2 k_2 m_{k1+1,k2-1} \\ & + \frac{\sigma^2}{2} k_2(k_2 - 1)\, m_{k1+2,k2-2} \end{aligned} \tag{x}$$

Equation (x) is the general differential equation of the response moments. From this equation one can write any set of moment equations of any desired order. Since the moment equations of any order are consistent, it is possible to examine their stability as will be shown in chapters 6 and 7.

Example 5.2: Introducing the inertia, damping, and stiffness non-linearities in the equation of motion of the column of example 5.1

$$\ddot{y} + 2\zeta\dot{y} + \{1 - W(\tau)\}y + \alpha y(y\,\ddot{y} + \dot{y}^2) + \beta y^2\,\dot{y} + \gamma y^3 = 0 \qquad \text{(i)}$$

where α, β, and γ are the non-linear coefficients of inertia, damping, and stiffness, respectively. $W(\tau)$ is the dimensionless random loading which has the same statistical properties of example 5.1. A dot denotes differentiation with respect to the time parameter τ, and $\tau=\omega_n t$. In order to write equation (i) in terms of a state Markov vector, the non-linear inertia term $\alpha y^2\ddot{y}$ must be removed by the following approximation

$$\ddot{y} = \{-2\zeta\dot{y} - y + W(\tau)y - \gamma y^3 - \beta y^2\dot{y} - \alpha y\dot{y}^2\}(1 + \alpha y^2)^{-1} \qquad \text{(ii)}$$

Expanding the last expression in a power series and retaining terms up to cubic order, equation (ii) can then be expressed in terms of the state vector equations

$$\begin{aligned} \dot{X}_1 &= X_2 \\ \dot{X}_2 &= \{-2\zeta X_2 - X_1 + (\alpha - \gamma)X_1^3 + (2\alpha\zeta - \beta)X_1^2\,X_2 - \alpha X_1 X_2^2\} \\ &\quad + (1 - \alpha X_1)\,X_1 W(\tau) \end{aligned} \qquad \text{(iii)}$$

The coefficients a_i and b_{ij} of the FPK equation are:

$$\begin{aligned} a_1 &= X_2 \\ a_2 &= -2\zeta X_2 - X_1 + (\alpha - \gamma)\,X_1^3 + (2\alpha\zeta - \beta)\,X_1^2\,X_2 - \alpha X_1 X_2^2 \\ b_{11} &= b_{12} = b_{21} = 0, \qquad b_{22} = \sigma^2\,X_1^2(1 - 2\alpha X_1^2) \end{aligned} \qquad \text{(iv)}$$

and the system FPK equation can be written in the form

$$\begin{aligned} \frac{\partial p}{\partial \tau} = &-\frac{\partial (X_2 p)}{\partial X_1} - \frac{\partial}{\partial X_2}\{(-2\zeta X_2 - X_1 + (\alpha - \gamma)\,X_1^3 + (2\alpha\zeta - \beta)\,X_1^2\,X_2 \\ &- \alpha X_1 X_2^2)p\} + \frac{\sigma^2}{2}\cdot\frac{\partial^2}{\partial X_2^2}\{X_1^2(1 - 2\,\alpha X_1^2)p\} \end{aligned} \qquad \text{(v)}$$

Following the same procedure of example 5.1, the general differential equation of moments of order $K = k_1 + k_2$ is

$$\dot{m}_{k1,k2} = k_1 m_{k1-1,k2+1} - 2\zeta k_2 m_{k1,k2} - k_2 m_{k1+1,k2-1}$$

$$+ (\alpha - \gamma) k_2 m_{k1+3,k2-1} + (2\alpha\zeta - \beta) k_2 m_{k1+2,k2}$$

$$- \alpha k_2 m_{k1+1,k2+1} + \frac{\sigma^2}{2} k_2 (k_2 - 1)(m_{k1+2,k2-2}$$

$$- 2\alpha m_{k1+4,k2-2} + \alpha m_{k1+6,k2-2}) \qquad \text{(vi)}$$

Equation (vi) shows that the moment equations of order K are coupled with moments of higher order. Thus the generating moment equations form a set of infinite hierarchy which is difficult to solve for the response and stability. In this chapter we will discuss a number of closure techniques which may serve to close the system moment equations to any desired order.

Example 5.3: Consider the column of example 5.1 subjected to a parametric excitation which is generated by passing a white noise loading W(t) through a second order linear system (shaping filter) described by the differential equation

$$\frac{d^2Z}{dt^2} + 2\zeta_z\omega_z \frac{dZ}{dt} + \omega_z^2 Z = W(t) \qquad \text{(i)}$$

where ζ_z and ω_z are the damping factor and the natural frequency of the linear filter. The lateral motion of the column is governed by the differential equation

$$\ddot{y} + 2\zeta\omega_n \dot{y} + \omega_n^2 (1 - \nu Z) y = 0 \qquad \text{(ii)}$$

where ν is a constant which depends on the filter characteristics. Equations (i) and (ii) are coupled through the term $-\omega_n^2 \nu Zy$. The filter response will have the characteristics of a narrow band process with hidden periodicity and possesses spectral density and autocorrelation given by the following expressions

$$S_z(\omega) = \left| H(\omega) \right|^2 S_o = S_o / \{(\omega_z^2 - \omega^2)^2 + (2\zeta_z\omega_z\omega)^2\} \qquad \text{(iii)}$$

$$R_z(\tau) = E[Z^2] \exp\{-\zeta_z\omega_z |\tau|\} \cdot \{\cos\sqrt{1 - \zeta^2}\, \omega_z^\tau + \frac{\zeta_z}{\sqrt{1 - \zeta^2}}$$

$$\sin\sqrt{1 - \zeta^2}\, \omega_n |\tau|\} \qquad \text{(iv)}$$

where $H(\omega)$ is the frequency response function and $S_o = \sigma^2/2\pi$.

$$E[Z^2] = \pi S_o(2\zeta_z\omega_z^3)$$

$$= \sigma^2/(4\zeta_z\omega_z^3) \qquad \text{(v)}$$

Introducing the coordinates transformation $X_1=y$, $X_2=\dot{y}$, $X_3=Z$ and $X_4=\dot{Z}$, equations (i) and (ii) can be written in the state Markov form

$$\dot{X}_1 = X_2$$

$$\dot{X}_2 = -2\zeta\omega_n X_2 - \omega_n^2(1 - \nu X_3)X_1 \qquad \text{(vi)}$$

$$\dot{X}_3 = X_4$$

$$\dot{X}_4 = -2\zeta_2\omega_z X_4 - \omega_z^2 X_3 - W(t)$$

The corresponding FPK equation is

$$\frac{\partial p}{\partial t} = -\frac{\partial(X_2 p)}{\partial X_1} + \frac{\partial}{\partial X_2}\{(2\zeta\omega_n X_2 + \omega_n^2 X_1 - \nu\omega_n^2 X_1 X_3)p\} - \frac{\partial(X_4 p)}{\partial X_3}$$

$$+ \frac{\partial}{\partial X_4}\{(2\zeta_z\omega_z X_4 + \omega_z^2 X_3)p\} + \frac{\sigma^2}{2}\cdot\frac{\partial^2 p}{\partial X_4^2} \qquad \text{(vii)}$$

The general differential equation of moments of order $K=k_1+k_2+k_3+k_4$ is

$$\dot{m}_{k1,k2,k3,k4} = k_1 m_{k1-1,k2+1,k3,k4} + k_3 m_{k1,k2,k3-1,k4+1}$$

$$- 2(\zeta\omega_n k_2 + \zeta_z\omega_z k_4)m_{k1,k2,k3,k4} - \omega_n^2 k_2 m_{k1+1,k2-1,k3,k4}$$

$$+ \nu\omega_n^2 k_2 m_{k1+1,k2-1,k3+1,k4} - \omega_z^2 k_4 m_{k1,k2,k3+1,k4-1}$$

$$+ \frac{\sigma^2}{2} k_4(k_4 - 1)m_{k1,k2,k3,k4-2} \qquad \text{(viii)}$$

Again equation (viii) constitutes an infinite coupled set of moment equations.

5.3 THE ITÔ FORMULA APPROACH

5.3.1 MOMENT DIFFERENTIAL EQUATIONS

If the function $F(\underset{\sim}{X},t)$ of the Itô stochastic differential (4.62) is replaced by $\Phi(\underset{\sim}{X})$, relation (4.62) becomes

$$d\Phi(\underset{\sim}{X}) = \{\frac{\partial\Phi(\underset{\sim}{X})}{\partial X_i}\}^T \{d\underset{\sim}{X}\} + \frac{1}{2}\,\text{Trace}\,\{\underline{G}\,\underline{Q}\,\underline{G}^T\underline{\Phi}_{XX}(\underset{\sim}{X})\}\,dt \tag{5.17}$$

Taking expectation of both sides and dividing by dt yields the following general form of the moment differential equation

$$\dot{m}_K = \frac{d}{dt}\,E[\Phi(\underset{\sim}{X})] = E[\{\frac{\partial\Phi(\underset{\sim}{X})}{\partial X_i}\}^T\{f_i(\underset{\sim}{X},t)\}] + \frac{1}{2}\,E[\text{Trace}\,\underline{G}\,\underline{Q}\,\underline{G}^T\underline{\Phi}_{XX}(\underset{\sim}{X})] \tag{5.18}$$

where $\underset{\sim}{f}(\underset{\sim}{X},t)$ and $\underline{G}(\underset{\sim}{X},t)$ are the matrices of the state vector equations (4.52). Relation (5.18) was derived by Cumming (1967) directly from the Itô differential equation (4.52). For processes with symmetrical distributions the moment equations may be written in terms of the central moments for which all odd central moments will vanish. Introducing the change of variables

$$y_i = X_i - E[X_i] \tag{5.19}$$

the function $\Phi(\underset{\sim}{y})$ takes the form

$$\Phi(\underset{\sim}{y}) = (y_1^{k_1}\, y_2^{k_2}\, \ldots\, y_n^{k_n}) \tag{5.20}$$

The Itô stochastic differential equation (4.52) can be written in terms of y_i as

$$d\underset{\sim}{y} = \{\underset{\sim}{f}(\underset{\sim}{y} + E[\underset{\sim}{X}],t) - E[\underset{\sim}{f}(\underset{\sim}{y} + E[\underset{\sim}{X}],t)]\}\,dt + \underline{G}(\underset{\sim}{y} + E[\underset{\sim}{X}],t)\,dB(t) \tag{5.21}$$

and the corresponding differential equation for central moments becomes

$$\frac{d}{dt}\,E[\Phi(\underset{\sim}{y})] = E[\{\frac{\partial\Phi}{\partial y_i}\}^T\{f_i(\underset{\sim}{y} + E[\underset{\sim}{X}],t)\}] - E[\{\frac{\partial\Phi}{\partial y_i}\}^T]\,E[f_i(\underset{\sim}{y}+E[\underset{\sim}{X}],t)]$$
$$+ \frac{1}{2}\,E[\text{Trace}\,\underline{G}\,\underline{Q}\,\underline{G}^T\Phi_{yy}] \tag{5.22}$$

5.3.2 SEMI-INVARIANT DIFFERENTIAL EQUATIONS

Recalling the definition of the characteristic equation (2.76)

$$F_{\underset{\sim}{x}}(\underset{\sim}{\theta}) = E[\exp(i\underset{\sim}{\theta}^T\underset{\sim}{X})] \qquad \text{here } i = \sqrt{-1}$$

$$= E[\Phi(\underset{\sim}{X})] \tag{5.23}$$

where $\Phi(\underset{\sim}{X}) = \exp(i\underset{\sim}{\theta}^T\underset{\sim}{X})$

The Itô differential equation in terms of the characteristic function becomes

$$\frac{d}{dt} F_{\underset{\sim}{x}}(\underset{\sim}{\theta}) = iE[\Phi(\underset{\sim}{X})\underset{\sim}{\theta}^T\underset{\sim}{f}(\underset{\sim}{X},t)] - \frac{1}{2} E[\Phi(\underset{\sim}{X})\underset{\sim}{\theta}^T\underset{=}{G}\ \underset{=}{Q}\ \underset{=}{G}^T\underset{\sim}{\theta}] \tag{5.24}$$

Introducing the logarithm of the characteristic function

$$\chi(\underset{\sim}{\theta}) = \ell n\ F_{\underset{\sim}{x}}(\underset{\sim}{\theta}) \tag{5.25}$$

Alternatively, the anti-log of (5.25) is

$$F_{\underset{\sim}{x}}(\underset{\sim}{\theta}) = \exp\big(\chi(\underset{\sim}{\theta})\big) \tag{5.26}$$

The differential equation for the log of characteristic function may be obtained by differentiating both sides of (5.25) with respect to t

$$\frac{d}{dt} \chi(\underset{\sim}{\theta}) = \frac{1}{F_{\underset{\sim}{x}}(\underset{\sim}{\theta})} \cdot \frac{dF_{\underset{\sim}{x}}(\underset{\sim}{\theta})}{dt} \tag{5.27}$$

Equation (5.24) can be written in terms of $\chi(\underset{\sim}{\theta})$ by using (5.26) and (5.27)

$$\frac{d}{dt} \chi(\underset{\sim}{\theta}) = \left\{iE[\underset{\sim}{\theta}^T\underset{\sim}{f}(\underset{\sim}{X},t)\Phi(\underset{\sim}{X})] - \frac{1}{2} E[\underset{\sim}{\theta}^T\underset{=}{G}\ \underset{=}{Q}\ \underset{=}{G}^T\underset{\sim}{\theta}\Phi(\underset{\sim}{X})]\right\}\exp\{-\chi(\theta)\} \tag{5.28}$$

Introducing the definition of the semi-invariant (cumulant), (2.81) of order $K = \sum_{j=1}^{n} k_j$,

$$\lambda_{k1,k2,\ldots,kn} = (-i)^K \frac{\partial^K \chi(\underset{\sim}{\theta})}{\partial\theta_1^{k_1}\cdots\partial\theta_n^{k_n}} \tag{5.29}$$

the differential equations of the semi-invariants of order K may be obtained by differentiating (5.28) repeatedly with respect to θ_i a total of K times and evaluating the result at $\underset{\sim}{\theta}=\underset{\sim}{0}$. This technique was introduced by Assaf and Zirkle (1976) for problems of non-linear closed-loop systems in automatic control. It is not difficult to show that the first four differential equations for the semi-invariants are:

First order equations

$$\dot{\lambda}_{k_i=1} = E[f_i(\underset{\sim}{X},t)] \tag{5.30}$$

Second order equations

$$\dot{\lambda}_{k_i,k_j=1} = E[y_j f_i(\underset{\sim}{X},t) + y_i f_j(\underset{\sim}{X},t)] + E[\beta_{ij}] \tag{5.31}$$

Third order equations

$$\dot{\lambda}_{k_i,k_j,k_\ell=1} = E[(y_j y_\ell - \lambda_{j\ell})f_i(\underset{\sim}{X},t) + (y_i y_\ell - \lambda_{i\ell})f_j(\underset{\sim}{X},t)$$

$$+ (y_i y_j - \lambda_{ij})f_\ell(\underset{\sim}{X},t)] + E[y_i\beta_{j\ell} + y_j\beta_{i\ell} + y_\ell\beta_{ij}] \tag{5.32}$$

Fourth order equations

$$\begin{aligned}\dot{\lambda}_{k_i=1, k_j=1, k_\ell=1, k_r=1} = {} & E[(y_j y_\ell y_r - y_j\lambda_{\ell r} - y_\ell\lambda_{jr} - y_r\lambda_{j\ell} - \lambda_{j\ell r})f_i(\underset{\sim}{X},t) \\ & + (y_i y_\ell y_r - y_i\lambda_{\ell r} - y_\ell\lambda_{ir} - y_r\lambda_{i\ell} - \lambda_{i\ell r})f_j(\underset{\sim}{X},t) \\ & + (y_i y_j y_r - y_i\lambda_{jr} - y_j\lambda_{ir} - y_r\lambda_{ij} - \lambda_{ijr})f_\ell(\underset{\sim}{X},t) \\ & + (y_i y_j y_\ell - y_i\lambda_{j\ell} - y_j\lambda_{i\ell} - y_\ell\lambda_{ij} - \lambda_{ij\ell})f_r(\underset{\sim}{X},t)] \\ & + E[(y_\ell y_r-\lambda_{\ell r})\beta_{ij} + (y_j y_r-\lambda_{jr})\beta_{i\ell} + (y_j y_\ell-\lambda_{j\ell})\beta_{ir} \\ & + (y_i y_r-\lambda_{ir})\beta_{j\ell} + (y_i y_\ell-\lambda_{i\ell})\beta_{jr} + (y_i y_j-\lambda_{ij})\beta_{\ell r}]\end{aligned} \tag{5.33}$$

where $\beta_{ij} = (\underline{G}\ \underline{Q}\ \underline{G}^T)_{ij}$,

$\lambda_{i\ell} = \lambda_{k_i=1,\ k_\ell=1}$ and so on.

Example 5.4: Consider a second order stochastic differential equation with random fluctuation in the damping and stiffness coefficients

$$\ddot{y} + \{2\zeta\omega_n + \xi_2(t)\}\dot{y} + \{\omega_n^2 - \xi_1(t)\}y = 0 \tag{i}$$

Derive the general differential equation of the response moments by using the Itô formula approach.

The parametric excitations $\xi_1(t)$ and $\xi_2(t)$ are assumed stationary wide band Gaussian processes. These physical random processes may be replaced by white noise processes $W_1(t)$ and $W_2(t)$, respectively, via the Wong-Zakai transformation. Both $W_1(t)$ and $W_2(t)$ may be expressed as the "formal" derivative of the Brownian motion process, $W_i(t)=dB_i(t)/dt$. The state vector equations are obtained through the coordinate transformation $y=X_1$, $\dot{y}=X_2$:

$$dX_1 = X_2 dt \tag{ii}$$

$$dX_2 = -\left\{(\omega_n^2 - \frac{\sigma_{12}^2}{2})X_1 + (2\zeta\omega_n - \frac{\sigma_{22}^2}{2})X_2\right\} + X_1 dB_1(t) - X_2 dB_2(t)$$

where $E[dB_i(t)\ dB_j(t)] = \sigma_{ij}^2 dt$

Introducing the scalar function $\Phi(\underset{\sim}{X})=(X_1^{k_1}\ X_2^{k_2})$, the following expressions are obtained:

$$\left\{\frac{\partial\Phi}{\partial X_i}\right\}^T = \left\{k_1 X_1^{k_1-1}\ X_2^{k_2} \qquad k_2 X_1^{k_1}\ X_2^{k_2-1}\right\}$$

$$\left[\frac{\partial^2\Phi}{\partial X_i \partial X_j}\right] = \begin{bmatrix} k_1(k_1-1)X_1^{k_1-2}\ X_2^{k_2} & k_1 k_2 X_1^{k_1-1}\ X_2^{k_2-1} \\ k_1 k_2 X_1^{k_1-1}\ X_2^{k_2-1} & k_2(k_2-1)X_1^{k_1}\ X_2^{k_2-1} \end{bmatrix}$$

$$\underset{\sim}{f} = \begin{bmatrix} 0 & 1 \\ -(\omega_n^2 - \frac{\sigma_{12}^2}{2}) & -(2\zeta\omega_n - \frac{\sigma_{22}^2}{2}) \end{bmatrix} \begin{Bmatrix} X_1 \\ X_2 \end{Bmatrix}$$

$$\underset{=}{G} = \begin{bmatrix} 0 & 0 \\ X_1 & -X_2 \end{bmatrix}, \qquad \underset{=}{Q}\ dt = \begin{bmatrix} \sigma_1^2 & \sigma_{12}^2 \\ \sigma_{12}^2 & \sigma_2^2 \end{bmatrix} dt$$

Substituting these expressions into the Itô formula (5.18), the following general differential equation of the response moments is obtained

$$\dot{m}_{k1,k2} = k_1 m_{k1-1,k2+1} + k_2\{\omega_n^2 - (k_2 - 0.5)\sigma_{12}^2\} m_{k1+1,k2-1}$$

$$+ \frac{1}{2} k_2(k_2\sigma_{22}^2 - 4\zeta\omega_n)m_{k1,k2} + \frac{1}{2} k_2(k_2 - 1)m_{k1+2,k2-2} \tag{iii}$$

5.4 THE INFINITE HIERARCHY PROBLEM

It has been shown in examples 5.2 and 5.3 that for linear systems with random coefficients generated from a linear (or non-linear) shaping white noise filter, or for non-linear systems driven by a white noise, the resulting moment equations are coupled with higher order moments. This infinite hierarchy creates difficulties in solving for the response or stability of such systems. The problem can be circumvented if a "proper" closure scheme is used to truncate the response statistical equations into a finite set. The scheme is said to be proper if it leads to results which do not violate the moment properties and other statistical conditions such as Schwarz's inequality. From the engineering point of view a truncation scheme is acceptable if the predicted response statistics are very close to the experimentally measured ones. Unfortunately, the literature lacks any correlation between the current existing methods and the experimental results.

The problem of infinite coupled moments may be described by the general equation

$$\frac{dM_i}{dt} = \psi_i(M_1, M_2, \ldots, M_i, M_{i+1}, \ldots) \qquad i=1,2,\ldots \tag{5.34}$$

with the initial conditions $M_i(t=0) = c_i$. The notation M_i is used to refer to the exact solution of the infinite set which is unknown. All truncation schemes reduce system (5.34) into a finite set of the form

$$\frac{dm_i}{dt} = \bar{\psi}_i(m_1, m_2, \ldots, m_i) \tag{5.35}$$

where $m_i(t=0) = b_i$. Here m_i are the approximate solution after truncation. A closure scheme is said to be valid if the error is very small, i.e.,

$$\left| M_i - m_i \right| < \varepsilon \tag{5.36}$$

where ε is a very small parameter which measures the deviation of the approximate solution m_i from the exact solution M_i. However, this condition is difficult to be verified since the exact solution is unknown. Alternatively, one may examine the validity of the solution from the properties of the moments. Among these properties the non-negativity of the mean squares and the Schwarz's inequality

$$(E[X])^2 \leq E[X^2] < \infty \tag{5.37}$$

are the most convenient criteria.

Bellman and Richardson (1968) and Wilcox and Bellman (1970) developed two lemmas which guarantee these properties for a limited class of one-dimensional systems. During the past twenty years a number of closure schemes have been developed for various applied mathematics and engineering problems. Some truncation assumptions were introduced by Kraichnan (1962) and Beran (1968) for the fluid turbulence theory, Keller (1964) and Haines (1967) for the stochastic eigenvalue of elastic structures, and Adomian (1971) and Adomian and Malakian (1979) for wave propagation in random media. Kistner (1977,1978) employed Lie algebra to develop a closure method applicable only to linear systems described by the first order differential equations

$$\dot{\underset{\sim}{X}} = \{\underline{A} + \underline{B}\, \underset{\sim}{\xi}(t)\}\underset{\sim}{X} \tag{5.38}$$

where $\underline{A}$ and $\underline{B}$ are square matrices and $\underset{\sim}{\xi}(t)$ is a vector of colored noise process with stationary Gaussian elements. The method depends on the structure of the Lie algebra generated by the matrices $\underline{A}$ and $\underline{B}$ (Belinfante, et al., 1966 and Sagirow, 1976). Closed moment equations have been given by Sagirow (1976) for the case of Abelian Lie algebra (with the property $\underline{AB} = \underline{BA}$). For the case of a general Lie algebra, Kistner (1977) outlined a scheme for deriving closed first and second order moment equations.

The basic linearized closure schemes based on a mean square approximation technique and Gaussian properties of the response have been established by Richardson (1964), Richardson and Levitt (1967), Bellman and Richardson (1968), and Wilcox and Bellman (1970). Sancho (1968,1969,1970a,b,c) applied the criteria of moment-preserving properties, established originally by Bellman, et al., to the moment equations of a special class of non-linear stochastic systems. Unfortunately, these techniques can be regarded as linearization procedures to the original non-linear systems, and the results obtained therein may not represent the true behavior of the system.

The available closure techniques may be classified into Gaussian and non-Gaussian schemes. The difference between the two classes is similar to a great extent to the difference between linear and non-linear analysis of deterministic mechanics. However, the closure schemes involve enormous difficulties which have not yet been well resolved.

5.4.1 GAUSSIAN CLOSURE SCHEMES

Three closure schemes have been developed based on the assumption that the response processes are "nearly" Gaussian. These include the cumulant truncation scheme, the central moment method, and the mean square closure technique. The cumulant closure scheme is regarded as the most powerful method for systems possessing small non-linearities. It has been indicated in section (2.3.8) that all semi-invariants of order greater than two vanish identically for Gaussian distributed processes. In this case the third and fourth order semi-invariants given by relations (2.82) may be set to zero, and one can express third and fourth order moments in terms of first and second order moments.

5.4.2 NON-GAUSSIAN CLOSURE METHODS

The concept of non-Gaussian probability density has been introduced in section 2.3.10 in terms of the Gram-Charlier and Edgeworth asymptotic expansions. The validity of these expansions depends on several factors such as their convergence and the positiveness of the probability density over the entire space domain. The basic ingredients of these expansions are the Gaussian joint probability density, which appears as the first term in the expansion, the quasi-moments, or the semi-invariants. The quasi-moments, semi-invariants, and moments are related by definite relationships given by Kendall and Stuart (1969). For non-Gaussian processes the semi-invariants of order greater than the second do not vanish. However, their contribution diminishes as their order increases if the process is slightly deviated from Gaussian. Thus one would expect that the first few terms will give an adequate measure of the deviation of the process from normality.

For dynamic systems with quadratic non-linearity, the differential equations of the response moments of order n will contain moments of order n+1. If cubic non-linearity is included, the n-th moment equations will contain moments of order n+2, and so on, for higher order non-linearities. Thus the first order non-Gaussian closure can be established by setting fifth order semi-invariants to zero for the case of quadratic non-linearity, or setting fifth and sixth order semi-invariants to zero for the case of cubic non-linearity, and so on, for higher order non-linearities. Second order approximation can be obtained by generating differential equations of the response moments up to fifth order and setting sixth order semi-invariants to zero for the case of quadratic non-linearity. For the case of cubic non-linearity, seventh and eighth semi-invariants are equated to zero. This procedure is referred to as the cumulant-neglect or cumulant-discard.

CHAPTER 6

Stochastic Averaging Methods

6.1 INTRODUCTION

The concept of the averaging principle, developed by Bogoliubov and Mitropol'skii (1961) for deterministic non-linear vibration, has been extended to solve stochastic differential equations. These equations include rapid fluctuations which are averaged to generate a set of simple equations for slowly fluctuating response coordinates. The stochastic averaging involves a procedure to take into account the effect of a random excitation multiplied by a correlated term. The basic method was established by Stratonovich (1963). Khas'miniskii (1966) developed a limit theorem, outlined in section 4.7.5, based on a rigorous physical treatment which justifies the application of the averaging principle to systems subjected to small intensity random excitations. Papanicolaou and Kohler (1974) provided another limit theorem which handles a wide class of random processes defined by non-linear stochastic differential equations.

The limit theorems require certain conditions concerning the random excitations and the system parameters such that, if these conditions are satisfied, the response process converges to a diffusion Markov process. Sethna and Orey (1980) have generalized the averaging method, developed originally by Sethna (1972,1973), which requires less restrictive hypotheses. They assumed that the system differential equations depend on two time scales; one is fast and the other is slow. Other versions of stochastic averaging and limit theorems have been developed by Chelpanov (1962), Weidenhammer (1964), Vrkoc (1966), Gikhman and Skorokhod (1972), and Howe (1974). The advantage of these methods is that they reduce the dimension of the response coordinates such that one can deal with the Fokker-Planck equation by considering the slowly varying amplitudes as independent coordinates. However, the effect of some terms such as cubic stiffness and special forms of non-linear inertia is lost during the averaging procedure. The effect of such non-linearities can be determined by performing a second order averaging which in most cases requires tedious mathematical manipulations. Alternatively, it is possible to perform the stochastic averaging on the energy envelope of "quasi-conservative" systems. In this case the energy envelope varies slowly during each period of oscillation and constitutes a Markov process. The probability density of the energy envelope can then be described by the FPK equation.

The versatility of the stochastic averaging schemes encouraged several researchers to investigate the random behavior of dynamic systems under random parametric excitations. However, the application of these methods to non-linear systems is very limited, partly because of the difficulty involved in solving the associated FPK equation.

In this chapter the basic concept of the stochastic averaging methods will be outlined and demonstrated by a number of examples. The recent results of dynamical systems treated by various stochastic averaging methods will be discussed.

6.2 BASIC CONCEPT

The stochastic averaging principle is identical to the Bogoliubov and Mitropol'skii procedure in every respect with one exceptional difference which provides a special treatment to stochastic terms. Basically, the method replaces random parametric terms by the sum of mean values and fluctuation components about the mean. The deterministic averaging method requires the equations of motion to be written in the standard form

$$\ddot{Y}_i + \omega_i^2 Y_i = \varepsilon F_i(Y_j, \dot{Y}_j, t) \qquad i,j = 1,2,\ldots,n \tag{6.1}$$

where Y_i are the normal coordinates, ε is a small parameter, and F_i are bounded non-linear functions in Y_j and $\dot{Y}_j$ and may contain forcing terms. The solution of (6.1) has a similar form to the linear solution ($\varepsilon = 0$), but both amplitudes and phase will be slowly varying time functions

$$Y_i(t) = a_i(t) \cos\phi_i(t) \tag{6.2}$$

where $\phi_i(t) = \omega_i t + \theta_i(t)$

$a_i(t)$ represent the full amplitudes and $\phi_i(t)$ are the full phase angles of the response.

In weakly non-linear systems the variations in a_i and θ_i are small and thus occur over a time interval slightly greater than the period of oscillation of the mode in question. Accordingly, the non-linear terms act as sources and sinks that generate or absorb very small amounts of energy during one cycle of oscillation. Substituting (6.2) into (6.1), such that

$$\dot{Y}_i(t) = - a_i\omega_i \sin\phi_i \tag{6.3}$$

transforms the n-differential equations into 2n-first order differential equations in the amplitudes and phases. The right-hand sides of the resulting equations are periodic, and the time derivative of the amplitudes and phases is of order ε which implies that they change very slowly during the time period $T_i = 2\pi/\omega_i$. By taking the time average over one complete interval, during which both a_i and θ_i are treated as constants, we obtain 2n-first order differential equations free from oscillatory terms. The solution of these equations can then be determined via analytical or numerical integration techniques. The resulting solution is regarded as the first order averaging solution. Higher order solutions can be developed; however, the procedure becomes involved mathematically.

6.3 STANDARD FORM OF STOCHASTIC AVERAGING

The basic concept of the deterministic averaging principle can be extended to the ordinary set of random differential equations

$$\ddot{Y}_i + \omega_i^2 Y_i = \varepsilon F_i\left(Y_j, \dot{Y}_j, \xi(t)\right), \qquad i=1,2,\ldots,n \tag{6.4}$$

where $\xi(t)$ is a stationary, measurable random process with zero mean and satisfies condition (4.84).

Introducing solution (6.2) subject to condition (6.3) and differentiating (6.2) with respect to the time gives

$$\dot{Y}_i(t) = \dot{a}_i(t)\cos\phi_i(t) - a_i(t)\{\omega_i + \dot{\theta}_i(t)\}\sin\phi_i(t) \tag{6.5}$$

Equating (6.3) and (6.5) yields

$$\dot{a}_i \cos\phi_i - a_i\dot{\theta}_i \sin\phi_i = 0 \tag{6.6}$$

Differentiating (6.3) with respect to t, gives

$$\ddot{Y}_i(t) = -\omega_i^2 a_i \cos\phi_i - \omega_i\dot{a}_i \sin\phi_i - \omega_i\dot{\theta}_i a_i \cos\phi_i \tag{6.7}$$

Substituting (6.7) into (6.4) and using (6.2) and (6.3) yields

$$\omega_i\dot{a}_i \sin\phi_i + \omega_i\dot{\theta}_i a_i \cos\phi_i = -\varepsilon F_i\{a_j \cos\phi_j, -\omega_j a_j \sin\phi_j, \xi(t)\} \tag{6.8}$$

Solving (6.6) and (6.8) for $\dot{a}_j$ and $\dot{\theta}_i$ results in the 2n-first order differential equations

$$\dot{a}_i = -\frac{\varepsilon}{\omega_i} F_i\{a_j \cos\phi_j, -\omega_j a_j \sin\phi_j, \xi(t)\} \sin\phi_i(t) \tag{6.9}$$

$$\dot{\theta}_i = -\frac{\varepsilon}{\omega_i a_i} F_i\{a_j \cos\phi_j, -\omega_j a_j \sin\phi_j, \xi(t)\} \cos\phi_i(t) \tag{6.10}$$

Equations (6.9) and (6.10) are referred to as the "standard form" equations and are equivalent to the system equations of motion (6.4). It is seen that both $\dot{a}_i$ and $\dot{\theta}_i$ are of order ε; hence, they are varying slowly with respect to time. Thus the changes in a_i and θ_i over the time interval $T_i = 2\pi/\omega_i$ are very small, i.e.,

$$\left| a_i(t + T_i) - a_i(t) \right| << a_i(t)$$

Accordingly, the right-hand sides of equations (6.9) and (6.10) may be replaced by their average values over one complete period of oscillation, where a_i and θ_i are treated as constants on the right-hand sides. If $\xi(t)$ are dropped, the method becomes identical to the deterministic procedure. All non-linear terms in F_i will generally contain products of sine and cosine functions. These products may be expanded into trigonometric functions at the multiple or combination phase angles ϕ_i. The functions which involve higher order multiple or combination phase angles represent high frequency oscillatory motions imposed on the slowly varying amplitudes and phases. If one is concerned with the stationary response, the high frequency oscillations may be eliminated during the averaging process over a long period of time.

At this stage it is convenient to replace the amplitude and phase vectors by the 2n-dimension vector $\underset{\sim}{X}$, i.e.,

$$\underset{\sim}{X} = \begin{Bmatrix} \underset{\sim}{a} \\ \underset{\sim}{\theta} \end{Bmatrix} \tag{6.11}$$

Equations (6.9) and (6.10) can be represented by the equivalent set of differential equations

$$\dot{X}_i = \varepsilon F_i\{X_j, \xi(t), t, \varepsilon\} \qquad i=1,2,\ldots,2n \tag{6.12}$$

If the time average of F_i exists

$$F_i^*(\underset{\sim}{X}^*) = \lim_{T\to\infty} \frac{1}{T}\int_0^T F_i\{\underset{\sim}{X}, \xi(t), t, \varepsilon\}\, dt < \infty \tag{6.13}$$

then solution of (6.12) will not differ significantly from the solution of the averaged equations

$$\dot{X}_i^* = \varepsilon F_i^*(\underset{\sim}{X}, S), \qquad \underset{\sim}{X}^*(t_o) = \underset{\sim}{X}_o \tag{6.14}$$

over a time interval of order $(1/\varepsilon)$. In (6.14) S represents the power spectral density of $\xi(t)$.

According to the limit theorems developed by Khas'miniskii (1966) and Papanicolaou and Kohler (1974) the solution process of (6.12) will converge in a certain sense to a Markov process as $\varepsilon \to 0$. These limit theorems require certain conditions to be satisfied regarding the functions F_i and the excitation vector $\xi(t)$ such as the smallness of the intensity of $\xi(t)$, the boundedness of the functions F_i, and their mixed partial derivatives with respect to $\underset{\sim}{X}$.

Based on the Khas'miniskii limit theorem, the system of differential equations (6.12) can be written in the form

$$\dot{X}_i = F_i^0(\underset{\sim}{X},\xi(t),t) + \varepsilon F_i^1(\underset{\sim}{X},\xi(t),t) \tag{6.15}$$

and the solution $X(\varepsilon^2 t)$ converges weakly as $\varepsilon \to 0$ to a Markov process $\underset{\sim}{X}^*$ which is continuous with probability one. Let $\tau = \varepsilon^2 t$; the solution is governed by the solution of the system of Itô's equations

$$dX_i^*(\tau) = f_i(\underset{\sim}{X}^*)\, d\tau + \sum_{k=1}^{2n} G_{ik}(\underset{\sim}{X}^*)\, dB_k(\tau) \tag{6.16}$$

where the drift f_i and diffusion G_{ik} coefficients are defined by relations (4.87) and (4.86). In performing the limits (4.86) precautions must be taken regarding the consistency of the dimensions of F_i^0, since half of the vector $\underset{\sim}{X}$ represents the amplitudes a_i and the other half are the phase angles.

Frequently equations (6.15) are written in the following general form

$$\dot{X}_i = \varepsilon^{2r} F_i^1(\underset{\sim}{X},t) + \varepsilon^r F_i^0\{\underset{\sim}{X},\xi(t),t\} \tag{6.17}$$

in most cases the power r is taken 0.5 or 1.

In terms of the amplitude and phase coordinates, equations (6.9) and (6.10) are written in the general form of (6.17)

$$\dot{a}_i = \{\varepsilon^{2r} \hat{F}_i^1(a_j,\theta_j,t) + \varepsilon^r \hat{F}_i^0(a_j,\theta_j,\xi(t),t)\} \sin\phi_i \tag{6.18}$$

$$\dot{\theta}_i = \frac{1}{a_i} \{\varepsilon^{2r} \hat{F}_i^1(a_j,\theta_j,t) + \varepsilon^r \hat{F}_i^0(a_j,\theta_j,\xi(t),t)\} \cos\phi_i \tag{6.19}$$

and the corresponding Itô equations are

$$da_i^* = \varepsilon^{2r} f_i(\underset{\sim}{a})\, dt + \varepsilon^r \sum_{j=1}^{n} G_{ij}(\underset{\sim}{a})\, dB_j(t) \tag{6.20}$$

$$d\theta_i^* = \varepsilon^{2r} \hat{f}_i(\underset{\sim}{a})\, dt + \varepsilon^r \sum_{j=1}^{n} \hat{G}_{ij}(\underset{\sim}{a})\, d\hat{B}_j(t) \tag{6.21}$$

The expressions of the drift and diffusion coefficients of (6.20) and (6.21) were expressed in detail by Ariaratnam and Srikantaiah (1978). It is observed that the averaged amplitude and phase equations are free from any oscillatory terms.

6.4 PAPANICOLAOU AND KOHLER ASYMPTOTIC THEOREM

Papanicolaou and Kohler (1974) developed an asymptotic theorem which guarantees the convergence of a wide class of random processes, defined by stochastic differential equations, to a diffusion Markov process. The stochastic equations do not necessarily have to be of the Itô type. The theorem and its conditions will be stated without proof.

Let $\underset{\sim}{F}(\tau,\underset{\sim}{X},t)$ be a random vector field, such that $\tau \in [0,T]$, $\underset{\sim}{X} \in R^{2n}$, and $t \in [0,\infty)$, that satisfies the following conditions:

i. F is jointly measurable with respect to its arguments.

ii. There is a constant C which is independent of the arguments of F such that:

$$\left| F_i(\tau,\underset{\sim}{X},t) \right| \leq C(1 + |X|) \tag{6.22a}$$

$$\left| \frac{\partial F_i(\tau,\underset{\sim}{X},t)}{\partial X_j} \right| \leq C \tag{6.22b}$$

$$\left| \frac{\partial^2 F_i(\tau,\underset{\sim}{X},t)}{\partial X_j \partial X_k} \right| \leq C(1 + |X|^q) \tag{6.22c}$$

$$\left| \frac{\partial^3 F_i(\tau,\underset{\sim}{X},t)}{\partial X_j \partial X_k \partial X_\ell} \right| \leq C(1 + |X|^q) \tag{6.22d}$$

$$\left| \frac{\partial^4 F_i(\tau,\underset{\sim}{X},t)}{\partial X_j \partial X_k \partial X_\ell \partial X_m} \right| \leq C(1 + |X|^q) \qquad i,j,k,\ell,m=1,2,\ldots,2n \tag{6.22e}$$

where the integer $q \geq 0$

iii. For the same constant C the following expectations exist

$$\begin{aligned} &E^{1/2}[\{F_i(s+h,\underset{\sim}{X},t) - F_i(s,\underset{\sim}{X},t)\}^2] \leq Ch(1 + |X|) \\ &E^{1/2}[\{\frac{\partial F_i(s+h,\underset{\sim}{X},t)}{\partial X_j} - \frac{\partial F_i(s,\underset{\sim}{X},t)}{\partial X_j}\}^2] \leq Ch \end{aligned} \tag{6.23}$$

Consider the stochastic differential equations

$$\frac{d\underset{\sim}{X}(t,s,X)}{dt} = \varepsilon \underset{\sim}{F}\big(\varepsilon^2 t, \underset{\sim}{X}(t,s,\underset{\sim}{X}),t\big) + \varepsilon^2 \underset{\sim}{F}^1\big(\varepsilon^2 t, \underset{\sim}{X}(t,s,\underset{\sim}{X}),t\big) \qquad t > s \tag{6.24}$$

where $\underset{\sim}{F}^1$ is a random vector field satisfying the same conditions as F, ε is a real parameter in (0,1]. In order to examine the behavior of the random process $\underset{\sim}{X}$ as $\varepsilon \to 0$ with $\varepsilon^2 t$ remaining fixed, the following scaled variables are introduced

$$\tau = \varepsilon^2 t, \quad s_1 = \varepsilon^2 s, \quad \underset{\sim}{X}^{\varepsilon}(\tau, s_1, \underset{\sim}{X}) = \underset{\sim}{X}\Big(\frac{\tau}{\varepsilon^2}, \frac{s_1}{\varepsilon^2}, \underset{\sim}{X}\Big) \tag{6.25}$$

equation (6.24) becomes

$$\frac{d\underset{\sim}{X}^{\varepsilon}(\tau,s_1,\underset{\sim}{X})}{d\tau} = \frac{1}{\varepsilon} \underset{\sim}{F}\Big(\tau, \underset{\sim}{X}^{\varepsilon}(\tau,s_1,\underset{\sim}{X}), \frac{\tau}{\varepsilon^2}\Big) + \underset{\sim}{F}^1\Big(\tau, \underset{\sim}{X}^{\varepsilon}(\tau,s_1,\underset{\sim}{X}), \frac{\tau}{\varepsilon^2}\Big) \qquad 0 \le s_1 < \tau \le T \tag{6.26}$$

It was shown that the process $\underset{\sim}{X}^{\varepsilon}$ converges weakly to a diffusion Markov process as $\varepsilon \to 0$. Because of the factor $1/\varepsilon$ in (6.26) it is clear that if a limit is to exist as $\varepsilon \to 0$, the random vector field $\underset{\sim}{F}$ must be centered. It is assumed that the mean value of $\underset{\sim}{F}$ is zero. The drift and diffusion coefficients are given by the following limiting expressions

$$f_i(\tau,\underset{\sim}{X}) = \lim_{\varepsilon\to 0} \frac{1}{\varepsilon^3} \int_{\tau}^{\tau+\varepsilon} \int_{\tau}^{s_1} \sum_{i=1}^{n} E\Big[F_i\Big(\tau,\underset{\sim}{X},\frac{s}{\varepsilon^2}\Big) \frac{\partial F_j(\tau,\underset{\sim}{X},s_1/\varepsilon^2)}{\partial X_i}\Big]\, ds\, ds_1$$

$$+ \lim_{\varepsilon\to 0} \frac{1}{\varepsilon} \int_{\tau}^{\tau+\varepsilon} E[F_j^1(\tau,\underset{\sim}{X},s/\varepsilon^2)]\, ds \tag{6.27}$$

$$G_{ij}(\tau,\underset{\sim}{X}) = \lim_{\varepsilon\to 0} \frac{1}{\varepsilon^3} \int_{\tau}^{\tau+\varepsilon} \int_{\tau}^{s_1} E[F_i(\tau,\underset{\sim}{X},s/\varepsilon^2)\, F_j(\tau,\underset{\sim}{X},s_1/\varepsilon^2)]\, ds\, ds_1 \qquad i,j = 1,2,\ldots,2n \tag{6.28}$$

The drift and diffusion coefficients possess the same regularity of $\underset{\sim}{F}$ and $\underset{\sim}{F}^1$. Other properties include the definiteness and non-negativeness of the diffusion matrix $[\underset{-}{G}(\tau,\underset{\sim}{X})]$.

Statement of the Theorem: Let $X^{\varepsilon}(\tau,s_1,\underset{\sim}{X})$ be a process defined by the stochastic differential equation (6.26), and the conditions (6.22) and (6.23) hold, then the processes $\underset{\sim}{X}$ converge weakly as $\varepsilon \to 0$ to a diffusion Markov process with the infinitesimal generator

$$\mathcal{L}\, p(\underset{\sim}{X}) = \sum_{j=1}^{2n} f_j(s_1,\underset{\sim}{X}) \frac{\partial p(\underset{\sim}{X})}{\partial X_j} + \sum_{i,j=1}^{2n} G_{ij}(s_1,\underset{\sim}{X}) \frac{\partial^2 p(\underset{\sim}{X})}{\partial X_i \partial X_j} \tag{6.29}$$

where $p(\underset{\sim}{X}) \in C^{4,r}(R^{2n})$, $r \geq 0$

Let $\bar{p}(s_1,\tau,\underset{\sim}{X})$ denote the solution of the partial differential equation

$$\partial \bar{p}(s_1,\tau,\underset{\sim}{X})/\partial s + \mathcal{L}\, \bar{p}(s_1,\tau,\underset{\sim}{X}) = 0 \tag{6.30}$$

$$0 < s < \tau < T$$

then there exists an integer $\bar{r} \geq r+4$ such that

$$\left| E\left[p\left(\underset{\sim}{X}(\tau,s_1,\underset{\sim}{X})\right) \middle| P_0^{s_1/\varepsilon^2} \right] - \bar{p}(s_1,\tau,\underset{\sim}{X}) \right| \leq \varepsilon C(p,T)\cdot(1+|X|^{\bar{r}}) \tag{6.31}$$

where $C(p,T)$ denotes a constant which depends on p and its derivatives up to order 4, on T and r but is independent of $\underset{\sim}{X}$ and ε.

This theorem differs from the Khas'miniskii limiting theorem (outlined in section 4.7.5) in that it does not deal with linear problems since the vector random fields $\underset{\sim}{F}$ and $\underset{\sim}{F}^1$ are uniformly bounded in $\underset{\sim}{X}$. Other properties have been pointed out by Papanicolaou and Kohler regarding the stationarity of $\underset{\sim}{F}$ and $\underset{\sim}{F}^1$, and the convergence of moments of the process X_i.

6.5 WEIDENHAMMER METHOD

Weidenhammer (1964) combined the concepts of the averaging principle and the perturbation technique in one procedure. However, the method depends on a critical assumption that the response processes are considered Gaussian when the parametric random excitations are Gaussian. The method follows exactly the same steps of section 6.3 to derive the standard equations of the amplitude and phase, (6.9) and (6.10), respectively. The solution of these differential equations may be written in the form

$$a(t) = a_0 \exp\{A(t)\} \tag{6.32}$$

$$\theta(t) = \theta_0 + \Theta(t) \tag{6.33}$$

where a_0 and θ_0 are constants. Both $A(t)$ and $\Theta(t)$ involve integrals of trigonometric and excitation functions. The amplitude and phase are expressed in power series in ε as

$$A(t) = \varepsilon A_1 + \varepsilon^2 A_2 + \ldots \tag{6.34}$$

$$\Theta(t) = \varepsilon\Theta_1 + \varepsilon^2\Theta_2 + \ldots \tag{6.35}$$

The argument of every trigonometric function in (6.32) and (6.33) will be replaced by (6.35) from which the sine and cosine functions can be expanded in terms of $(n\omega t + n\theta_0)$ and $n\,\Theta$. In the analysis $n\,\Theta$ are assumed very small such that

$$\sin n\,\Theta_i = n\,\Theta_i \quad \text{and} \quad \cos n\,\Theta_i = 1.$$

Based on the assumption that $a(t)$ and $\theta(t)$ are "approximately" Gaussian, it is possible to express their joint probability density $p(a,\theta)$ in a Gaussian form. Weidenhammer employed the method to determine the stochastic stability of linear systems with random coefficients which have specific forms of spectral density. These include excitations with exponential correlation function, delta correlation function and limited-band white noise. Graefe (1966) applied the method to random parametric excitations having arbitrary power spectrum. Example 6.2 (page 126) will demonstrate the method.

6.6 STOCHASTIC AVERAGING OF THE RESPONSE ENERGY ENVELOPE

This approach is very useful for "quasi-conservative" systems with non-linear restoring and inertia forces. The method is essentially based on constructing the FPK equation for the average energy envelope of the system response. It was first introduced by Stratonovich (1963) and has been extensively applied for non-linear systems under external excitations [see, for example, J. B. Roberts (1978) and Spanos (1978)]. For dynamic systems subjected to random parametric excitations, the method has recently been employed by Dimentberg and Menyailov (1979), Dimentberg (1980a), and Zhu (1983a,b). Unlike the standard averaging method, the average of the energy envelope takes into account the influence of a wide class of non-linearities. For small random excitation and light linear damping the energy envelope varies slowly during each period of oscillation, and this energy envelope can be treated as a one-dimensional Markov process. Thus the probability of the average energy envelope can be described by the FPK equation.

The method will be described with reference to random parametric vibration as outlined by Zhu (1983a). He considered a general non-linear system with random coefficients described by the differential equation of motion

$$\ddot{Y} + g(Y) = -\,\varepsilon F(Y,\dot{Y}) + \varepsilon^{1/2} \sum_{i=1}^{\ell} h_i(Y,\dot{Y})\xi_i(t) \tag{6.36}$$

where $g(Y)$ is an odd monotonically increasing continuous function representing the system elastic restoring force, ε is a small positive parameter, the function $F(Y,Y)$ includes linear and non-linear damping and inertia terms, and the functions $h_i(Y,Y)$ contain those stiffness and damping components associated with the random fluctuations $\xi_i(t)$. The parametric excitations $\xi_i(t)$ are assumed physically white noise independent processes in the Stratonovich sense. The effects of the mean squares of the random excitations and the damping terms over one period of oscillation are assumed small (of order ε).

The energy envelope V of the conservative system (obtained by setting ε to zero in equation (6.36)) is

$$V = \frac{1}{2}\dot{Y}^2 + G(Y) \tag{6.37}$$

where the first term represents the kinetic energy of the system and the second term is the potential energy given by the expression

$$G(Y) = \int_0^Y g(y)\, dy \tag{6.38}$$

From relation (6.37) the velocity can be written in the form

$$\dot{Y} = \pm\sqrt{2\{V - G(Y)\}} \tag{6.39}$$

It is seen that the sign of $\dot{Y}$ is not uniquely determined by the values of V and Y. J. B. Roberts (1978) indicated that this feature may lead to considerable complications which may be resolved by expressing the system displacement in terms of V and a phase process. A differential equation for the energy envelope can be derived by multiplying both sides of equation (6.36) by $\dot{Y}$ and differentiating equation (6.37) with respect to time. Combining the two results gives

$$\dot{V} = \dot{Y}\left\{\varepsilon^{1/2}\sum_{i=1}^{\ell} h_i(Y,\dot{Y})\xi_i(t) - \varepsilon F(Y,\dot{Y})\right\} \tag{6.40}$$

This equation represents the net power of the system, where the first expression is the power input due to the parametric excitation, and the second term represents the power dissipated by the damping mechanism. Since ε is assumed to be very small, the energy envelope V(t) is a slowly varying process while the displacement Y is a rapidly varying process.

Equations (6.39) and (6.40) can be transformed into stochastic equations of the Itô type:

$$dY = \{2[V-G(Y)]\}^{1/2}dt$$

$$dV = \varepsilon\{-[2(V-G(Y))]^{1/2}F(Y,[2(V-G(Y))]^{1/2}) + H(Y,[2(V-G(Y))]^{1/2})\}\,dt$$

$$+ \varepsilon^{1/2}[2(V-G(Y))]^{1/2}\sum_{i=1}^{\ell} h_i(Y,[2(V-G(Y))]^{1/2})\, dB_i(t) \tag{6.41}$$

where

$$H(Y,[2(V-G(Y))]^{1/2}) = \frac{1}{2}\sum_{i=1}^{\ell}[2(V-G(Y))]^{1/2}h_i(Y,[2(V-G(Y))]^{1/2}).$$

$$\frac{\partial}{\partial V}\{[2(V-G(Y))]^{1/2}h_i(Y,[2(V-G(Y))]^{1/2})\}$$

(6.42)

Taking the average of both sides of the second equation of (6.41) with respect to time, and using the first equation, the following equation is obtained

$$d\bar{V} = \varepsilon A(\bar{V})\ dt + \varepsilon^{1/2}\sum_{i=1}^{\ell} b_i(\bar{V})\ dB_i(t) \tag{6.43}$$

where

$$A(\bar{V}) = \frac{1}{2\Phi(\bar{V})}\int_R [V-G(y)]^{-1/2}\{-[2(V-G(y))]^{1/2}F(y,[2(V-G(y))]^{1/2})$$

$$+ H(y,[2(V-G(y))]^{1/2})\}\ dy \tag{6.44}$$

$$b_i(\bar{V}) = \{\frac{1}{\Phi(\bar{V})}\int_R [V-G(y)]^{1/2}\ h_i^2(y,[2(V-G(y))]^{1/2})\ dy\}^{1/2} \tag{6.45}$$

$$\Phi(\bar{V}) = \frac{1}{2}\int_R dy/[V-G(y)]^{1/2} \tag{6.46}$$

and the domain of integration of R includes all values of y for which $V \geq G(y)$. Thus the average energy envelope $\bar{V}$ constitutes a Markov process which possesses a probability density given by the solution of the FPK equation

$$\frac{\partial p(\bar{V},t)}{\partial t} = -\varepsilon\frac{\partial}{\partial\bar{V}}\{A(\bar{V})\ p(\bar{V},t)\} + \frac{\varepsilon}{2}\frac{\partial^2}{\partial\bar{V}^2}\{\sum_{i=1}^{\ell} b_i^2(\bar{V})\ p(\bar{V},t)\} \tag{6.47}$$

If the stiffness function $g(Y)$ has a power-law form and $F(Y,\dot{Y})$ and $h_i(Y,\dot{Y})$ are polynomials in Y and $\dot{Y}$, one may obtain an analytical solution for (6.47). Once $p(\bar{V},t)$ is determined, it is not difficult to evaluate the joint probability density $p(Y,\dot{Y};t)$ of the system displacement and velocity by using the well known property of the conditional probability density

$$p(Y,\bar{V};t) = p_1(Y;t\mid\bar{V})\ p(\bar{V};t) \tag{6.48}$$

where $p_1(Y;t \mid \bar{V})$ is the conditional probability density that the system displacement is Y at time t, provided that its energy envelope V is known. Stratonovich (1963, p. 116) established an expression for $p_1(Y;t \mid \bar{V})$ based on the fact that the energy $\bar{V}$ is "nearly" conserved during a large number of oscillations, and the time which Y spends is proportional to the velocity $\dot{Y} = [2(\bar{V} - G(Y))]^{1/2}$, that is

$$p_1(Y;t \mid \bar{V}) = \begin{cases} C/[\bar{V} - G(Y)]^{1/2} & \text{for } G(Y) \leq V \\ 0 & \text{for } G(Y) \geq V \end{cases} \tag{6.49}$$

where the constant C can be obtained from the normalized condition

$$\int_{-\infty}^{\infty} p_1 \, dy = 1$$

and by using (6.46) it can be written in terms of $\Phi(\bar{V})$ as

$$C = \frac{1}{2\Phi(\bar{V})}$$

Thus the joint probability density (6.48) becomes

$$p(Y,\bar{V};t) = p(\bar{V};t)/\{2\Phi(\bar{V})[\bar{V} - G(Y)]^{1/2}\} \tag{6.50}$$

The joint probability of the state vector Y and $\dot{Y}$ can be obtained from (6.50), based on the transformation (6.37), in the form

$$p(Y,\dot{Y};t) = \frac{p(V;t)}{\sqrt{2}\ \Phi(\bar{V})} \cong \frac{p(\bar{V};t)}{\sqrt{2}\ \Phi(\bar{V})} \tag{6.51}$$

This result indicates that a constant contour of the energy envelope in the phase plane corresponds to a contour of constant probability density.

The solution of equation (6.47) is usually difficult except for certain special cases. Dimentberg (1980a) considered the special case

$$\ddot{Y} + g(Y) = - F(Y,\dot{Y}) + Y\xi_1(t) + \xi_2(t) \tag{6.52}$$

where the parameter ε is absorbed in the function F and the excitations $\xi_1(t)$ and $\xi_2(t)$. The excitations are assumed uncorrelated stationary white noise processes with spectral densities $2D_1$ and $2D_2$, respectively. g(Y) is represented by a continuous odd function. The function $F(Y,\dot{Y})$ is even in Y and odd in $\dot{Y}$. The average stochastic equation of the energy envelope of (6.52) is, (Dimentberg, 1980a)

$$d\bar{V} = \left\{- \frac{Q(\bar{V})}{S_o'(\bar{V})} + \frac{D_1 S_1(\bar{V}) + D_2 S_o'(\bar{V})}{S_o'(\bar{V})}\right\} dt + h_1 dB_1(t) + h_2 dB_2(t) \tag{6.53}$$

where $B_1(t)$ and $B_2(t)$ are two Brownian motion processes of unit intensity. The other parameters are defined by the following expressions:

$$h_1^2 = \frac{2D_1 S_1(\bar{V})}{S_o'(\bar{V})} \quad , \quad h_2 = \frac{2D_2 S_o(\bar{V})}{S_o'(\bar{V})}$$

$$Q(\bar{V}) = \int_0^{Y_o} F\left(Y, \sqrt{2[V-G(Y)]}\ \right) dY, \quad S_o(\bar{V}) = \int_0^{Y_o} \sqrt{2[V-G(y)]}\ dy$$

$$S_o'(\bar{V}) = \int_0^{Y_o} \left(1/\sqrt{2[V-G(y)]}\ \right) dy \quad , \quad S_1(\bar{V}) = \int_0^{Y_o} y^2\sqrt{2[V-G(y)]}\ dy$$

Y_o is the root of the equation $V-G(Y_o)=0$, and $G(y)$ is defined by relation (6.38). The stationary solution of the FPK equation of the system is

$$p(\bar{V}) = p_o S_o'(\bar{V}) \exp\left\{- \int_o^{\bar{V}} \frac{Q(v)}{D_1 S_1(v) + D_2(v)} dv\right\} \tag{6.54}$$

where p_o is a constant which can be obtained from the normalized condition. Solution (6.54) was obtained by Landa and Stratonovich (1962) and was rigorously justified by Khas'miniskii (1964).

The stationary probability density can be expressed in terms of Y_o instead of $\bar{V}$, both are related by the implicit relationship $V=G(Y_o)$. In view of the assumption that $g(Y)$ is monotonic increasing with respect to Y, the inverse dependence of Y on V will be single-valued. In addition, the variable Y will be a slowly varying function like the energy envelope $\bar{V}$. Thus $Y_o(t)$ can be interpreted as the amplitude of oscillation of the dynamic system (6.52). Introducing the expression $\left| dV/dY_o \right| = g(Y_o)$, the probability density of Y_o can be obtained from expression (6.54)

$$p(Y_o) = p_o S_o'\left(G(y_o)\right) g(Y_o) \exp\left\{- \int_o^{Y_o} \frac{Q\left(G(y)\right) g(y)}{D_1 S_1\left(G(y)\right) + D_2 S_o\left(G(y)\right)} dy\right\} \tag{6.55}$$

The integration in (6.55) is generally complicated. Since the nonlinearity of the restoring force is assumed very small, it can be treated by approximate analysis. The functions Q, S_o, and S_1 can be evaluated in explicit forms by replacing the function $2\{G(Y_o) - G(Y)\}$ in their integrands by the quadratic relation $\Lambda^2(Y_o)(Y_o^2 - Y^2)$, where $\Lambda^2(Y_o)$ is the linearization factor. This process is identical to the harmonic linearization technique used in deterministic mechanics. In the present case g(y) is linearized and Dimentberg obtained $\Lambda^2(Y_0)$ in the form

$$\Lambda^2(Y_0) = \frac{4}{\pi Y_0^2} \int_0^{Y_0} \{ yg(y)/[Y_0^2 - y^2]^{1/2} \} dy \tag{6.56}$$

6.7 STOCHASTIC BEHAVIOR OF LINEAR SYSTEMS

6.7.1 SINGLE DEGREE-OF-FREEDOM SYSTEMS

The applications of the stochastic averaging to study the behavior of linear dynamic systems with random coefficients are extensive. The general equation of motion of single degree-of-freedom systems has the following form

$$\ddot{Y} + \{2\zeta\omega + \mu_1\xi_1(t)\} \dot{Y} + \omega^2\{1 + \mu_2\xi_2(t)\} Y = \mu_3\xi_3(t) \tag{6.57}$$

where $\xi_1(t)$ and $\xi_2(t)$ are random parametric excitations and they essentially govern the state of the system response stability regardless of the external random excitation $\xi_3(t)$. The excitation $\xi_3(t)$ constitutes a non-homogeneous term in (6.57) and, hence, governs the level of the system response statistics. The coefficients μ_i and ζ are, in general, very small.

In the absence of $\xi_1(t)$ and $\xi_3(t)$ Stratonovich and Romanovskii (1965) and Stratonovich (1967) obtained the well known stability condition which states that the system becomes unstable if the damping factor ζ is less than a critical value ζ_{cr}. ζ_{cr} depends on the magnitude of the spectral intensity of $\xi_2(t)$ at a frequency which is twice the natural frequency of the system. The effect of simultaneous occurrence of harmonic and random parametric excitations was also considered. The results showed that the effect of the random component is to displace the region of parametric instability.

Ariaratnam and Tam (1976) extended the analysis of Stratonovich and Romanovskii, and included a harmonic component to the random excitation $\xi_2(t)$. They derived stability conditions of the first and second moments of the response in the neighborhood of the parametric resonance condition $\Omega = 2\omega$, where Ω is the frequency of the harmonic parametric excitation. When the random excitation $\xi_2(t)$ is replaced by a white noise, they showed that the white noise does not have any effect on the first moment of the response, but has a destabilizing effect on the second moment. For colored noise of exponential autocorrelation, the stability boundaries of the first and second moments exhibit the typical V shape of deterministic stability boundaries.

Dynamic systems described by equation (6.57) were further examined by Ariaratnam and Tam (1979). Conditions of moment stability were derived in terms of the damping factor and the spectral densities of $\xi_1(t)$ and $\xi_2(t)$. The probability density of the stationary response was derived in a closed analytical form. J. B. Roberts (1982) determined the effect of random parametric excitation on ship rolling motion in the presence of random waves. The roll angle motion is governed by equation (6.57) with $\xi_1(t)=0$. For the case of linear damping Roberts obtained the same stability conditions derived earlier by Statonovich and Romanovskii. The stationary response was found to be identical to the result of Ariaratnam and Tam (1979).

Weidenhammer (1964) applied his averaging method to determine the stability of dynamic systems with stiffness uncertainties and considered special forms of the spectral density of $\xi_2(t)$. Graefe (1966) and Ariaratnam (1967) employed Weidenhammer's method to the case when $\xi_2(t)$ has arbitrary spectrum. Chelpanov (1962) used a similar method and obtained a stability condition of the response mean square for colored noise $\xi_2(t)$.

The effect of atmospheric turbulence on the stability of helicopter rotor blades experiencing uncoupled flapping motion in hovering flight was examined by Lin, et al., (1979). The study was extended to the case of uncoupled flapping motion in forward flight by Fujimori (1978) and Fujimori, et al., (1979). The resulting moment differential equations of the flapping angle include periodic coefficients which are functions of the azimuth angle of the blade. The eigenvalues of the resulting Floquet transition matrix were obtained numerically. Various aspects of helicopter blade dynamics will be treated in chapter 8.

The classical problem of the inverted pendulum stability has received an extensive number of analytical and experimental investigations. This particular system has negative restoring moment, i.e., $\omega^2 < 0$. Bogdanoff and Citron (1965a,b) pointed out that a pendulum which is stabilized in the upward position with a discrete high frequency motion of its support could be destabilized by adding a continuous spectrum of Gaussian random noise. Hemp and Sethna (1968) showed that the effect of the support motion at two frequencies close to each other can destabilize the pendulum. Sethna (1972) indicated that it is not possible to stabilize an inverted pendulum unless the power spectrum of the parametric fluctuations is zero in the neighborhood of the origin. He (1973) extended the averaging method and obtained a criterion for the stability of the pendulum subjected to arbitrary rapid support motion. By including the low frequency support motion components he obtained results identical to those found by Bogdanoff and Citron. Mitchell (1972) employed the method of averaging to examine the stability of the inverted pendulum subjected to deterministic and stochastic support motions. He found additional regions of instability to exist when the difference between two frequencies of an almost periodic motion is small. Mitchell extended his analysis to include second order approximation which resulted in an additional instability region that takes place when one excitation frequency is about twice the other. Analog computer simulation revealed that the pendulum can be stabilized when the damping in the support is sufficiently large and the support motion is a stochastic process with a high-pass power spectral density.

Howe (1974) confirmed the results of Bogdanoff and Citron and derived another condition involving the power spectrum of the effective support excitations at twice the frequency of the induced stabilized oscillation of the pendulum about the upward vertical. Howe's analysis is a combination of the averaging method and an extension of the energy distribution concept used to study wave propagation in random media. He obtained an integro-differential equation that describes the energy content of each frequency component of the smoothed variable pendulum motions. The equation indicates that an interchange of energy occurs among the various frequency components of the oscillation, and is caused by interactions (scattering) with the support motions. Another important feature of the equation is that it accounts for resonance interactions between the various frequency components of the pendulum motion and the Fourier components of the support motion. The Fourier components have twice the frequency of the oscillation of the inverted pendulum.

The results of the analog computer simulation reported by Mitchell (1972) motivated Sethna and Orey (1980) to examine the possibility of stabilizing the pendulum by allowing the support motions to be sample functions of a stochastic process with a continuous spectrum. Sethna and Orey applied the averaging method and showed that the pendulum can indeed be stable in the straight up position when the support motion is stationary in the wide sense and has continuous spectrum. This result has recently been confirmed by Prussing (1981) who proved that the inverted pendulum can be stabilized in terms of the first and second moments if the support motion is a physical white noise. However, if the excitation is a mathematical white noise, the pendulum cannot be stabilized as proved by Nevel'son and Khas'miniskii (1966b) and Nakamizo and Sawaragi (1972).

6.7.2 MULTI-DEGREE-OF-FREEDOM SYSTEMS

The study of the random behavior of linear multi-degree-of-freedom systems has been confined mainly to two-freedom systems. In general, the differential equations of motion of multi-freedom systems with random fluctuations in the stiffness matrix are given in terms of the generalized coordinates q_i

$$\ddot{q}_i + \varepsilon^2 \sum_{j=1}^{n} \hat{C}_{ij}\dot{q}_j + \sum_{j=1}^{n} \{\omega_{ij}^2 + \varepsilon\xi(t)\,\hat{K}_{ij}\}\, q_j = \mu_i\xi_i(t) \tag{6.58}$$

The second term is referred to as the velocity-dependent forces. The matrices $\hat{\underline{C}}$, $\underline{\omega}^2$, and $\hat{\underline{K}}$ may be symmetric or non-symmetric. The velocity-dependent forces are commonly termed dissipative viscous forces when the corresponding quadratic form $\hat{C}_{ij}\dot{q}_i\dot{q}_j$ is positive for all $\dot{q}_i$ with non-zero norm $||\dot{q}||$. They are called gyroscopic when the matrix $\hat{\underline{C}}$ is a skew symmetric.

The analysis of (6.58) can be simplified if the equations of motion are written in terms of the principal coordinates Y_i. In this case the matrix $\underline{\omega}^2$ is reduced to a diagonal matrix and system (6.58) becomes

$$\ddot{Y}_i + \omega_i^2 Y_i + \varepsilon^2 \sum_{j=1}^{n} C_{ij}\dot{Y}_j + \varepsilon\xi(t) \sum_{j=1}^{n} K_{ij}Y_j = \bar{\mu}_i\xi_i(t) \tag{6.59}$$

Ariaratnam (1972), Ariaratnam and Srikantaiah (1978), and Nemat-Nasser (1972) examined the second moment stability of systems under parametric excitation only (when $\mu_i=0$). Their results, while differing in form, indicate that the second moment stability depends on the excitation spectral densities at frequencies $2\omega_i$ or $|\omega_i \pm \omega_j|$. Fujimori (1978) and Fujimori, et al., (1979) extended the analysis of the stability of helicopter rotor blades during forward flight to include the coupled flapping-torsional motion. The resulting equations of motion involve periodic and random coefficients. Numerical solutions for the second moment stability boundaries show substantial deviations from the stability boundaries of the non-turbulent case. Prussing and Lin (1982,1983) determined the moment stability of coupled flap-lag motion of a rotor blade in hover and in forward flight. Closed form conditions for first moment stability showed that the lead-lag damping decreases quadratically with the blade angle of attack and, hence, with the blade lift. In addition, for non-zero spectral density of the vertical turbulent flow, the critical value of lift to cause instability is higher than the corresponding value in the absence of turbulence.

Dimentberg and Isikov (1977) analyzed system (6.59) when $\xi(t)$ is harmonic and $\xi_i(t)$ are random. They obtained the joint probability density and the response second moments in the neighborhood of the principal and combination parametric resonance conditions. For the case of narrow band excitations $\xi_i(t)$ generated by a second order linear filter, they showed that a reduction in the bandwidth of the excitation spectrum results in an increase in the irregularity of the phase distribution associated with an increase in the mean square of the response amplitude.

6.7.3 APPLICATIONS

Classical Dynamic Systems: The random behavior of classical systems shown in fig. (1.1) will be studied by applying the stochastic averaging methods. The following examples are considered.

Example 6.1: Consider the motion of the liquid free surface under vertical random acceleration $\ddot{\xi}(t)$. The linear description of the liquid sloshing is given by the differential equation

$$\ddot{Y} + 2\zeta\omega\dot{Y} + \omega^2\{1 + \varepsilon\ddot{\xi}(t)\}\, Y = 0 \tag{i}$$

where Y is the amplitude of the sloshing mode in question, $\varepsilon=1/g$, g is the gravitational acceleration, and $\ddot{\xi}(t)$ is the random vertical acceleration which is assumed stationary with zero mean. In addition, $\ddot{\xi}(t)$ has a small intensity and small correlation time. The damping factor ζ is also very small and can be written in terms of the small parameter ε as $\zeta = \varepsilon^2\hat{\zeta}$.

Introducing solution (6.2), the standard differential equations of the amplitude and phase are:

$$\dot{a} = \left\{-2\varepsilon^2\omega\hat{\zeta}a\sin\phi + \varepsilon\omega\ddot{\xi}(t)a\cos\phi\right\}\sin\phi \tag{ii}$$

$$\dot{\theta} = \frac{1}{a}\left\{-2\varepsilon^2\omega\hat{\zeta}a\sin\phi + \varepsilon\omega\ddot{\xi}(t)a\cos\phi\right\}\cos\phi \tag{iii}$$

According to the Khas'miniskii limit theorem, equations (ii) and (iii) converge weakly to a Markov process governed by a couple of Itô stochastic differential equations. The drift and diffusion coefficients (4.87) and (4.86) are evaluated as follows

$$f_1(a) = -\frac{1}{T}\int_0^T E[2\omega\hat{\zeta}a\sin^2\phi]\,dt$$

$$+\frac{1}{T}\int_0^T ds\Big\{\int_{-\infty}^0 \{E[(\omega\ddot{\xi}(s)\cos\phi)(\omega\ddot{\xi}(s+\tau)a\cos(\phi+\omega\tau))]\sin\phi\sin(\phi+\omega\tau)$$

$$+\frac{1}{a}E[(-\omega\ddot{\xi}(s)a\sin\phi)(\omega\ddot{\xi}(s+\tau)a\cos(\phi+\omega\tau))]\sin\phi\cos(\phi+\omega\tau)\}\,d\tau$$

$$+\frac{1}{a}\int_{-\infty}^0 E[(\omega\ddot{\xi}(s)a\cos\phi)(\omega\ddot{\xi}(s+\tau)a\cos(\phi+\omega\tau))]\cos\phi\cos(\phi+\omega\tau)d\tau\Big\}$$

$$= -\omega\hat{\zeta}a + \frac{3}{16}\omega^2 a\,S(2\omega) \tag{iv}$$

where $S(\Omega) = \int_{-\infty}^{\infty} R(\tau)\cos\Omega\tau d\tau$, $\quad R(\tau) = E[\ddot{\xi}(t)\ddot{\xi}(t+\tau)]$

$$\hat{f}_i(a) = -\frac{1}{T}\int_0^T \frac{1}{a}E[2\omega\hat{\zeta}a\sin\phi\cos\phi]\,d\tau$$

$$+\frac{1}{T}\int_0^T ds\Big\{\int_{-\infty}^0 \{\frac{1}{a}E[(\omega\ddot{\xi}(s)\cos\phi)(\omega\ddot{\xi}(s+\tau)a\cos(\phi+\omega\tau))]\cos\phi\sin(\phi+\omega\tau)$$

$$+\frac{1}{a^2}E[(-\omega\ddot{\xi}(s)a\sin\phi)(\omega\ddot{\xi}(s+\tau)a\cos(\phi+\omega\tau))]\cos\phi\cos(\phi+\omega\tau)\}\,d\tau$$

$$- \frac{1}{a^2} \int_{-\infty}^{0} E[(\omega\ddot{\xi}(s)a\ \cos\phi)(\omega\ddot{\xi}(s+\tau)a\ \cos(\phi+\omega\tau))](\cos\phi\ \sin(\phi+\omega\tau)$$

$$+ \sin\phi\ \cos(\phi+\omega\tau))\ d\tau\}$$

$$= \frac{\omega^2}{4} \int_{-\infty}^{0} R(\tau)\ \sin 2\omega\tau \cdot d\tau \qquad \text{(v)}$$

Stratonovich (1967, p. 288) obtained the same expression but with negative sign. It is not difficult to show that the above result is the correct one.

$$(\underline{G}(a)\ \underline{G}(a)^T)_{11} = \frac{1}{T} \int_{0}^{T} ds \int_{-\infty}^{\infty} E[(\omega\ddot{\xi}(s)a\ \cos\zeta)(\omega\ddot{\xi}(s+\tau)a\ \cos(\phi+\omega\tau))].$$

$$\sin\phi\ \sin(\phi+\omega\tau)\ d\tau$$

$$= \omega^2 a^2 S(2\omega)/8 \qquad \text{(vi)}$$

$$(\hat{\underline{G}}(a)\ \hat{\underline{G}}(a)^T)_{11} = \frac{1}{T} \int_{0}^{T} ds \int_{-\infty}^{\infty} \frac{1}{a^2} E[(\omega\ddot{\xi}(s)a\ \cos\phi)(\omega\ddot{\xi}(s+\tau)a\ \cos(\phi+\omega\tau))].$$

$$\cos\phi\ \cos(\phi+\omega\tau)\ d\tau$$

$$= \frac{\omega^2}{4} \{S(0) + \frac{1}{2} S(2\omega)\} \qquad \text{(vii)}$$

The Itô stochastic differential equations of the amplitude and phase are:

$$da = \varepsilon^2\{-\omega\hat{\zeta}a + \frac{3}{16} \omega^2 a S(2\omega)\}\ dt + \varepsilon\omega(\frac{S(2\omega)}{8})^{1/2}\ a\ dB(t) \qquad \text{(viii)}$$

$$d\theta = \varepsilon^2\{\frac{\omega^2}{4} \int_{-\infty}^{0} R(\tau)\sin 2\omega\tau\ d\tau\}\ dt + \varepsilon\ \frac{\omega}{2}\ \{S(0) + \frac{1}{2} S(2\omega)\}^{1/2} d\hat{B}(t) \qquad \text{(ix)}$$

These two equations indicate that the amplitude and phase are both independent. In most cases the amplitude behavior is of more concern than the phase. The rest of the analysis will be confined to examine the properties of the amplitude probability density which is described by the FPK equation

$$\frac{\partial p(a,t)}{\partial t} = - \frac{\partial}{\partial a} \{\varepsilon^2(-\omega\hat{\zeta}\ a + \frac{3}{16} \omega^2 a S(2\omega))p(a,t)\}$$

$$+ \frac{\varepsilon^2\omega^2}{16} S(2\omega) \frac{\partial^2}{\partial a^2} \{a^2 p(a,t)\} \qquad \text{(x)}$$

The solution of this equation was obtained by Stratonovich (1967, p. 303) for the initial condition

$$p(a,t) \to \delta(a - a_0) \qquad \text{as } t \to t_0 \tag{xi}$$

with the result

$$p(a,t) = \frac{1}{a\sqrt{2\pi\mu t}} \exp\{-\frac{1}{2\mu t} \ell n^2(\frac{a}{a_0} \exp(-\gamma t))\} \tag{xii}$$

where $\mu = \varepsilon^2\omega^2 S(2\omega)/8, \quad \gamma = \varepsilon^2\{\frac{1}{8}\omega^2 S(2\omega) - \omega\hat{\zeta}\}$

This solution shows that as $t \to \infty$ the probability that the amplitude a is greater than a given value, say a_1, approaches one if $\gamma>0$, and zero if $\gamma<0$. This means that a stationary response of the liquid free surface as described by equation (i) cannot be achieved since the oscillations of the free surface would either increase without limit or diminish to zero. Thus one can draw the stability condition

$$\gamma < 0, \quad \text{or} \quad \hat{\zeta} > \omega S(2\omega)/8$$

in terms of the original damping ratio this condition becomes

$$\zeta > \varepsilon^2\omega S(2\omega)/8 \tag{xiii}$$

The stability of the amplitude moments can be determined by generating the differential equations of the moments from equation (x). The differential equations of the first and second moments are

$$\dot{m}_1 = \varepsilon^2\{\frac{3}{16}\omega^2 S(2\omega) - \omega\hat{\zeta}\} m_1 \tag{xiv}$$

$$\dot{m}_2 = 2\varepsilon^2\{-\omega^2 S(2\omega) - \omega\hat{\zeta}\} m_2, \qquad m_k = E[a^k] \tag{xv}$$

It is evident that the stability conditions of the first and second moments are, respectively:

$$\hat{\zeta} > \frac{3}{16}\omega S(2\omega), \text{ and} \tag{xvi}$$

$$\hat{\zeta} > \frac{1}{4}\omega S(2\omega) \tag{xvii}$$

Example 6.2: Consider one of the classical systems of fig. (1.1) to examine the mean square stability by employing the Weidenhammer averaging method. The equation of motion of these systems is

$$\ddot{Y} + 2\zeta\omega_n\dot{Y} + \{\omega_n^2 \pm \varepsilon_1\xi(t)\} Y = 0 \tag{i}$$

where the negative sign is associated with the equation of motion of an elastic column under random loading. The sign of the excitation term has no effect on the final result since the stability conditions depend on the spectral density of $\xi(t)$. The parameter ε_1 is very small indicating that the random fluctuation is small. $\xi(t)$ is assumed a stationary Gaussian process with zero mean.

Introducing the transformation

$$Y(t) = Z(t)\exp(-\zeta\omega_n t) \tag{ii}$$

equation (i) becomes

$$\ddot{Z} + \omega_d^2\{1 - \varepsilon\xi(t)\}\, Z = 0 \tag{iii}$$

$$\text{where } \omega_d = \omega_n\sqrt{1 - \zeta^2}, \qquad \varepsilon = \varepsilon_1/(1 - \zeta^2) \tag{iv}$$

The solution of (iii) may be expressed in terms of slowly varying amplitude and phase in the form

$$Z(t) = a(t)\cos\phi(t), \qquad \phi(t) = \omega_d t + \theta(t) \tag{v}$$

such that

$$\dot{Z}(t) = -\,\omega_d a(t)\sin\phi(t) \tag{vi}$$

which requires

$$\dot{a}(t)\cos\phi(t) - a(t)\dot{\theta}\sin\phi(t) = 0 \tag{vii}$$

Substituting (v) and (vi) into (iii) gives

$$\dot{a}\sin\phi + \dot{\theta}a\cos\phi = \omega_d\varepsilon\xi(t)\, a\cos\phi \tag{viii}$$

solving (vii) and (viii) simultaneously for $\dot{a}$ and $\dot{\theta}$ yields the two first order differential equations

$$\dot{a}/a = -\frac{1}{2}\varepsilon\omega_d\xi(t)\sin 2\phi \tag{ix}$$

$$\dot{\theta} = -\frac{1}{2}\varepsilon\omega_d\xi(t)\{1 + \cos 2\phi\} \tag{x}$$

The solutions of these equations may be written in the form

$$a(t) = a_0\exp\{A(t)\} \tag{xi}$$

$$\theta(t) = \theta_0 + \Theta(t) \tag{xii}$$

where a_0 and θ_0 are constants of integration, and

$$A(t) = -\frac{1}{2}\varepsilon\omega_d \int_0^t \xi(t)\ \sin\{2\omega_d\tau + 2\theta(\tau)\}\ d\tau \qquad \text{(xiii)}$$

$$\Theta(t) = -\frac{1}{2}\varepsilon\omega_d \int_0^t \xi(t)\{1 + \cos[2\omega_d\tau + 2\theta(\tau)]\}\ d\tau \qquad \text{(xiv)}$$

Expanding A(t) and $\Theta(t)$ in power series in the small parameter ε

$$A(t) = \varepsilon A_1(t) + \varepsilon^2 A_2(t) + \varepsilon^3 A_3(t) + \ldots \qquad \text{(xv)}$$

$$\Theta(t) = \varepsilon\Theta_1(t) + \varepsilon^2\Theta_2(t) + \varepsilon^3\Theta_3(t) + \ldots \qquad \text{(xvi)}$$

also, expanding $\cos 2\phi(t)$ and $\sin 2\phi(t)$ in a Taylor series about $2\phi_0(t) = 2\omega_d t + \theta_0(t)$:

$$\cos 2\phi(t) = \cos\{2\phi_0(t) + 2\Theta(t)\}$$

$$= \cos 2\phi_0(t) - 2\varepsilon\Theta_1(t)\sin 2\phi_0(t) + O(\varepsilon^2) \qquad \text{(xvii)}$$

$$\sin 2\phi(t) = \sin 2\phi_0(t) + 2\varepsilon\Theta_1(t)\cos 2\phi_0(t) + O(\varepsilon^2) \qquad \text{(xviii)}$$

and substituting (xvi), (xvii), and (xviii) into (xiii) and (xiv) and equating coefficients of equal powers in ε yields:

$$\Theta_1(t) = -\frac{1}{2}\omega_d \int_0^t \xi(\tau)\{1 + \cos(2\omega_d\tau + 2\theta_0(\tau))\}\ d\tau \qquad \text{(xixa)}$$

$$\Theta_2(t) = \omega_d \int_0^t \xi(\tau)\ \sin(2\omega_d\tau + 2\theta_0(\tau))\Theta_1(\tau)\ d\tau \qquad \text{(xixb)}$$

$$A_1(t) = -\frac{1}{2}\omega_d \int_0^t \xi(\tau)\ \sin(2\omega_d\tau + 2\theta_0(\tau))\ d\tau \qquad \text{(xixc)}$$

$$A_2(t) = -\omega_d \int_0^t \xi(\tau)\ \cos(2\omega_d\tau + 2\theta_0(\tau))\Theta_1(\tau)\ d\tau \qquad \text{(xixd)}$$

In order to obtain a condition for asymptotic stability of second moment of the system displacement, the response mean square will be evaluated first

$$E[Y^2(t)] = E[Z^2(t)\exp(-2\zeta\omega_n t)]$$

$$= \frac{1}{2} a_o^2 \exp(-2\zeta\omega_n t)\{E[\exp(2A(t))]$$

$$+ \cos(2\omega_d t + 2\theta_o)\cdot E[\exp(2A(t))\cdot\cos 2\Theta(t)]$$

$$- \sin(2\omega_d t + 2\theta_o)\cdot E[\exp(2A(t))\cdot\sin 2\Theta(t)]\} \qquad \text{(xx)}$$

Since $\xi(t)$ is Gaussian and linear operations on Gaussian processes result in Gaussian processes, it is evident that both $A_1(t)$ and $\Theta_1(t)$ will be Gaussian, but $A_2(t)$ and $\Theta_2(t)$ and all other higher terms are not. However, for very small ε, one may assume that $A(t)$ and $\Theta(t)$ are approximately Gaussian. To determine the expected values in (xx) the joint Gaussian probability will be used

$$p(A,\Theta) = \frac{1}{2\pi\sigma_A\sigma_\theta\sqrt{1-\rho^2}} \exp\{- \frac{1}{2(1-\rho^2)} [\frac{(A - m_A)^2}{\sigma_A^2} + \frac{(\Theta - m_\theta)^2}{\sigma_\theta^2}$$

$$- \frac{2\rho(A - m_A)(\Theta - m_\theta)}{\sigma_A\sigma_\theta}]\} \qquad \text{(xxi)}$$

where $m_A = E[A]$, $\qquad \sigma_A^2 = E[(A - m_A)^2]$

$$m_\theta = E[\Theta], \qquad \sigma_\theta^2 = E[(\Theta - m_\theta)^2] \qquad \text{(xxii)}$$

$$\rho = E[A - m_A]\cdot E[\Theta - m_\theta]/(\sigma_A\sigma_\theta)$$

The expected values in (xx) are evaluated as follows

$$E[\exp(2A)] = \int_0^{2\pi}\int_0^{\infty} \exp(2A)\ p(A,\Theta)\ dA\ d\Theta = \exp\{2(m_A + \sigma^2)\} \qquad \text{(xxiiia)}$$

$$E[\cos 2\Theta\ \exp(2A)] = \cos 2(m_\theta + 2\rho\sigma_A\sigma_\theta)\cdot\exp\{2(m_A + \sigma_A^2 - \sigma_\theta^2)\} \qquad \text{(xxiiib)}$$

$$E[\sin 2\Theta\ \exp(2A)] = \sin 2(m_\theta + 2\rho\sigma_A\sigma_\theta)\cdot\exp\{2(m_A + \sigma_A^2 - \sigma_\theta^2)\} \qquad \text{(xxiiic)}$$

Substituting (xxiii) into (xx) gives

$$E[Y^2(t)] = \frac{1}{2} a_o^2 \exp[2(-\zeta\omega_n t + m_A + \sigma_A^2)]\{1 +$$

$$\cos 2\phi_o(t)\cdot\cos 2(m_\theta + 2\rho\sigma_A\sigma_\theta)\cdot\exp(-2\sigma_\theta^2) -$$

$$\sin 2\phi_o(t)\cdot\sin 2(m_\theta + 2\rho\sigma_A\sigma_\theta)\cdot\exp(-2\sigma_\theta^2)\} \qquad \text{(xxiv)}$$

The mean square is asymptotically stable if

$$-\zeta\omega_n + \lim_{t\to\infty} \frac{1}{t}(m_A + \sigma_A^2) < 0 \qquad \text{(xxv)}$$

Since $E[\xi(t)]=0$, then $E[A_1(t)]=0$, and

$$m_A = \varepsilon^2 E[A_2(t)] + O(\varepsilon^3)$$

$$\sigma_A^2 = \varepsilon^2 E[A_1^2(t)] + O(\varepsilon^3)$$

Keeping terms of quadratic order in ε, condition (xxv) becomes

$$-\zeta\omega_n + \lim_{t\to\infty} \frac{\varepsilon^2}{t}\{E[A_2] + E[A_1^2]\} < 0 \qquad \text{(xxvi)}$$

Introducing (xix) we write

$$E[A_2] = \frac{\omega_d^2}{4}\int_0^t\int_0^{\tau_2} R(\tau_1-\tau_2)\{2\cos(2\omega_d\tau_d + 2\theta_0) + \cos[2\omega_d(\tau_1-\tau_2)]$$

$$+ \cos[2\omega_d(\tau_1+\tau_2) + 4\theta_0]\}\, d\tau_1 d\tau_2 \qquad \text{(xxvii)}$$

$$E[A_1^2] = \frac{\omega_d^2}{8}\int_0^t\int_0^t R(\tau_1-\tau_2)\{\cos 2\omega_d(\tau_1-\tau_2) - \cos[2\omega_d(\tau_1+\tau_2)+4\theta_0]\}\, d\tau_1 d\tau_2$$

(xxviii)

where $R(\tau_1-\tau_2) = E[\xi(\tau_1)\xi(\tau_2)]$ is the autocorrelation function of $\xi(t)$ and is related to the spectral density $S(\omega)$ through the Fourier transform

$$R(\tau_1-\tau_2) = \frac{1}{2\pi}\int_{-\infty}^{\infty} S(\omega)\cos\omega(\tau_1-\tau_2)\, d\omega \qquad \text{(xxix)}$$

Substituting (xxix) into (xxvii) and (xxviii), and assuming $S(\omega)$ converges to zero fast enough as $\omega \to \infty$ so that the order of integration can be interchanged, the following five integrals will be obtained

$$I_1 = \frac{1}{2\pi}\int_{-\infty}^{\infty}\int_0^t\int_0^{\tau_2} S(\omega)\cos\omega(\tau_1-\tau_2)\cos(2\omega_d\tau_2+2\theta_0)\, d\tau_1 d\tau_2\, d\omega$$

$$I_2 = \frac{1}{2\pi}\int_{-\infty}^{\infty}\int_0^t\int_0^{\tau_2} S(\omega)\cos\omega(\tau_1-\tau_2)\cos[2\omega_d(\tau_1-\tau_2)]\, d\tau_1 d\tau_2\, d\omega$$

$$I_3 = \frac{1}{2\pi}\int_{-\infty}^{\infty}\int_0^t\int_0^{\tau_2} S(\omega)\cos\omega(\tau_1-\tau_2)\cos[2\omega_d(\tau_1+\tau_2) + 4\theta_0]\, d\tau_1 d\tau_2\, d\omega$$

$$I_4 = \frac{1}{4\pi}\int_{-\infty}^{\infty}\int_0^t\int_0^t S(\omega)\cos\omega(\tau_1-\tau_2)\cos[2\omega_d(\tau_1-\tau_2)]\, d\tau_1 d\tau_2\, d\omega = 2I_2$$

$$I_5 = \frac{1}{2\pi}\int_{-\infty}^{\infty}\int_0^t\int_0^t S(\omega)\cos\omega(\tau_1-\tau_2)\cos[2\omega_d(\tau_1+\tau_2) + 4\theta_0]\, d\tau_1 d\tau_2\, d\omega = 2I_3$$

Evaluating these integrals and using the following relation (Laning and Battin, 1956)

$$\lim_{t\to\infty}\left\{\frac{1}{2\pi t}\,\frac{\sin^2\omega t}{\omega^2}\right\} = \delta(2\omega)$$

the following results are obtained:

$$\lim_{t\to\infty}\frac{1}{t}I_1 = \lim_{t\to\infty}\frac{1}{t}I_3 = \lim_{t\to\infty}\frac{1}{t}I_5 = 0$$

$$\lim_{t\to\infty}\frac{1}{t}I_2 = \frac{1}{4}\int_{-\infty}^{\infty} S(\omega)\lim_{t\to\infty}\frac{1}{2\pi t}\left\{\frac{\sin^2(\frac{1}{2}\omega+\omega_d)}{(\frac{1}{2}\omega+\omega_d)^2} + \frac{\sin^2(\frac{1}{2}\omega-\omega_d)}{(\frac{1}{2}\omega-\omega_d)^2}\right\}$$

$$= \frac{1}{4}\int_{-\infty}^{\infty} S(\omega)\{\delta(\omega + 2\omega_d) + \delta(\omega - 2\omega_d)\}\, d\omega$$

$$= \frac{1}{2} S(2\omega_d) \qquad \text{(xxx)}$$

Substituting the above results into the stability condition gives

$$-\zeta\omega_n + \frac{\varepsilon^2}{4}\omega_d^2 S(2\omega_d) < 0 \qquad \text{(xxxi)}$$

This condition can be written in terms of the original system parameters as

$$4\zeta(1-\zeta^2) > \varepsilon_1^2\,\omega_n S(2\omega_n\sqrt{1-\zeta^2}) \qquad \text{(xxxii)}$$

where the power spectral density is evaluated at twice the damped natural frequency of the system. For small damping ratio condition (xxxii) becomes

$$4\zeta > \varepsilon_1^2\,\omega_n S(2\omega_n) \qquad \text{(xxxiii)}$$

It is interesting to note that this condition, although approximate, is exactly identical to the second moment stability condition obtained in example (6.1).

Ship Roll Dynamics in Random Sea Waves:

In their Probabilistic Theory of Ship Dynamics, Price and Bishop (1974) stated "... waves have a profound effect on hull resistance, structural dynamics and on handling characteristics. In short, the dynamics of ships in waves is one of those fields where it is necessary to make the best of a bad job and to try to make the best better and better." This view is true since sea waves are not harmonic and may cause severe or dangerous ship motions. The study of ship motions was first reported by Froude (1863) who observed that ships have undesirable roll characteristics when the frequency of a small, free oscillation in pitch is twice the frequency of a small, free oscillation in roll. It was Paulling and Rosenberg (1959) who formulated the analytical modeling of the coupled roll-pitch motion [see fig. (6.1) for ship motion nomenclature]. This coupled motion is described by a set of non-linear equations. If the non-linear effect of roll is neglected, the pitch equation of motion is reduced to a linear differential equation which is free from roll motion terms. When the pitch equation is solved, its response appears as a coefficient to the restoring moment of the roll motion, and the roll equation of motion is reduced to a Mathieu equation.

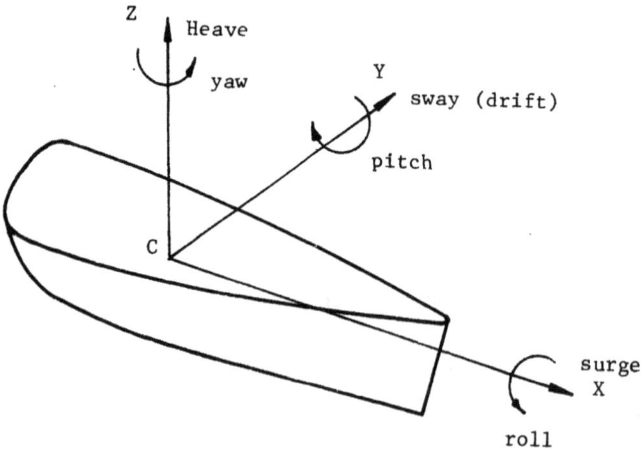

FIG. (6.1) Body axes and rotational motions of a ship (Price and Bishop, 1974)

The analysis of the ship roll stability in an irregular sea was treated by Price (1975), Haddara (1975), and Muhuri (1980). They assumed purely white noise of the restoring moment. J. B. Roberts (1982) extended the analysis and included non-white restoring moment and the effect of non-linear damping. He employed the stochastic averaging method to determine the random roll motion of a ship in irregular sea waves.

Example 6.3 (J. B. Roberts, 1982): The uncoupled roll motion of a ship in a rough sea is described by the stochastic non-homogeneous differential equation

$$\ddot{\Phi} + 2\zeta\omega\dot{\Phi} + \omega^2\{1 + \varepsilon\Theta(t)\}\Phi = \varepsilon\xi(t) \tag{i}$$

where Φ is the roll angle. Originally, $\Theta(t)$ represents the pitch angle which is assumed to be a random stationary process $\Theta(t)$. $\xi(t)$ represents the wave random excitation.

Introducing the solution

$$\Phi(t) = a(t)\cos\phi(t), \qquad \phi(t) = \omega t + \theta(t) \tag{ii}$$

the standard equations of the amplitude and phase shift are

$$\dot{a} = \left\{-2\zeta\omega\, a\sin\phi + \varepsilon\omega a\,\Theta(t)\cos\phi - \frac{\varepsilon}{\omega}\xi(t)\right\}\sin\phi \tag{iii}$$

$$\dot{\theta} = \frac{1}{a}\left\{-2\zeta\omega\, a\sin\phi + \varepsilon\omega a\,\Theta(t)\cos\phi - \frac{\varepsilon}{\omega}\xi(t)\right\}\cos\phi \tag{iv}$$

It is assumed that $\Theta(t)$ and $\xi(t)$ are stationary and correlated random processes. However, the cross-correlation characteristics of these two processes will not enter the analysis since the ship reacts only at one frequency component to each excitation. Thus the ship will respond at ω under $\xi(t)$ and 2ω to $\Theta(t)$ as will be shown in this example. In addition, both $\Theta(t)$ and $\xi(t)$ have small intensities. According to the limit theorems, the state space coordinates of the response converge weakly, as $\varepsilon \to 0$, to a Markov process governed by the Itô equations

$$da = (-\alpha a + \beta/2a)\,dt + (Ta^2 + \beta)^{1/2}dB(t) \tag{va}$$

$$d\theta = \eta dt + (\delta + \beta/a^2)^{1/2}d\hat{B}(t) \tag{vb}$$

where

$$\alpha = \zeta\omega - \frac{3}{16}\varepsilon^2\omega^2 S_\Theta(2\omega), \qquad \delta = \frac{1}{4}\varepsilon^2\omega^2\{S_\Theta(0) + \frac{1}{2}S_\Theta(2\omega)\}$$

$$\beta = \frac{\varepsilon^2}{2\omega^2}S_\xi(\omega), \qquad \eta = \frac{1}{4}\varepsilon^2\omega^2\int_{-\infty}^{0} R_\Theta(\tau)\sin 2\omega\tau\, d\tau$$

$$T = \frac{1}{8}\varepsilon^2\omega^2 S_\Theta(2\omega), \qquad S_\xi(\Omega) = \int_{-\infty}^{\infty} R_\xi(\tau)\cos\Omega\tau \, d\tau$$

$$R_\xi(\tau) = E[\xi(t)\xi(t+\tau)], \; R_\Theta(\tau) = E[\Theta(t)\Theta(t+\tau)] \tag{vi}$$

In view of equations (v), the probability density of the amplitude is governed by the FPK equation

$$\frac{\partial p(a,t)}{\partial t} = -\frac{\partial}{\partial a}\{(-\alpha a + \beta/2a)p\} + \frac{1}{2}\frac{\partial^2}{\partial a^2}\{(Ta^2 + \beta)p\} \tag{vii}$$

The stationary solution of (vii) is

$$p(a) = p_0 a T^{1+\nu}/(\beta + Ta^2)^{1+\nu} \tag{viii}$$

where $\nu = \frac{1}{2} + \alpha/T$

and p_0 is the constant of integration which is determined from the normalized condition $\int_0^\infty p(a)\,da = 1$. This definite integral may be found in tables of integration (see, for example, section 3.251, formula no. 5 in Gradshteyn and Ryzhik (1980)), with the result

$$p_0 = 2\nu(\beta/T)^\nu \tag{ix}$$

The complete expression of the stationary probability density is

$$p(a) = 2T\nu\beta^\nu \, a/(\beta + Ta^2)^{1+\nu} \tag{x}$$

which exists if $\nu > 0$, or

$$\zeta > \varepsilon^2\omega S_\Theta(2\omega)/8 \tag{xi}$$

This inequality defines the condition of sample stability of the roll angle amplitude. Moreover, it reveals that the onset of instability is not affected by the forcing excitation $\xi(t)$ of the sea waves. If this excitation is removed, the probability density will degenerate into a delta function at zero amplitude.

The n-th moment of the amplitude is

$$E[a^n] = \int_0^\infty a^n p(a)\,da$$

$$= \nu(\beta/T)^{n/2}\,\Gamma(1 + \tfrac{n}{2})\,\Gamma(\nu - \tfrac{n}{2})\,\Gamma^{-1}(\nu + 1) \tag{xii}$$

which exists if

$\nu > n/2$

Thus the stability conditions of the first and second moments are, respectively

$$\zeta > \frac{3}{16}\varepsilon^2 \omega S_\Theta(2\omega) \qquad n=1 \qquad \text{(xiii)}$$

$$\zeta > \frac{1}{4}\varepsilon^2 \omega S_\Theta(2\omega) \qquad n=2 \qquad \text{(xiv)}$$

It may be noticed that p(a) is not of Rayleigh type distribution for the amplitude. However, in the limiting case as $T \to 0$ (i.e., as $\Theta(t)$ approaches zero) the probability density (x) approaches the Rayleigh formula

$$p(a) = \frac{a}{\sigma^2}\exp(-\frac{a^2}{2\sigma^2}) = \frac{A}{\sigma}\exp(-\frac{1}{2}A^2) \qquad \text{(xv)}$$

where $\sigma^2 = \frac{\beta}{2\zeta\omega} = \frac{\varepsilon^2 S_\xi(\omega)}{4\zeta\omega^3}$, and $A = a/\sigma$

Equation (xv) is obtained by using the standard formula

$$\lim_{\varepsilon \to 0} (1 + \varepsilon X)^{1/\varepsilon} = \exp(X)$$

In terms of A the probability density takes the form

$$p(A) = \nu A\,(1+\nu)^{-1}\{1 + \frac{A^2}{2(1+\nu)}\}^{-1-\nu} \qquad \text{(xvi)}$$

The dependence of p(A) upon ν is shown in fig. (6.2). The case of the Rayleigh distribution ($\nu \to \infty$) is also included.

It has been shown in section (4.4) that the amplitude and phase are two independent random processes for which the joint probability density $p(a,\theta)$ is written in the form

$$p(a,\theta) = p(a)\,p(\theta) = p(a)/2\pi \qquad \text{(xvii)}$$

where p(a) is given by relation (v).

Following the transformation described in section (2.3.9), the amplitude probability (x) can also be written in terms of the roll

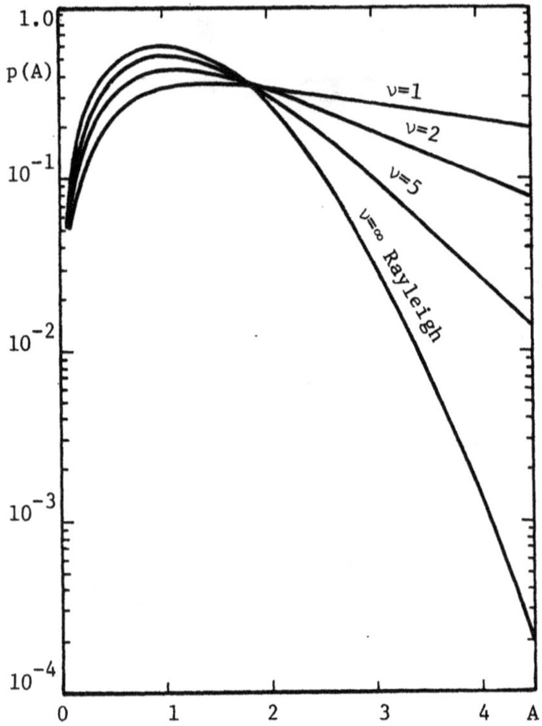

FIG. (6.2) Probability density function of the roll angle amplitude parameter for various values of parametric excitation levels ν (J. B. Roberts, 1982)

angle Φ and its time rate $\dot{\Phi}$ in the form

$$p(\Phi,\dot{\Phi}) = \frac{T\nu\beta^{\nu}}{\pi\omega}\{\beta + T(\Phi^2 + \dot{\Phi}^2/\omega^2)\}^{-1-\nu} \qquad \text{(xviii)}$$

as $T \to 0$, formula (xviii) is reduced to the bivariate Gaussian density function for a non-parametric excitation

$$p(\Phi,\dot{\Phi}) = \frac{1}{2\pi\sigma^2}\exp\{-\frac{1}{2\sigma^2}(\Phi^2 + (\dot{\Phi}/\omega)^2)\} \qquad \text{(xix)}$$

Integrating (xviii) with respect to $\dot{\Phi}$ yields the stationary density for Φ

$$P(\Phi) = (T/\pi)^{1/2}\,\nu\beta^{\nu}\Gamma(\tfrac{1}{2} + \nu)\Gamma^{-1}(1+\nu)(\beta+T\Phi^2)^{-1/2-\nu} \qquad \text{(xx)}$$

as $T \to 0$, this relation is reduced to the Gaussian form

$$p(\Phi) = \frac{1}{\sqrt{2\pi}\,\sigma} \exp\left(-\frac{\Phi^2}{2\sigma^2}\right) \tag{xxi}$$

Let $\Psi = \Phi/\sigma$, equation (xv) becomes

$$p(\Psi) = \frac{\nu}{\sqrt{2\pi(1+\nu)}} \frac{\Gamma(\frac{1}{2}+\nu)}{\Gamma(1+\nu)} \left\{1 + \frac{\Psi^2}{2(1+\nu)}\right\}^{-1/2-\nu} \tag{xxii}$$

Figure (6.3) shows the dependence of $p(\Psi)$ on ν including the limiting case $\nu \to \infty$.

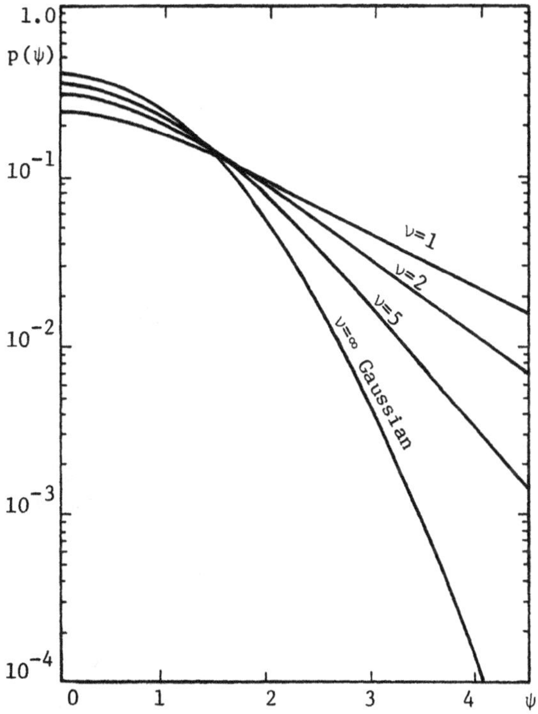

FIG. (6.3) **Probability density function of the roll angle $\psi = \Phi/\sigma$ for various values of parametric excitation levels ν (J. B. Roberts, 1982)**

The n-th moment of the roll angle $E[\Phi^n]$ can be evaluated by using the probability density (xx)

$$E[\Phi^n] = \begin{cases} \dfrac{\nu}{\sqrt{\pi}} (\beta/T)^{n/2} \dfrac{\Gamma(\frac{1+n}{2})\Gamma(\nu-n/2)}{\Gamma(1+\nu)} & n \text{ even} \\ 0 & n \text{ odd} \end{cases} \tag{xxiii}$$

The even order moments exist if $\nu > n/2$

The mean square of the roll angle is

$$E[\Phi^2] = \sigma^2 = \beta/\{2T(\nu - 1)\}$$

$$= \frac{\varepsilon^2 S_\xi(\omega)}{\omega^3\{4\zeta-\varepsilon^2\omega S_\Theta(2\omega)\}} \tag{xxiv}$$

The level of parametric amplification of oscillation caused by the external random excitation ξ can be determined from the ratio

$$\frac{E[\phi^2]}{E[\phi^2]_{\theta=0}} = \left\{1 - \frac{\varepsilon^2\omega}{4\zeta} S_\Theta(2\omega)\right\}^{-1} \tag{xxv}$$

provided that condition (xiv) is satisfied.

The problem of evaluating the level of parametric amplification was treated in detail by Dimentberg and Isikov (1977) and Dimentberg and Sidorenko (1978).

Example 6.4: Flexural-Torsional Random Vibration of Beams

It has been indicated in chapter 1 that parametric instability of combination resonance can take place in structural systems such as beams and plates. Nemat-Nasser (1972) and Ariaratnam and Srikantaiah (1978), who employed different versions of stochastic averaging, examined the flexural-torsional stability of a simply supported thin beam subjected to random end couple M(t) as shown in fig. (6.4). The end moments are assumed stationary with spectral density which covers a frequency range well above the fundamental flexural and torsional frequencies of the beam. For the full derivation of the equations of motion the reader may refer to Bolotin (1964) or to one of the references cited in the review article by Ibrahim and Barr (1978a). Considering the fundamental modes in bending u(z,t) and torsion $\Theta(z,t)$

$$\begin{aligned} u(z,t) &= k\, Y_1(t) \sin(\pi z/L) \\ \Theta(z,t) &= Y_2(t) \sin(\pi z/L) \end{aligned} \tag{i}$$

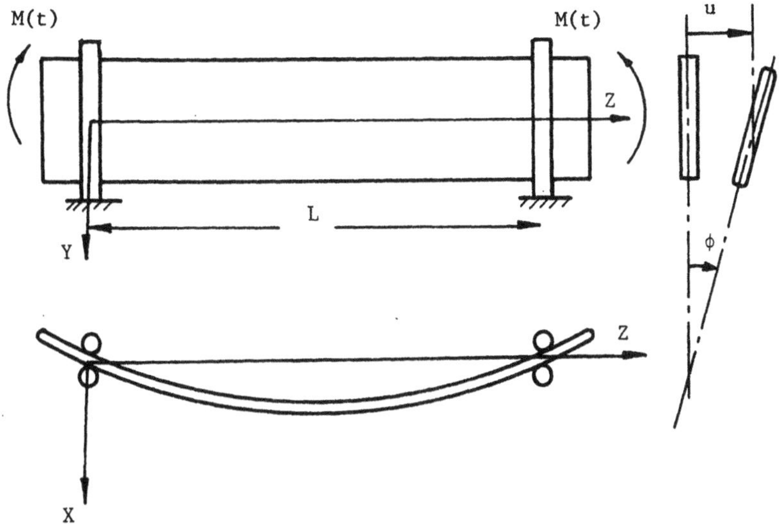

FIG. (6.4) Elastic beam under random end moments

where k is the polar radius of gyration, $Y_1(t)$ and $Y_2(t)$ are the generalized coordinates (dimensionless). The equations of motion are

$$\ddot{Y}_1 + \omega_1^2 Y_1 + 2\zeta_1\omega_1\dot{Y}_1 - \omega_1\omega_2\xi(t)Y_2 = 0$$
$$\ddot{Y}_2 + \omega_2^2 Y_2 + 2\zeta_2\omega_2\dot{Y}_2 - \omega_1\omega_2\xi(t)Y_1 = 0 \qquad \text{(ii)}$$

where $\omega_1^2 = \pi^4 EI_y/(mL^4)$, $\qquad \omega_2^2 = \pi^2 GJ/(mL^2k^2)$

$\xi(t) = M(t)/M_{cr}$, $\qquad M_{cr} = (EI_yGJ)^{1/2}/L$

The critical moment M_{cr} is the couple at which static buckling begins to occur. EI_y, and GJ are the flexural and torsional rigidities of the cross section, respectively. m is the mass per unit length, ζ_1 and ζ_2 are the damping ratios.

Introducing the solutions

$$Y_i = a_i(t)\cos\phi_i(t), \qquad \phi_i(t) = \omega_i t + \theta_i(t), \qquad i=1,2 \qquad \text{(iii)}$$

The standard equations of the amplitudes and phases are

$$\dot{a}_1 = -\{2\zeta_1\omega_1 a_1 \sin\phi_1 + \omega_2\xi(t)a_2 \cos\phi_2\} \sin\phi_1$$

$$\dot{\theta}_1 = -\frac{1}{a_1}\{2\zeta_1\omega_1 a_1 \sin\phi_1 + \omega_2\xi(t)\, a_2 \cos\phi_2\} \cos\phi_1$$

$$\dot{a}_2 = -\{2\zeta_2\omega_2 a_2 \sin\phi_2 + \omega_1\xi(t)\, a_2 \cos\phi_2\} \sin\phi_2 \tag{iv}$$

$$\dot{\theta}_2 = -\frac{1}{a_2}\{2\zeta_2\omega_2 a_2 \sin\phi_2 + \omega_1\xi(t)\, a_1 \cos\phi_1\} \cos\phi_2$$

Following the same steps of the previous examples, the Itô equations of the amplitudes are

$$da_1 = \{-\zeta_1\omega_1 a_1 + \frac{1}{16}(2\omega_1\omega_2 a_1 S_{12}^- + \omega_2^2 a_2^2 S_{12}^+/a_1)\}\, dt$$

$$+ \omega_2 a_2\sqrt{S_{12}^+/8}\, dB_1(t) + (\omega_1\omega_2 a_1 a_2 S_{12}^-/8)^{1/2} dB_2(t)$$

$$da_2 = \{-\zeta_2\omega_2 a_2 + \frac{1}{16}(2\omega_1\omega_2 a_2 S_{21}^- + \omega_1^2 a_1^2 S_{12}^+/a_2)\}\, dt \tag{v}$$

$$+ \omega_2 a_1\sqrt{S_{21}^+/8}\, d\hat{B}_1(t) + (\omega_1\omega_2 a_1 a_2 S_{21}^-/8)^{1/2} d\hat{B}_2(t)$$

where $S_{ij}^{\pm} = S(\omega_i + \omega_j) \pm S(\omega_i - \omega_j)$, $\quad S(\omega) = \int_{-\infty}^{\infty} R_\xi(\tau) \cos\omega\tau \, d\tau$

In order to determine the mean square stability of the amplitude, the differential equations of the moments of the amplitudes are generated by using the Itô formula for stochastic differential which gives

$$\begin{Bmatrix} \frac{d}{dt} E[a_1^2] \\ \frac{d}{dt} E[a_2^2] \end{Bmatrix} = \begin{bmatrix} -2\zeta_1\omega_1 + \alpha_{11} & \alpha_{12} \\ \alpha_{21} & -2\zeta_2\omega_2 + \alpha_{22} \end{bmatrix} \begin{Bmatrix} E[a_1^2] \\ E[a_2^2] \end{Bmatrix} \tag{vi}$$

where $\alpha_{11} = (\omega_1\omega_2 S_{12}^-/4) + \omega_1^2 S(2\omega_1)/2$, $\quad \alpha_{12} = \omega_2^2 S_{12}^+/4$

$$\alpha_{22} = (\omega_1\omega_2 S_{12}^-/4) + \omega_2^2 S(2\omega_2)/2 , \quad \alpha_{21} = \omega_1^2 S_{12}^+/4$$

The necessary and sufficient conditions for mean square stability are that all the eigenvalues of the coefficient matrix of $E[a_i^2]$ have negative real parts. These conditions may be obtained readily by applying the Routh-Hurwitz criteria to the coefficient matrix. The following conditions are obtained

$$2\zeta_1\omega_1 > \alpha_{11}, \quad 2\zeta_2\omega_2 > \alpha_{22}, \quad \text{and}$$

$$(2\omega_1\zeta_1 - \alpha_{11})(2\omega_2\zeta_2 - \alpha_{22}) > \alpha_{12}\alpha_{21} \tag{vii}$$

These conditions were also obtained by Wedig (1969) and Kolovskii and Troitskaya (1972) who used different approaches such as the perturbation technique and moment functions method.

Ariaratnam and Srikantaiah (1978) considered a band-pass process excitation with a frequency bandwidth $\Delta\omega$ and central frequency ω_0. When this excitation is applied to two degree-of-freedom systems described by equations (6.58), in the absence of any gyroscopic terms, the following stability conditions are obtained

Sum Frequency Case:

When $\omega_0 = \omega_r + \omega_s$, $(r \neq s)$, and setting $C_{ii} = 2\zeta_{ii}\omega_i$ in equation (6.58), the second moments are stable if

$$\frac{\zeta_{rr}\zeta_{ss}\omega_r^2\omega_s^2}{\zeta_{rr}\omega_r+\zeta_{ss}\omega_s} > K_{rs}K_{sr}\, S(\omega_r + \omega_s)/8 \tag{6.60}$$

Difference Frequency Case:

If $\omega_0 = \left| \omega_r - \omega_s \right|$, the system (6.58) is stable in the second moments if

$$\frac{\zeta_{rr}\zeta_{ss}\omega_r^2\omega_s^2}{\zeta_{rr}\omega_r+\zeta_{ss}\omega_s} > - K_{rs}K_{sr}S(\omega_r - \omega_s)/8 \tag{6.61}$$

Double Frequency Case:

If $\omega_0 = 2\omega_r$, the following well known condition is obtained

$$\zeta_{rr} > K_{rr}^2S(2\omega_r)/(4\omega_r^3) \tag{6.62}$$

which is identical to condition (xvii) of example (6.1) if one replaces K_{rr} by ω_r^2.

6.8 STOCHASTIC BEHAVIOR OF NON-LINEAR SYSTEMS

6.8.1 SINGLE DEGREE-OF-FREEDOM SYSTEMS

The linear analysis of vibrating systems under random parametric excitations (in the absence of non-parametric external excitations) predicts the stability boundaries of the equilibrium configuration. It has been shown in example (6.1) that bounded stationary response statistics cannot be reached within the scope of linear analysis. If the system is brought into the regions of stochastic instability, the system response statistics will grow without limit. Physically this is not always true and the inherent non-linearities of the system bring the system into a bounded limit cycle.

It has been indicated in chapter 1 that there are three types of non-linearities: stiffness, damping, and inertia. Within the first order approximation the inclusion of cubic stiffness non-linearity does not show any influence upon the response behavior of the averaged solution. In this case, second order averaging or the averaged energy envelope method may be used to determine the influence of stiffness non-linearity.

The early treatment of non-linear systems is believed to be due to Stratonovich and Romanovskii (1965) and Stratonovich (1967). They considered a system with non-linear damping, and random stiffness coefficient. The stationary solution indicates two main features: the first is the limiting growth of the response amplitude, and the second is the presence of small fluctuations of a non-parametric nature under stable conditions. Ariaratnam (1980) analyzed the same system and showed a discrepancy in the results obtained by the stochastic averaging and the Gaussian closure scheme. The stationary responses of ships, in rough sea waves, involving linear plus quadratic damping or linear plus cubic damping have been determined by J.B. Roberts (1982).

Kolomiets (1967,1972) applied the averaging method to derive the stationary solutions for a number of non-linear systems which include sources of self-excited vibrations and periodic damping coefficients. Dimentberg (1967) obtained the stationary probability density of the response of a non-linear damped system subjected to periodic parametric and random forced excitations. The stationary solution was then used to solve diagnostic problems such as the estimation of the threshold of parametric instability and identification of oscillations caused by random forced and harmonic parametric disturbances. The identification of self-excited vibration when both excitations are random was studied by Gorbunov and Dimentberg (1974) and Dimentberg and Gorbunov (1975). For quasi-conservative systems with hard non-linear stiffness, Dimentberg (1980a) adopted the total energy envelope and the displacement as new state variables. The right-hand side of the energy equation was averaged in accordance with a scheme outlined by Landa and Stratonovich (1962). The probability density of the response amplitude indicated that the growth of the response amplitude, within the stochastic instability region of the corresponding linear system, can be limited by introducing hard stiffness non-linearity into the system equation of motion. For the case of systems with non-linear damping Dimentberg, et al., (1981) employed the Stratonovich-Khas'miniskii averaging method. Within the limits of the instability region of the linear system the mean square response was found to increase linearly as the linear damping coefficient decreases. Their analysis led to the conclusion that the low-frequency component of the random parametric excitation can exert a strong effect on the intensity of the response process. Zhu (1983a) considered a single degree-of-freedom oscillator possessing the three common types of non-linearities. He used the energy envelope averaging method and showed that when the destabilizing excitation was applied to the stiffness parameter, the response process would be limited by sufficiently strong non-linearities in one of three main categories: the first includes non-linear stiffness with linear damping and inertia, the second includes non-linear damping with linear stiffness and inertia, and the third involves non-linear inertia with linear stiffness and damping. When the excitation was applied to the damping term, however, the response could be limited in the second and third cases, but it could not be limited by any form of stiffness non-linearity coupled with linear damping and inertia. Ibrahim (1982) examined the simultaneous occurrence of self-excited and random parametric vibrations in water lubricated bearings of submarines. It was found that the inclusion of a positive non-linear damping brings the system response into a stable limit cycle. The non-linear behavior of liquid free surface sloshing in moving containers was investigated by Ibrahim and Soundararajan (1983). It was concluded that the stochastic averaging solution is very close to the experimental results obtained by Dalzell (1967). Recently, Chen and Huang (1984) determined the response of a string to random parametric excitation. Their results showed that the introduction of non-linear fluid damping brings the originally unstable string into a bounded stable limit cycle.

Baxter (1971) examined the non-linear response of a simply supported elastic column when the parametric excitation is narrow band, harmonic plus random, and wide band. He also outlined the second order averaging procedure. Schmidt (1976,1981) analyzed structural systems with quadratic and cubic restoring forces, and cubic damping forces, subjected to white noise random parametric excitation. To the first order averaging approximation the effect of non-linear restoring forces did not appear in the stationary probability density of the response. However, Schmidt (1981) extended the analysis to include a second order averaging which took into account the influence of non-linear stiffness terms.

6.8.2 MULTI-DEGREE-OF-FREEDOM SYSTEMS

The random analysis of multi-degree-of-freedom systems involves tedious mathematical manipulations. Unlike deterministic non-linear analysis, which has been well examined by several investigators (see, for example, the review of Ibrahim and Barr (1978b)), the stochastic analysis of these systems is well behind the deterministic progress. Schmidt (1977a) and Schmidt and Schulz (1983) employed the averaging method to determine the joint probability density of a non-linear two degree-of-freedom system. Their solution is considered incomplete due to the difficulty in determining the constant of integration. However, they provided qualitative description of the main features of the response probability density of the amplitudes which reflect the main characteristics of autoparametric coupling obtained by Ibrahim and Roberts (1976).

The stochastic stability and response characteristics of wind-loaded structures were examined by Lin and Holmes (1978). The dynamic interaction of two orthogonal modes of the elastic structure (the along-wind mode and the cross-wind mode) was described by two non-linear differential equations. The random wind loads enter the equations of motion in the form of non-homogeneous and parametric inputs. Because the equation of motion of the structure along the wind is independent of the perpendicular motion, the stationary probability density of the along-wind mode was obtained. The response analysis of the structure perpendicular to the wind direction was found difficult; however, Lin and Holmes obtained the stability condition of the first order moment of the response. They disclosed that their analysis is incomplete especially for the cross-wind mode.

The lack of investigations of systems with non-linear coupling is attributed to the difficulty of solving the system FPK equation and the difficulty of closing the infinite hierarchy of the moment equations. In chapter 8 the response and stability of non-linear systems will be examined by non-Gaussian closure schemes.

6.8.3 APPLICATIONS

Example 6.5: Non-Linear Liquid Sloshing Under Vertical Random Excitation

The stationary response of liquid free surface in a partially filled circular container subjected to a wide band random excitation was examined by Ibrahim and Soundararajan (1983). The differential equation of motion of the liquid surface amplitude is

$$\ddot{A}_{mn} + 2\zeta_{mn}\dot{A}_{mn} + \{1 + W(\tau)\}A_{mn}(1 - K_1 A_{mn} - K_2 A_{mn}^2) + K_3\dot{A}^2 + K_4 A_{mn}\dot{A}_m$$

$$+ K_5 A_{mn}\dot{A}_{mn}^2 + K_6 A_{mn}^2\ddot{A}_{mn} = 0 \qquad (1)$$

where $A_{mn} = Y_{mn}/a$ is the dimensionless free surface amplitude of the sloshing mode mn, a is the tank radius, and Y_{mn} is the free surface amplitude. The dot denotes differentiation with respect to the dimensionless time $\tau = \omega_{mn}t$, ω_{mn} is the sloshing natural frequency of the mode mn. The excitation $W(\tau) = \ddot{\xi}(\tau)/g$; $\ddot{\xi}(\tau)$ is the vertical acceleration which is assumed to be a stationary wide band random process with zero mean and smooth spectral density 2D. In addition, the correlation time of $W(\tau)$ is considered very short compared with the characteristic time of the sloshing mode mn. With this assumption it is possible to treat the response as a Markov process and $W(\tau)$ can be approximated as the formal derivative of the Brownian motion, $W(\tau) = dB(\tau)/d\tau$. The constant K_1 is associated with the symmetric sloshing modes while K_2 appears with the asymmetric modes.

In order to generate the Markov vector, the double derivative $\ddot{A}_{mn}$ associated with the non-linear terms must be removed by successive elimination. Retaining terms up to cubic order and setting $A_{mn} = X_1$ and $\dot{X}_1 = X_2$, equation (i) may be written in the state vector form

$$\dot{X}_1 = X_2$$

$$\dot{X}_2 = -2\zeta X_2 + C_1 X_1 X_2^2 + C_2 X_2^2 + C_3 \zeta X_1^2 X_2 + C_4 X_1 X_2 \qquad \text{(ii)}$$

$$-[1 + W(\tau)](X_1 - C_6 X_1^2 - C_5 X_1^3)$$

where the new constants C_i are related to K_i. ζ is the damping ratio of the mode in question. Introducing the solution

$$X_1 = a(\tau)\cos\phi(\tau), \qquad X_2 = -a(\tau)\sin\phi(\tau) \qquad \text{(iii)}$$

where $\phi(\tau) = \tau + \theta(\tau)$

The standard equation of the amplitude is

$$da(\tau) = \{(-2\zeta a \sin\phi - C_1 a^3 \sin\phi \cos\phi - C_2 a^2 \sin^2\phi + C_3\zeta a^3 \cos\phi \sin\phi$$

$$+ C_4\zeta a^2 \cos\phi \sin\phi - C_5 a^3 \cos^3\phi - C_6 a^2 \cos^2\phi)d\tau$$

$$+ \sigma(-a\cos\phi + C_5 a^3 \cos^3\phi + C_6 a^2 \cos^2\phi)dB(\tau)\} \sin\phi \qquad \text{(iv)}$$

where $\sigma^2 = 2D$, $W(\tau) = dB(\tau)/d\tau$

Carrying out the stochastic averaging to equation (iv) and introducing the Wong-Zakai correction term gives the Itô equation

$$da = \{-\zeta + \frac{1}{8} C_3\zeta a^2 + \frac{1}{16}\sigma^2[3 + \frac{5}{2}(C_6^2 - 2C_5)a^2 + \frac{35}{16} C_5^2 a^4]\} a\, d\tau$$

$$+ \frac{5}{\sqrt{8}}\{1 + \frac{1}{2}(C_6^2 - 2C_5)a^2 + \frac{5}{16} C_5^2 a^4\}^{1/2} a\, dB(\tau) \qquad \text{(v)}$$

The stationary Fokker-Planck equation of the amplitude is

$$-\frac{\partial}{\partial a}\{[-\zeta + \frac{1}{8} C_3 \zeta a^2 + \frac{D}{8}(3 + 5C_7 a^2 + 7C_8 a^4)]a\, p(a)\}$$

$$+ \frac{D}{8}\frac{\partial^2}{\partial a^2}[(1 + C_7 a^2 + C_8 a^4)a^2\, p(a)] = 0 \qquad \text{(vi)}$$

where $C_7 = \frac{1}{2}(C_6^2 - 2C_5)$, and $C_8 = \frac{5}{16} C_5^2$

The stationary solution of (vi) is

$$p(a) = p_0 a^{n_1} \exp(-n_2 a^2 - n_3 a^4) \qquad \text{(vii)}$$

where $n_1 = 1 - 8\zeta/D$, $n_2 = \left|\frac{1}{3} C_3 + 4C_7\right| \zeta/D$

$$n_3 = \frac{3}{4} C_7^2 - \frac{\zeta}{4D}(8C_8 - C_3 C_7)$$

and p_0 is the constant of integration which is determined from the normalized condition

$$\int_0^\infty p(a)\, da = 1$$

This integral can be transformed into one of the standard formulae listed in the tables of integration by Gradshteyn and Ryzhik (1980), (formula 3.462). The following result is obtained

$$p_0 = 2(2n_3)^{\nu/2}/\{\Gamma(\nu)\exp(n_2^2/8n_3)\, \underline{D}_{-\nu}(n_2/\sqrt{2n_3})\} \qquad \text{(viii)}$$

provided $n_3 > 0$ and $\nu > 0$, where $\nu = \frac{1}{2}(n_1 + 1)$,

and $\underline{D}_\nu(\ )$ is the parabolic cylindrical function known as Whittaker's function which is tabulated in Abramowitz and Stegen (1972). The full expression of the stationary probability density of the response is

$$p(a) = \frac{2(2n_3)^{\nu/2}}{\Gamma(\nu)} \exp\left(\frac{-n_2^2}{8n_3}\right) \underline{D}_{-\nu}^{-1}\left(\frac{n_2}{\sqrt{2n_3}}\right) a^{n_1} \exp\{-n_2 a^2 - n_3 a^4\} \qquad \text{(ix)}$$

It is clear that the shape of the probability density is determined basically by the value of the excitation parameter n_1 and the damping ratio. The characteristic behavior of the fluid free surface may be classified into three regimes:

i. Zero-Motion Response. The fluid free surface remains plane with identical zero motion if the excitation level is in the range

$$n_1 \leq -1, \quad \text{i.e.,} \quad 0 \leq D/2\zeta \leq 2 \qquad \text{and} \quad \nu < 0 \tag{x}$$

Under this condition the probability density function at zero amplitude becomes delta function and the normalization condition is invalid since $\nu < 0$. Physically this means that the excitation level is not strong enough to overcome the fluid damping force. Another important feature is that condition (x) corresponds to the sample stability boundary obtained by Mitchell and Kozin (1974) for linear systems.

ii. Undeveloped Sloshing. This regime is characterized by very small motions of the liquid free surface if

$$-1 \leq n_1 \leq 0, \quad \text{or} \quad 2 \leq D/2\zeta \leq 4, \text{ and} \quad \nu > 0 \tag{xi}$$

Here the normalization is valid and solution (ix) shows that p(a) is infinite at a = 0.

iii. Partially Developed Sloshing. Finite random motion of the liquid surface takes place in the excitation range

$$0 \leq n_1 \leq 1: \text{ i.e.,} \quad 4 \leq D/2\zeta < \infty \tag{xii}$$

The predicted normalized probability density function $\bar{p}(a_{mn}$, $\{\bar{p}(a_{mn}) = p(a_{mn})/p_{max}(a_{mn})\}$ of the sloshing modes mn = 11 and 01-03 is shown by the solid curves for three excitation levels $D/2\zeta = 5$, 10 and 20 in figs. (6.5). It is seen that as the excitation level increases the probability of small amplitudes ($a_{mn} \approx 0$) decreases. The peak location of these density curves is the value of the most probable amplitude a^*_{mn}. The dotted curves are obtained from the least square fitting of the experimental results of Dalzell (1967).

Statistical Response Parameters. The most probable value of the liquid amplitude a* occurs at the point of maximum probability for which $\partial p/\partial a \big|_{a^*} = 0$. This amplitude is given by the real root of the quadratic equation in a^{*2},

$$a^{*4} + \left(\frac{n_2}{2n_3}\right)a^{*2} + \left(\frac{n_1}{4n_3}\right) = 0 \tag{xiii}$$

as $D/2\zeta \to \infty$, a* is found to have the asymptotic value

$$a^* = (3)^{-1/4}/\sqrt{|c_7|} \tag{xiv}$$

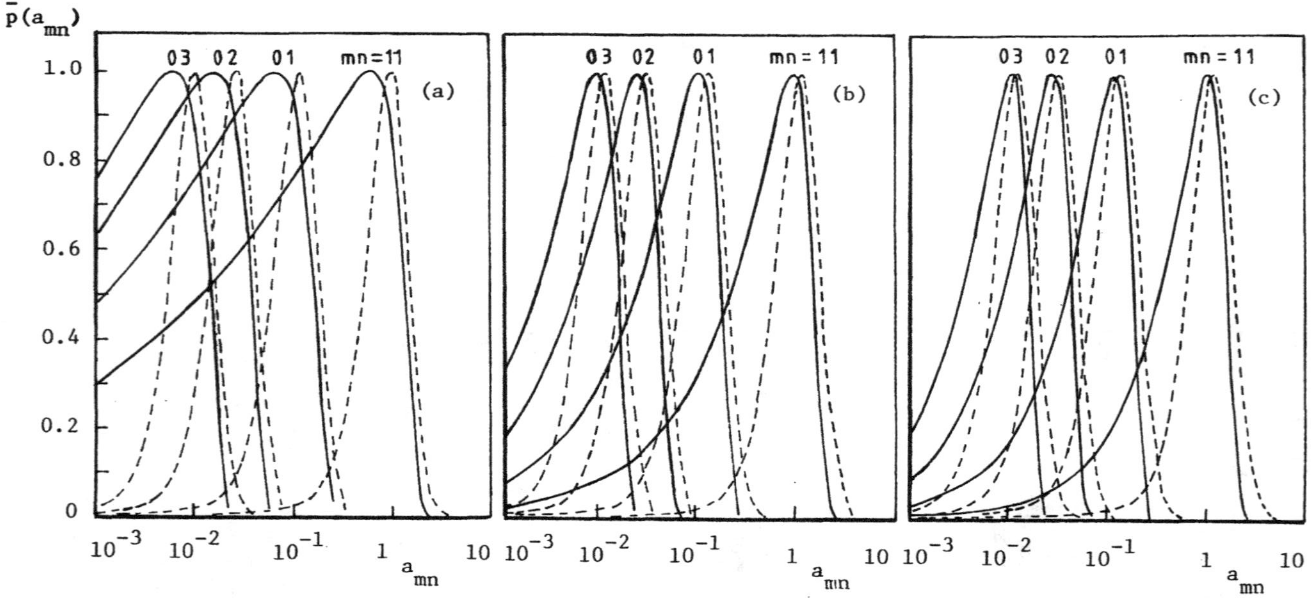

FIG. (6.5) Normalized probability for symmetric sloshing modes 01–03 and asymmetric mode 11

——— stochastic averaging

- - - - - Dalzell's experimental fitting

(a) $D/2\zeta = 5.0$, (b) $D/2\zeta = 10.0$ and (c) $D/2\zeta = 20.0$

The m-th order moment of the amplitude is

$$E[a^m] = \int_0^\infty a^m\, p(a)\, da$$

$$= \bar{D}\,(2n_3)^{-m/4}\, \Gamma(\mu)/\Gamma(\nu) \qquad \text{(xv)}$$

where $\bar{D} = D_{-\mu}(n_2/\sqrt{2n_3})/D_{-\nu}(n_2/\sqrt{2n_3})$, $\mu = (m + 2\nu)/2$

From equation (xv) the mean and mean square of a are obtained

$$E[a] = \{\Gamma(\nu + \tfrac{1}{2})/\Gamma(\nu)\}(2n_3)^{-1/4}\,\bar{D}, \qquad E[a^2] = \nu\bar{D}/\sqrt{2n_3} \qquad \text{(xvi)}$$

6.9 SECOND ORDER AVERAGING

The general characteristics of linear and non-linear systems can fairly be defined within the framework of the first order stochastic averaging. However, if the mathematical model involves cubic stiffness or inertia non-linearities of special forms, the first order averaging fails to predict their influence upon the response statistics. In order to determine the effect of such non-linearities, the averaging procedure must be extended to include higher order approximation.

In this section the second order averaging methods developed by Baxter (1971) and Schmidt (1979) will be outlined. Let the amplitude and phase differential equations (6.9) and (6.10) be written in a power series in ε

$$\begin{aligned} \dot{a}(t) &= \varepsilon G(a,\phi,\xi(t)) = \varepsilon(G_1 + \varepsilon G_2 + \varepsilon^2 G_3 + \ldots) \\ \dot{\theta}(t) &= \varepsilon H(a,\phi,\xi(t)) = \varepsilon(H_1 + \varepsilon H_2 + \varepsilon^2 H_3 + \ldots) \end{aligned} \qquad (6.63)$$

These equations are called the "standard asymptotic equations." Each component of G_i and H_i is defined by comparing (6.63) with (6.9) and (6.10).

Introducing the following transformation of variables to define the non-oscillatory amplitude a*, phase angle ϕ*, and phase shift θ*

$$\begin{aligned} a(t) &= a^*(t) + \varepsilon u(a^*,\phi^*) \\ \theta(t) &= \theta^*(t) + \varepsilon v(a^*,\phi^*) \end{aligned} \qquad (6.64)$$

where a* and θ* should satisfy a new set of differential equations

$$\begin{aligned} \dot{a}^* &= \varepsilon G^*(a^*,\theta^*,\xi(t)) = \varepsilon(G_1^* + \varepsilon G_2^* + \ldots) \\ \dot{\theta}^* &= \varepsilon H^*(a^*,\theta^*,\xi(t)) = \varepsilon(H_1^* + \varepsilon H_2^* + \ldots) \end{aligned} \qquad (6.65)$$

These equations are referred to as the "non-oscillatory asymptotic equations" and are similar in form to equations (6.63). The oscillatory terms involved in (6.63) are replaced by the functions u and v. These oscillatory terms may be represented by the expansions

$$u(a^*,\phi^*) = u_1 + \varepsilon u_2 + \varepsilon^2 u_3 + \dots$$
$$v(a^*,\phi^*) = v_1 + \varepsilon v_2 + \varepsilon^2 v_3 + \dots \tag{6.66}$$

In order to simplify the analysis, the procedure will be divided into two stages. The first stage does not include the oscillatory terms associated with the random excitation, while the second stage deals with the effect of these oscillatory terms. Setting $\xi(t)=0$ in equations (6.63) and (6.65) and differentiating (6.64) gives

$$\dot{a}(t) = \dot{a}^* + \varepsilon\{\dot{a}^*\partial u/\partial a^* + (\omega + \dot{\theta}^*)\partial u/\partial\phi^*\}$$
$$\dot{\theta}(t) = \dot{\theta}^* + \varepsilon\{\dot{a}^*\partial v/\partial a^* + (\omega + \dot{\theta}^*)\partial v/\partial\phi^*\} \tag{6.67}$$

Substituting (6.65) into (6.57), with $\xi(t)=0$, gives

$$\dot{a}(t) = \varepsilon G^*(a^*,\theta^*,0) + \varepsilon\{\varepsilon G^*\partial u/\partial a^* + (\omega + \varepsilon H^*)\partial u/\partial\phi^*\}$$
$$= \varepsilon(G_1^* + \varepsilon G_2^* + \dots) + \varepsilon\{\varepsilon(G_1^* + \varepsilon G_2^* + \dots)\frac{\partial}{\partial a^*}(u_1 + \varepsilon u_2 + \dots)$$
$$+ [\omega + \varepsilon(H_1^* + \varepsilon H_2^* + \dots)]\frac{\partial}{\partial\phi^*}(u_1 + \varepsilon u_2 + \dots)\}$$

$$\dot{\theta}(t) = \varepsilon H^*(a^*,\theta^*,0) + \varepsilon\{\varepsilon G^*\frac{\partial v}{\partial a^*} + (\omega + \varepsilon H^*)\frac{\partial v}{\partial\phi^*}\} \tag{6.68}$$
$$= \varepsilon(H_1^* + \varepsilon H_2^* + \dots) + \varepsilon\{\varepsilon(G_1^* + \varepsilon G_2^* + \dots)\frac{\partial}{\partial a^*}(v_1 + \varepsilon v_2 + \dots)$$
$$+ [\omega + \varepsilon(H_1^* + \varepsilon H_2^* + \dots)]\frac{\partial}{\partial\phi^*}(v_1 + \varepsilon v_2 + \dots)\}$$

In parallel to the above expansion, we expand the original equations (6.63), with $\xi(t)=0$, into a Taylor series

$$\dot{a}(t) = \varepsilon G(a^* + \varepsilon u,\phi^* + \varepsilon v,0)$$
$$= \varepsilon\{G(a^*,\phi^*,0) + \varepsilon u\left.\frac{\partial G}{\partial a}\right|^* + \varepsilon v\left.\frac{\partial G}{\partial\phi}\right|^* + \dots\}$$
$$\dot{\theta}(t) = \varepsilon H(a^* + \varepsilon u,\phi^* + \varepsilon v,0) \tag{6.69}$$
$$= \varepsilon\{H(a^*,\phi^*,0) + \varepsilon u\left.\frac{\partial H}{\partial a}\right|^* + \varepsilon v\left.\frac{\partial H}{\partial\phi}\right|^* + \dots\}$$

Equating (6.68) and (6.69) and equating the coefficients of equal powers in ε, the following relations are obtained

i. First order equations (coefficients of ε) are

$$G_1^*(a^*,\theta^*,0) + \omega \frac{\partial u_1}{\partial \phi^*}(a^*,\phi^*) = G_1(a^*,\phi^*,0)$$

$$H_1^*(a^*,\theta^*,0) + \omega \frac{\partial v_1}{\partial \phi^*}(a^*,\phi^*) = H_1(a^*,\phi^*,0) \qquad (6.70)$$

ii. The second order equations (coefficients of ε^2) are

$$G_2^* + \omega \frac{\partial u_2}{\partial \phi^*} + G_1^* \frac{\partial u_1}{\partial a^*} + H_1^* \frac{\partial u_1}{\partial \phi^*} = G_2\Big|^* + u_1 \frac{\partial G_1}{\partial a}\Big|^* + v_1 \frac{\partial G_1}{\partial \phi}\Big|^*$$

$$(6.71)$$

$$H_2^* + \omega \frac{\partial v_2}{\partial \phi^*} + G_1^* \frac{\partial v_1}{\partial a^*} + H_1^* \frac{\partial v_1}{\partial \phi^*} = H_2\Big|^* + u_1 \frac{\partial H_1}{\partial a}\Big|^* + v_1 \frac{\partial H_1}{\partial \phi}\Big|^*$$

In these equations the functions $G_1^*(a^*,\theta^*,0)$ and $H^*(a^*,\theta^*,0)$ are equated to the non-oscillatory terms. The differential terms of the type $\omega\partial u_i/\partial\phi^*$ and $\omega\partial v_i/\partial\phi^*$ are equated to the oscillatory terms appearing in each equation. The value of the oscillatory functions $u_i(a^*,\phi^*)$ and $v_i(a^*,\phi^*)$ are obtained by integrating $\partial u_1/\partial\phi^*$ and $\partial v_1/\partial\phi^*$. Thus from (6.64) we have

$$G_1^*(a^*,\theta^*,0) = \text{non-oscillatory terms of } G_1(a^*,\phi^*,0)$$

$$(6.72)$$

$$H_1^*(a^*,\theta^*,0) = \text{non-oscillatory terms of } H_1(a^*,\phi^*,0)$$

$$u_1(a^*,\phi^*) = \frac{1}{\Omega}\int \{\text{oscillatory terms of } G_1(a^*,\phi^*,0)\}\, d\phi^* + g_1(a^*)$$

$$v_1(a^*,\phi^*) = \frac{1}{\Omega}\int \{\text{oscillatory terms of } H_1(a^*,\phi^*,0)\}\, d\phi^* + h_1(a^*)$$

$$(6.73)$$

where $g_1(a^*)$ and $h_1(a^*)$ are constants of integration. These constants are determined such that the non-oscillatory amplitude will be the full amplitude of the response at the fundamental harmonic $\cos\phi^*$. This is guaranteed by substituting the oscillatory functions $u_1(a^*,\phi^*)$ and $v_1(a^*,\phi^*)$ into equation (6.2) of the total response of the system

$$Y(t) = a(t)\cos\phi(t)$$

$$= (a^* + \varepsilon u_1 + \ldots)\cos(\phi^* + \varepsilon v_1 + \ldots)$$

$$\cong a^*\cos\phi^* + \varepsilon(u_1\cos\phi^* - v_1 a^*\sin\phi^*) + \ldots \tag{6.74}$$

The higher order non-oscillatory functions $G_i^*(a^*,\phi^*,0)$ and $H_i^*(a^*,\phi^*,0)$, and the higher order oscillatory components $u_i(a^*,\phi^*)$ and $v_i(a^*,\phi^*)$, $(i>1)$ are determined in the same manner. Note that the expressions on the left side of (6.71) depend on the previously calculated values from (6.70), or more precisely on (6.72) and (6.73). Terms on the right-hand side of (6.71) are evaluated from the coefficients of the standard asymptotic equations (6.64) with $\xi(t)=0$.

The above outlined procedure constitutes the first stage where $\xi(t)=0$. The second stage takes into account the effect of the excitation terms. This will be performed by substituting the transformation (6.64) into the excitation terms and expanding each term into a power series in ε. The resulting excitation components are then added to the corresponding components of $G_i^*(a^*,\theta^*,0)$ and $H_i^*(a^*,\theta^*,0)$ of the same power in ε.

Example 6.6: Baxter (1971) studied the non-linear behavior of an elastic column subjected to random parametric loading as shown in fig. (6.6). The differential equation of the lateral motion of the column is

$$\ddot{Y} + \omega^2 Y = \varepsilon\{-2\zeta\omega\dot{Y} + 2\mu_1\omega^2\xi(t)Y - \mu_2\omega^2 Y^3 - \mu_3(\dot{Y}^2 + Y\ddot{Y})Y\}$$

$$+ \varepsilon^2\{-2\zeta\omega\delta Y^2\dot{Y} + \frac{3}{4}\mu_1\omega^2\xi(t)Y^3 - \mu_6\omega^2 Y^5 - \mu_4\dot{Y}^2 Y^3$$

$$- \mu_5(\dot{Y}^2 + Y\ddot{Y})Y^3\} \tag{i}$$

where Y is a dimensionless amplitude of the lateral vibration, defined by the ratio

$$\varepsilon^{1/2}Y(t) = \pi y_0(t)/L,$$

$y_0(t)$ is the amplitude of vibration at the midpoint of the column,
ω is the flexural natural frequency of the column $= \lambda\{\frac{P}{m}(1 - \frac{KZ_0}{P})\}^{1/2}$,

$$\lambda = \pi/L, \qquad P = \lambda^2 EI + (M + \frac{1}{2}mL)g,$$

m is the mass of the column per unit length,
K is the spring stiffness, Z_0 is the static spring deflection,
μ_1 is the excitation parameter $= K\sigma/\big(2(P - KZ_0)\big)$,

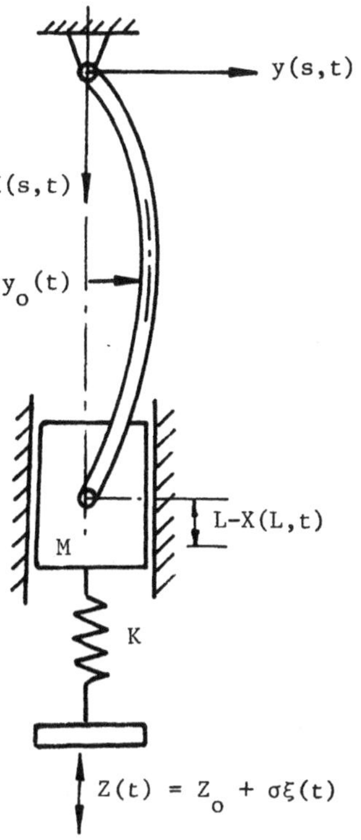

FIG. (6.6) Elastic column under random transmitted force (Baxter, 1971)

σ^2 is the variance of Z about the mean Z_0,
$\sigma\xi(t)$ is the random component of Z(t),
δ is the non-linear damping parameter,
μ_2, μ_6 are the non-linear spring parameters,
μ_3, μ_4, and μ_5 are non-linear inertia parameters.

Introducing the solution (6.2) and eliminating the double derivative Y from the non-linear terms, the standard equations for the amplitude and phase may be written in terms of sine and cosine functions of the multiple phase angle of the form (6.63), where

$$G_1\left(a,\phi,\xi(t)\right) = -\zeta\omega a(1 - \cos2\phi) + \frac{1}{8}\mu_2\omega a^3(2\sin2\phi + \sin4\phi)$$

$$- \frac{1}{4}\mu_3\omega a \sin4\phi - \underline{\mu_1\omega\xi(t)a \sin2\phi}$$

$$H_1\big(a,\phi,\xi(t)\big) = -\zeta\omega\ \sin 2\phi + \frac{1}{8}\mu_2\omega a^2(3 + 4\cos 2\phi + \cos 4\phi)$$

$$- \frac{1}{4}\mu_3\omega a^2(1 + 2\cos 2\phi + \cos 4\phi) - \underline{\mu_1\omega\xi(t)(1 + \cos 2\phi)}$$

(ii)

$$G_2\big(a,\phi,\xi(t)\big) = -\frac{1}{4}\zeta\omega\delta a^3(1 - \cos 4\phi) + \frac{1}{32}\mu_6\omega a^5(5\sin 2\phi + 4\sin 4\phi$$

$$+ \sin 6\phi) + \frac{1}{32}\mu_4\omega a^5(3\sin 2\phi - \sin 6\phi)$$

$$- \frac{1}{16}\mu_5\omega a^5(\sin 2\phi + 2\sin 4\phi + \sin 6\phi)$$

$$+ \frac{1}{32}\mu_3 a^2\{8\zeta\omega a(1 - \cos 4\phi) - \mu_2\omega a^3(5\sin 2\phi + 4\sin 4\phi$$

$$+ \sin 6\phi) + 2\mu_3\omega a^3(\sin 2\phi + 2\sin 4\phi + \sin 6\phi)\}$$

$$- \underline{\frac{1}{32}\mu_1\omega(3 - 8\mu_3)\xi(t)\ a^3(2\sin 2\phi + \sin 4\phi)}$$

$$H_2\big(a,\zeta,\xi(t)\big) = -\frac{1}{4}\zeta\omega\delta a^2(2\sin 2\phi + \sin 4\phi) + \frac{1}{32}\mu_6\omega a^4(10 + 15\cos 2\phi$$

$$+ 6\cos 4\phi + \cos 6\phi) + \frac{1}{32}\mu_4\omega a^4(2 + \cos 2\phi - 2\cos 4\phi$$

$$- \cos 6\phi) - \frac{1}{16}\mu_5\omega a^4(4 + 7\cos 2\phi + 4\cos 4\phi + \cos 6\phi)$$

$$+ \frac{1}{32}\mu_3 a^2\{8\zeta\omega(2\sin 2\phi + \sin 4\phi) - \mu_2\omega a^2(10 + 15\cos 2\phi$$

$$+ 6\cos 4\phi + \cos 6\phi) + 2\mu_2\omega a^2(4 + 7\cos 2\phi + 4\cos 4\phi$$

$$+ \cos 6\phi)\} - \underline{\frac{1}{32}\mu_1\omega(3 - 8\mu_3)\xi(t)a^2(3+4\cos 2\phi+\cos 4\phi)}$$

(iii)

Notice that the underlined terms involve the random excitation $\xi(t)$ and the remaining terms constitute the functions $G_i(a,\phi,0)$ and $H_i(a,\phi,0)$. The non-oscillatory terms in G_1 and H_1 are equated to G^* and H^* which are given by the expressions

$$G_1^*(a^*,0,0) = -\zeta\omega a^*$$

$$H_1^*(a^*,0,0) = \frac{1}{8}\omega a^{*2}(3\mu_2 - 2\mu_3)$$

(iv)

The oscillatory terms in (ii) are equated to $\omega\partial u_1/\partial\phi^*$ and $\omega\partial v_1/\partial\phi^*$ and then integrated with respect to ϕ to give

$$u_1(a^*,\phi^*) = \frac{1}{2}\zeta a^*\sin 2\phi^* - \frac{1}{8}\mu_2 a^{*3}\cos 2\phi^* - \frac{a^{*3}}{32}(\mu_2 - 2\mu_3)\cos 4\phi^* + g_1(a^*)$$

$$v_1(a^*,\phi^*) = \frac{1}{2}\zeta\cos 2\phi^* + \frac{1}{4}a^{*2}(\mu_2 - \mu_3)\sin 2\phi^* + \frac{a^{*2}}{32}(\mu_2 - 2\mu_3)\sin 4\phi^* + h_1(a^*) \qquad \text{(v)}$$

The constants of integration $g_1(a^*)$ and $h_1(a^*)$ are determined in such a manner that the non-oscillatory amplitude a^* will be the full average amplitude of the response at the fundamental harmonic $\cos\phi^*$. This can be obtained by substituting the oscillatory functions $u_1(a^*,\phi^*)$ and $v_1(a^*,\phi^*)$ into equation (6.74) such that $Y(t)$ will have no terms of order ε or higher that contain the fundamental harmonic. The procedure results in the following identity

$$(u_1\cos\phi^* - v_1 a^*\sin\phi^*) = (\frac{1}{2}\zeta a^* - a^*h_1)\sin\phi^* - \frac{1}{16}(-3\mu_2 a^{*3} + 2\mu_3 a^{*2} + g_1)\cos\phi^* + \frac{1}{32}a^{*3}(\mu_2 - 2\mu_3)\cos 3\phi^* - \frac{1}{64}a^* \cdot (\mu_2 - 2\mu_3)\cos 5\phi^* + \frac{1}{32}a^{*3}(\mu_2 - 2\mu_3)\sin 5\phi^* \qquad \text{(vi)}$$

Thus $g(a^*)$ and $h(a^*)$ must have the values

$$g_1(a^*) = a^{*3}(3\mu_2 - 2\mu_3)$$
$$h_1(a^*) = \zeta/2 \qquad \text{(vii)}$$

The second order correction terms $G_2^*(a^*,0,0)$ and $H_2^*(a^*,0,0)$ are evaluated from equations (6.71) as follows

$$G_2^*(a^*,0,0) = -\frac{1}{8}\zeta\omega a^{*3}(2\delta - 49\mu_2 + 30\mu_3)$$

$$H_2^*(a^*,0,0) = -\frac{1}{2}\zeta\omega + \frac{1}{16}\omega a^{*4}(\mu_4 - 4\mu_5 + 5\mu_6 - 3\mu_2\mu_3 - \frac{3}{16}\mu_2^2 + \frac{9}{4}\mu_2^3) \qquad \text{(viii)}$$

Adding the underlined oscillatory terms in (ii) and (iii) to the corresponding terms yields the first and second order approximations

$$G_1^*(a^*,\theta^*,\xi(t)) = -\zeta\omega a^* - \mu_1\omega\xi(t)a^*\sin 2\phi^*$$

$$H_1^*(a^*,\theta^*,\xi(t)) = \frac{1}{8}\omega a^{*2}(3\mu_2 - 2\mu_3) - \mu_1\omega\xi(t)(1 + \cos 2\phi^*) \qquad \text{(ix)}$$

$$G_2^*(a^*,\theta^*,\xi(t)) = -\frac{1}{8}\zeta\omega a^*(2\delta - 48\mu_2 + 28\mu_3)$$

$$-\frac{1}{32}\mu_1\omega(3 - 8\mu_3)\xi(t)a^{*3}(2\sin 2\phi^* + \sin 4\phi^*)$$

$$H_2^*(a^*,\theta^*,\xi(t)) = -\frac{1}{2}\zeta\omega + \frac{\omega a^{*4}}{16}(\mu_4 - 4\mu_5 + 5\mu_6 - 3\mu_2\mu_3 - \frac{3}{16}\mu_2^2$$

$$+\frac{9}{4}\mu_3^2) - \frac{1}{32}\mu_1\omega(3 - 8\mu_3)\xi(t)a^{*3}(3 + 4\cos 2\phi^* + \cos 4\phi^*)$$

(x)

Equations (ix) and (x) may be combined to form the Stratonovich equation for the amplitude and phase shift

$$\dot{a}^*(t) = \varepsilon(G_1^* + G_2^*)$$
(xi)
$$\dot{\theta}^*(t) = \varepsilon(H_1^* + H_2^*)$$

Confine the analysis to the amplitude characteristics, the Itô equation of the amplitude is

$$da^*(t) = \varepsilon\{-\zeta\omega a^* - \frac{1}{8}\varepsilon\zeta a^{*3}(2\delta - 49\mu_2 + 30\mu_3) + \frac{3}{4}\varepsilon\mu_1^2\omega^2 a^* S(2\omega)\}\,dt$$

$$+ \varepsilon\mu_1\omega\sqrt{\frac{1}{2}S(2\omega)}\;a^*\,dB(t) \qquad \text{(xii)}$$

where $S(\omega) = \int_{-\infty}^{\infty} E[\xi(t)\xi(t+\tau)]\cos\omega\tau\,d\tau$

The stationary Fokker-Planck equation of the amplitude is

$$-\frac{\partial}{\partial a^*}\{[-\varepsilon\zeta\omega a^* - \frac{1}{8}\varepsilon^2\zeta\omega a^{*3}(2\delta - 49\mu_2 + 30\mu_3) + \frac{3}{4}\varepsilon^2\mu_1^2\omega^2 a^* S(2\omega)]p\}$$

$$+\frac{1}{4}\varepsilon^2\mu_1^2\omega^2 S(2\omega)\frac{\partial^2}{\partial a^{*2}}(a^{*2}p) = 0 \qquad \text{(xiii)}$$

The solution of equation (xiii) is

$$p(a^*) = p_0 a^\lambda \exp(-\nu a^{*2}) \qquad \text{(xiv)}$$

where $\lambda = 1 - 2\zeta\omega/\varepsilon S_0$, $\qquad S_0 = \mu_1^2\omega^2 S(2\omega)/2$

$$\nu = \zeta\omega(2\delta - 49\mu_2 + 30\mu_3)/8S_0$$

and p_0 is the constant of integration which is determined from the normalized condition with the result

$$p_0 = 2\nu^{(\lambda+1)/2}\ \Gamma^{-1}\ (\frac{\lambda+1}{2}) \qquad \text{(xv)}$$

provided that $\nu > 0$ or

$$\delta > 24\mu_2 - 14\mu_3 \qquad \text{(xvi)}$$

Thus the probability density of the non-oscillatory amplitude is

$$p(a^*) = 2\nu^{(\lambda+1)2}\ \Gamma^{-1}(\frac{\lambda+1}{2})a^{*\lambda}\ \exp(-\nu a^{*2}) \qquad \text{(xvii)}$$

Example 6.7: Schmidt (1979,1981) extended the second order averaging method developed by Baxter (1971). The main feature of the Schmidt method is that it does not require the estimation of the integration functions g(a*) and h(a*). The method is outlined through an example of a non-linear elastic structure described by the differential equation

$$\ddot{Y} + Y = -2\zeta\dot{Y} - \beta Y^2\dot{Y} + \delta Y^3 + \mu Y(\dot{Y}^2 + Y\ddot{Y}) + \varepsilon\xi(t)Y \qquad \text{(i)}$$

Removing $\ddot{Y}$ from the non-linear term $Y\ddot{Y}$ by successive elimination and introducing solution (6.2), the following Itô equations are obtained for amplitude and phase shift

$$da(t) = f\ dt + G\ dB(t)$$

$$d\theta(t) = \hat{f}\ dt + \hat{G}\ d\hat{B}(t) \qquad \text{(ii)}$$

where $B(t)$ and $\hat{B}(t)$ are two independent Brownian motion processes, each has a unit variance, and

$$f = -\zeta a(1 - \cos2\phi) - \frac{\beta}{8}a^3(1 - \cos4\phi) - \frac{\delta}{8}a^3(2\sin2\phi + \sin4\phi)$$

$$+ \frac{1}{4}\mu a^3 \sin4\phi + \frac{\varepsilon^2 a}{16}(3 + 4\cos2\phi + \cos4\phi)$$

$$G = -\frac{\varepsilon}{2} a \sin2\phi$$

$$\hat{f} = -\frac{1}{2}\zeta\sin2\phi - \frac{\beta}{8}a^2(2\sin2\phi + \sin4\phi) - \frac{\delta}{8}a^2(3 + 4\cos2\phi + \cos4\phi)$$

$$+ \frac{1}{4}\mu a^2(2\cos2\phi + \cos4\phi) - \frac{\varepsilon^2}{4}(2\sin2\phi + \sin4\phi)$$

$$\hat{G} = -\frac{\varepsilon}{2}(1 + \cos2\phi) \qquad \text{(iii)}$$

Introducing the following transformations

$$a(t) = a^*(t) + u(a^*,\phi^*)$$
$$\phi(t) = \phi^*(t) + v(a^*,\phi^*) \qquad \text{(iv)}$$

where a^* and ϕ^* are the non-oscillatory amplitude and phase respectively, u and v are called the auxiliary correction functions. For the phase transformation we write

$$\phi^*(t) = t + \Theta^*, \quad \text{and} \quad \theta = \Theta^* + v(a^*,\phi^*) \qquad \text{(v)}$$

The stochastic differentials of the non-oscillatory processes a^* and Θ^* may be expressed in the form of the Itô equations

$$da^* = R\,dt + Q\,dB_1(t)$$
$$d\Theta^* = \hat{R}\,dt + \hat{Q}\,d\hat{B}_1(t) \qquad \text{(vi)}$$

where $R = \sum_{i=1}^{\ell} R_i$, $\quad Q = \sum_{i=1}^{\ell} Q_i$

$$\hat{R} = \sum_{i=1}^{\ell} \hat{R}_i, \text{ and } \quad \hat{Q} = \sum_{i=1}^{\ell} \hat{Q}_i \qquad \text{(vii)}$$

ℓ stands for the order of approximation.

The stochastic differential of the oscillatory function $u = u(a^*,\phi^*)$ can be written in the Itô form

$$du = \frac{\partial u}{\partial a^*}\,da^* + \frac{\partial u}{\partial \phi^*}\,d\phi^* + \left(\frac{1}{2}\frac{\partial^2 u}{\partial a^{*2}} Q^2 + \frac{\partial^2 u}{\partial a^*\partial \phi^*}\hat{Q}Q + \frac{1}{2}\frac{\partial^2 u}{\partial \phi^{*2}}\hat{Q}^2\right) dt \qquad \text{(viii)}$$

Alternatively, the differential da^* can be written from (iv) and using (ii,v,vi)

$$da^* = da - du = \big(f\,dt + G\,dB(t)\big) - \text{right-hand side of (viii)}$$

$$= \left\{-\frac{\partial u}{\partial \phi^*} - \frac{\partial u}{\partial a^*}R - \frac{\partial u}{\partial \phi^*}\hat{R} - \frac{1}{2}\frac{\partial^2 u}{\partial a^{*2}}Q^2 - \frac{\partial^2 u}{\partial a^*\partial \phi^*}\hat{Q}Q - \frac{1}{2}\frac{\partial^2 u}{\partial \phi^{*2}}\hat{Q}^2\right.$$

$$\left. + f(a^*+u,\phi^*+v)\right\} dt + \left\{-\frac{\partial u}{\partial a^*}Q - \frac{\partial u}{\partial \phi^*}\hat{Q} + G(a^*+u,\phi^*+v)\right\} dB(t) \qquad \text{(ix)}$$

Similarly, the differential of phase is

$$d\theta^* = \left\{-\frac{\partial v}{\partial \phi^*} - \frac{\partial v}{\partial a^*} R - \frac{\partial v}{\partial \phi^*} \hat{R} - \frac{1}{2}\frac{\partial^2 v}{\partial a^{*2}} Q^2 - \frac{\partial^2 v}{\partial a^* \partial \phi^*} Q\hat{Q} - \frac{1}{2}\frac{\partial^2 v}{\partial \phi^{*2}} \hat{Q}^2\right.$$

$$\left. + \hat{f}(a^*+u, \phi^*+v)\right\} dt + \left\{-\frac{\partial v}{\partial a^*} Q - \frac{\partial v}{\partial \phi^*} \hat{Q} + \hat{G}(a^*+u, \phi^*+v)\right\} dB(t) \tag{x}$$

<u>First Order Averaging</u>: The first order averaging solution is obtained by setting $u=v=0$ and considering only the non-oscillatory terms $f, \hat{f}$, G^2, $G\hat{G}$, and $\hat{G}^2$. The corresponding Fokker-Planck equation for the stationary probability density of the amplitude a^* is

$$\frac{\partial\{Rp(a^*)\}}{\partial a^*} - \frac{1}{2}\frac{\partial^2}{\partial a^{*2}}\{Q^2 p(a^*)\} = 0 \tag{xi}$$

where $R = R_1 = \frac{1}{2} a^*(-2\zeta + \frac{3}{8}\frac{\varepsilon^2}{\omega^2}) - \frac{\beta}{8} a^{*3}$

$$Q^2 = Q_1^2 = \varepsilon^2 a^{*2}/8$$

For abbreviation sake let

$$J = (2R - \partial Q^2/\partial a^*)/Q^2$$

The stationary solution of (xi) may be written in the form

$$p = p_0 \exp(\int J \, da^*)$$

$$= p_0 a^{*(1-16\zeta/\varepsilon^2)} \exp(-\beta a^{*2}/\varepsilon^2) \tag{xii}$$

It is not difficult to show, from the normalized condition and Gradshteyn and Ryznik (1980) tables (formula 3.381, 4), that p_0 is given in the form

$$p_0 = (\beta/\varepsilon^2)^{1-8\zeta/\varepsilon^2} \Gamma^{-1}(1 - \frac{8\zeta}{\varepsilon^2})$$

provided $\varepsilon^2 > 8\zeta$ (xiii)

It is clear that the solution reveals the influence of the parametric excitation, the linear and non-linear damping coefficients on the response probability density of the amplitude. However, the solution does not take into account the effect of the stiffness and inertia non-linearities which can only be determined by considering the second order approximation.

Second Order Averaging: Let $R = R_1 + R_2$ and $Q = Q_1 + Q_2$.

The value of u in equation (ix) will be determined in such a way that it compensates the expressions involving oscillatory terms. Notice that the maximum value of $u = -\frac{\partial u}{\partial \phi^*}$ is

$$-\frac{\partial u}{\partial \phi^*} = \text{oscillatory terms in } f(a^*,\phi^*) \qquad \text{(xiv)}$$

Integrating (xiv) and setting the integration constant (which is a function of the amplitude) to zero, gives

$$u(a^*,\phi^*) = \frac{\zeta a^*}{2}\sin 2\phi^* + \frac{\beta_1}{32}a^{*3}\sin 4\phi^* + \frac{\delta a^*}{32}(4\cos 2\phi^* + \cos 4\phi^*)$$

$$-\frac{\mu}{16}a^{*3}\cos 4\phi^* + \frac{\varepsilon^2}{64}a^*(8\sin 2\phi^* + \sin 4\phi^*) \qquad \text{(xv)}$$

Similarly, the oscillatory terms in R are equated to $-\partial v/\partial\phi^*$

$$-\partial v/\partial\phi^* = \text{oscillatory terms in } \hat{f}(a^*,\phi^*) \qquad \text{(xvi)}$$

Integrating (xvi) gives

$$v(a^*,\phi^*) = \frac{\zeta}{2}\cos 2\phi^* + \frac{\beta a^{*2}}{32}(4\cos 2\phi^* + \cos 4\phi^*) - \frac{\delta a^{*2}}{32}(8\sin 2\phi^*$$

$$+ \sin 4\phi^*) + \frac{\mu a^{*2}}{16}(4\sin 2\phi^* + \sin 4\phi^*) + \frac{\varepsilon^2}{32}(4\cos 2\phi^*$$

$$+ \cos 4\phi^*) \qquad \text{(xvii)}$$

The expression in the first braces in equation (ix) consists of $R_1 + R_2$. Substituting R_1, $\hat{R}_1$, Q_1, and $\hat{Q}_1$ for R, $\hat{R}$, Q and $\hat{Q}$ into equation (ix), and simplifying the terms of the function $f(a^*+u,\phi^*+v)$ according to the approximation

$$(a^* + u)^3\cos(4\phi^* + 4v) \cong a^{*3}\cos 4\phi^* + 3a^{*3}u\cos 4\phi^* - 4a^{*3}v\sin 4\phi^*$$

results in R_2,

$$R_2 = -\frac{\partial u}{\partial a^*} R_1 - \frac{\partial u}{\partial \phi^*} \hat{R}_1 - \frac{1}{2}\frac{\partial^2 u}{\partial a^{*2}} Q_1^2 - \frac{\partial^2 u}{\partial a^* \partial \phi^*} Q_1 \hat{Q}_1 - \frac{1}{2}\frac{\partial^2 u}{\partial \phi^{*2}} \hat{Q}_1^2$$

$$- \zeta u(1 - \cos 2\phi^*) - 2\zeta a^* v \sin 2\phi^* - \frac{3}{8}\beta a^{*2} u(1 - \cos 4\phi^*)$$

$$- \frac{1}{2}\beta a^{*3} v \sin 4\phi^* - \frac{3\delta}{8} a^{*2} u(2\sin 2\phi^* + \sin 4\phi^*) - \frac{\delta}{2} a^{*3} v(\cos 2\phi^*$$

$$+ \cos 4\phi^*) + \frac{3}{4}\mu a^{*2} u \sin 4\phi^* + \mu a^{*3} v \cos 4\phi^* + \frac{\varepsilon^2 u}{16\omega^2}(3 + 4\cos 2\phi^*$$

$$+ \cos 4\phi^*) - \frac{\varepsilon^2 a^*}{4} v(2\sin 2\phi^* + \sin 4\phi^*) \qquad \text{(xviii)}$$

Substituting (xv,xvii) into (xviii) and extracting the non-oscillatory terms gives

$$R_2 \text{ (non-osc.)} = -\frac{\zeta}{4}\mu a^{*3} + \frac{25\delta\varepsilon^2}{128} a^{*3} - \frac{94\varepsilon^2}{64} a^{*3} - \frac{\beta\delta}{32} a^{*5} \qquad \text{(xix)}$$

The total expression of $R = R_1 + R_2$ becomes

$$R = \frac{1}{2} a^*(-2\zeta + \frac{3\varepsilon^2}{8}) - \frac{a^{*3}}{128}(16\beta + 32\zeta\mu - 25\delta\varepsilon^2 + 18\mu\varepsilon^2)$$

$$- \frac{\beta\delta}{32} a^{*5} \qquad \text{(xx)}$$

Similarly, the second braces of equation (ix) give Q_2

$$Q_2 = -\frac{\partial u}{\partial a^*} Q_1 - \frac{\partial u}{\partial \phi^*} \hat{Q}_1 - \frac{\varepsilon u}{2} \sin 2\phi^* - \varepsilon v a^* \cos 2\phi^* \qquad \text{(xxi)}$$

Substituting (xv,xvii) into (xxi) to determine the non-oscillatory terms of $Q^2 = (Q_1 + Q_2)^2$ gives

$$Q^2 = \frac{\varepsilon^2}{64} a^{*2}(8 + 5\delta a^{*2} - 2\mu a^{*2}) \qquad \text{(xxii)}$$

Introducing R and Q^2 in the J expression yields

$$J = (2R - \partial Q^2/\partial a)/Q^2$$

$$= \{2(C_1 - C_4) - 2(C_2 + 2C_5)a^{*2} - 2C_3 a^{*4}\}/\{a(C_4 + C_5 a^2)\}$$

$$= -2M_1 a^* + \frac{M_2}{a^*} + \frac{M_3 a^*}{(C_4 + C_5 a^{*2})} \qquad \text{(xxiii)}$$

where $C_1 = \frac{1}{2}(-2\zeta + \frac{3\varepsilon^2}{8})$, $C_3 = \frac{\beta}{32}$, $C_4 = \frac{\varepsilon^2}{8}$, $C_5 = \frac{\varepsilon^2}{64}(5\delta - 2\mu)$

$$C_2 = \frac{1}{128}(16\beta + 32\zeta\mu - 25\delta\varepsilon^2 + 18\mu\varepsilon^2), \quad M_1 = C_3/C_5,$$

$$M_2 = 2(\frac{C_1}{C_4} - 1), \quad M_3 = 2\frac{C_3C_4}{C_5} - 2(\frac{C_1}{C_4} - 1)C_5 - 2(C_2 + 2C_5)$$

The stationary probability density is given by using (xxiii) in the stationary Fokker-Planck equation with the result

$$p = p_0 a^{*M_2}(C_4 + C_5 a^{*2})^{M_3/2C_5}\exp(-M_1 a^{*2}) \tag{xxiv}$$

provided $M_2 > 0$ or $\varepsilon^2 > 8\zeta$.

The constant of integration p_0 is determined numerically from the normalized condition.

CHAPTER 7

Parametric Stochastic Stability

7.1 INTRODUCTION AND HISTORICAL REVIEW

Stochastic stability provides an important criterion in studying the behavior of dynamic and control systems under random parametric excitation. In deterministic parametric vibration it is well known that the stability properties are determined from the Mathieu equation together with the corresponding Ince-Strutt diagram. The Ince-Strutt diagram delineates the regions of dynamic stability in terms of the excitation amplitude and frequency. The first two instability regions are situated near the harmonic and twice the natural frequency of the system. If the parametric excitation becomes random, the stability boundaries of the system are expressed in terms of the statistical properties of the excitation and the system physical parameters.

The concept of stochastic stability is essentially based upon the stochastic modes of convergence of the system response. The stochastic stability may thus be studied in terms of convergence in probability, convergence in mean square, and almost sure convergence. With regard to these modes, Kozin (1969) indicated that "not all definitions of stochastic stability will be of interest simply because they may be too weak to be of practical significance." In attempting to determine stability boundaries from experimental or simulation records it is possible, from a simple observation of the system response, to judge whether or not the system is stable. However, the stability decision is, in general, not an easy task experimentally and theoretically. Kozin and Sugimoto (1977) discussed cases of sample solutions that appear to decay during the period of observation and yet are generated from unstable systems. On the other hand, one can observe samples that appear to grow during the period of observation, but are generated from stable systems. Kozin and Sugimoto proposed some measures concerning the response sample processing and its time duration, which is required to make a "reasonable" decision for its asymptotic behavior.

The subject of stochastic stability is a result of several theorems established over the last few decades by investigators who were motivated by the development and design of adaptive control systems and structural dynamic systems. The first work is believed to be due to Andronov, et al., (1933). They studied the probabilistic properties of the sample response functions of one- and two-dimensional nonlinear systems driven by a Gaussian white noise. Twenty years later, Rosenbloom, et al., (1955) and Tikhonov (1958) examined the statistical properties of first order differential equations with correlated external and parametric excitations. Samuels and Eringen (1959) introduced the concept of the mean square stability of systems excited parametrically by a white noise process. Since then several techniques have been developed to examine the stochastic parametric stability. These techniques include the stochastic averaging, the Liapunov direct method, the moment function method, the Khas'miniskii method, the perturbation techniques, and algebraic and numerical methods. The stochastic averaging methods have already been treated in chapter 6. The present chapter deals with other methods used frequently in mechanical vibrations. This section provides a historical review of the methods and results pertaining to parametric stochastic stability. The review is classified into five categories which may have some overlapping.

7.1.1 THE LIAPUNOV DIRECT METHOD

The second method of Liapunov has promoted the study of stochastic stability in automatic control and mechanical vibrations. Originally the Liapunov second (or direct) method was used by mathematicians who established stability theorems in theoretical mechanics. Basically, the method establishes stability conditions of the system equilibrium without solving the system differential equations of motion. The method is based on constructing a scalar function defined in the phase plane ($\underset{\sim}{X}$) or in the motion space ($\underset{\sim}{X}$,t). These functions are called "Liapunov functions" and they are regarded as an extension of the energy principle which states that the equilibrium position is stable if the energy of the system is always decreasing as the equilibrium state is approached. The sign of the Liapunov function and its time derivative, evaluated along the trajectory of the system equations of motion, thus determines whether or not the system is stable. Techniques for constructing Liapunov functions of linear systems are well documented by La Salle and Lefschetz (1961) and Hahn (1963). However, there is no unique procedure for establishing a Liapunov function for a general dynamic system. In addition, there is no unique Liapunov function for a given system. Kushner (1965) suggested a method for constructing stochastic Liapunov functions. In his approach the form $LV(\underset{\sim}{X})<0$ is assumed and $V(\underset{\sim}{x})$ is evaluated; $L(\cdot)$ is a Liapunov operator, and $V(x)$ is the Liapunov function. The computed $V(\underset{\sim}{x})$ is then checked for positive-definiteness for values of $\underset{\sim}{x}$ in the stability domain.

Bertram and Sarachik (1959) employed the second method of Liapunov to determine the stochastic stability in the mean square. Kats and

Krasovskii (1960), Krasovskii (1961) and Kats (1964) derived a number of criteria for asymptotic stability in the probability sense of a system of differential equations with random coefficients which resemble a homogeneous Markov chain with a finite number of states. Khas'miniskii (1962) obtained a necessary and sufficient condition for the stability of a Markov process trajectory for which the coefficients of diffusion and drift become zero at the origin. Khas'miniskii showed that as the number of equations of motion of the deterministic system increases, the introduction of a sufficiently large diffusion process will reduce the stability of the system. Palmer (1966) extended the work of Kats and Krasovskii (1960) and applied the ergodic properties of Markov chains to study linear systems with coefficients described by a stationary positive recurrent Markov chain. Palmer determined sufficient conditions which guarantee almost sure asymptotic stability of the equilibrium solution.

Samuels (1960) developed a general theory for the mean square stability of linear systems with white noise coefficients. The results of Samuels attracted the attention of Caughey (1960) who found a number of errors which invalidated the results of the given examples. Caughey pointed out that it is not possible to stabilize the unstable system (with negative damping) by means of a random parametric noise. Another important fact is that the mean square stability is a necessary but not sufficient condition for the system stability. In order to ensure stability, all the moments must be stable. Rabotnikov (1964) treated the same problem and confirmed Caughey's remarks. Kozin (1969) added that the sample equations may possess asymptotic stability with probability one, yet all moments become unstable. Valeev (1971) and Valeev and Dolya (1974) used the Liapunov function and the method of averaging to examine the stability of the inverted pendulum under different types of parametric excitations. Samuels (1961) determined the second order moments of linear systems with Gaussian random coefficients. He obtained conditions for stability in mean and mean square, and found that the mean square stability condition depends only upon the values of the auto- and cross-correlation functions of the excitation at the origin and not on the detailed structure of these functions. It was concluded that the mean of the response is not related to the random excitation. However, this conclusion was questioned by Bogdanoff and Kozin (1962) who showed that the mean motion is influenced by the random coefficients. Necessary and sufficient conditions guaranteeing the mean square stability were obtained by Sawaragi, et al., (1967). The so called frozen system (which is obtained by setting the random coefficients to zero) was introduced and the stability criteria were established based on the assumption that the frozen system is asymptotically stable. Nakamizo and Sawaragi (1972) generalized the results to n-th order linear equations with random coefficients.

The behavior of linear dynamic systems with a single random parameter was investigated by Caughey and Dienes (1962). For a non-white parametric excitation the moments of the response diminish asymptotically and the system becomes stable if the power spectral density of the excitation does not contain frequency components. If the excitation, on the other hand, contains zero frequency components, then the moments of sufficiently higher order will be unstable. The question of moments stability was also discussed for an n-th order system with random coefficients by Leibowitz (1963). Leibowitz showed that Gaussian noise coefficients act as destabilizing sources to the system. Nevel'son and Khas'miniskii (1966a) proved that for asymptotic stability in the mean square of a linear system, it is necessary and sufficient that the Liapunov operator performed on a positive definite quadratic Liapunov function results in a preassigned negative definite form. This theorem led to algebraic criteria of asymptotic stability in the mean square. However, these criteria involve cumbersome analysis and computations. Nevel'son and Khas'miniskii (1966b) established another technique which involves determinants of order (n+1), where n is the order of the differential equation of the system. They showed that if the equilibrium solution of a deterministic second order system is not asymptotically stable, then the n-th moments of the equilibrium solution of the same system under white noise parametric excitation cannot be asymptotically stable. If the excitation is a wide band random process, the system will possess regions of stable second moments and almost sure sample stability as indicated by Mitchell (1970).

Nevel'son (1966) provided an interpretation for stability criteria employed in controlled circuit theory. For small intensity parametric excitation he (1967) confirmed some results obtained earlier by Khas'miniskii (1962) that if the deterministic system is unstable in the Liapunov sense, it remains almost sure unstable by adding a small diffusion disturbance.

7.1.2 THE ALMOST SURE STOCHASTIC STABILITY

The almost sure stochastic stability of linear systems with non-white random coefficients has received an extensive number of investigations. Kozin (1963) employed the fundamental Gronwall-Bellman lemma and the strong law of large numbers for strictly stationary stochastic processes. For second order systems, Kozin's results were close to those obtained by Samuels (1961). Caughey and Gray (1965) indicated that Kozin's results are, however, too conservative since one condition yields zero standard deviation of the parametric excitation when the system is nearly critically damped. They obtained sharper conditions sufficient to guarantee the almost sure stability of linear systems with stochastic coefficients. Kozin (1966) subsequently questioned their results concerning the almost sure stability of a class of non-linear systems. Mehr and Wang (1966) established additional theorems to the results of Caughey and Gray and provided a comparison between Kozin's result (1963) and the results of Caughey and Gray. The result of Kozin (1963) was further examined by Gray (1967) who obtained a less conservative result when the excitation is not Gaussian.

Gray (1967), Caughey and Dickerson (1967), and Hsu and Lee (1969) obtained almost sure stability conditions for linear systems with narrow band parametric coefficients. Kozin (1965a,b) investigated the implication of the almost sure asymptotic Liapunov stability "in the large" that are contained in the moment properties of the system response. Kozin (1966) presented some of the basic properties of the concept of Liapunov stability for stochastic systems. Wang (1965,1966) generalized the most important results of Kozin (1963) and Caughey and Gray (1965) to particular classes of linear distributed-parameter and time-lag dynamical systems described by a set of linear partial-integral equations with stochastic parameters.

Nevel'son and Khas'miniskii (1966a), Kozin (1972) and Pinsky (1974) gave various definitions and properties of stochastic stability of ordinary differential equations with white noise coefficients of intensity which depends on the state of the system. Wonham (1966) obtained sufficient stability conditions for a diffusion process defined by stochastic differential equations of the Itô type. The conditions require the existence of functions which resemble Liapunov functions. The "weak" stability of a stochastic system was obtained by formulating a Liapunov function for the Lagrange stability (La Salle and Lefschetz, 1961) of a corresponding deterministic system. Kushner (1967) and Khas'miniskii (1980) provided a number of stability theorems for stochastic linear differential equations. Dickerson and Caughey (1969) developed a scheme based on a Liapunov-type approach to determine sufficient conditions guaranteeing the asymptotic stability of a class of dynamic systems with random coefficients. They showed that a parametrically excited system will be stable if the system is stable in the absence of the excitation.

Akhmetkaliev (1965,1966) established a relationship between the stability of the solution of stochastic differential equations and the corresponding difference equations. He obtained stability conditions of stochastic difference systems under parametric excitations. The conditions of exponential stability and stability in probability of the equilibrium solution of stochastic difference equations with Markov parametric excitations were treated by Konstantinov (1970), Yudaev (1979) and Danilin and Yadykin (1981).

The almost sure asymptotic stability conditions of linear systems with piecewise constant coefficients were obtained by Morozan (1967b,c). He (1967a,1968) developed a number of stability theorems for linear systems described by the Itô type equations. Lepore and Stoltz (1971b) indicated that for the case of uncorrelated excitation and response, the excitation covariance increases without limit as the damping increases.

A significant step was taken by Infante (1968) who developed a theorem which gives sharp boundaries of almost sure stability of dynamic systems with real random parametric coefficients. Basically, Infante employed the theory of pencils of quadratic forms (Gantmacher, 1959a,b) to determine an optimum norm. The results showed that the variance of the random coefficient process approaches infinity as the damping coefficient approaches infinity, a fact which confirmed a conjecture mentioned by Mehr and Wang (1966). Bunke (1970,1971) extended the results of Infante and developed a number of stability theorems dealing with the Itô equations and other types of random differential equations.

Khas'miniskii (1962) established a theorem to determine the almost sure stability of the static equilibrium of linear systems excited parametrically by a Gaussian white noise or by a random process approximated by a Gaussian white noise in the sense of Stratonovich. Prodromou (1970) and Kozin and Prodromou (1971) employed Khas'miniskii's theorem and calculated the stability boundaries of three different dynamic systems. Kozin (1972) correlated the results obtained in terms of Itô's equations with those derived by Infante for systems with real noise coefficients. As the real noise coefficient approaches white noise, the system equations of motion should be modified by including the Wong-Zakai correction terms described in section 4.7.4.

For the Itô stochastic differential equations, Kozin (1965a) obtained sufficient conditions via the properties of second moments. Mitchell (1970) and Kozin and Prodromou (1971) derived necessary and sufficient conditions in terms of the intensity level (power spectral height) of the Gaussian white parametric excitation. For linear stochastic differential equations with physical noise coefficients, Kozin and Wu (1973) obtained sample stability conditions in terms of the noise variance. Mitchell and Kozin (1974) showed that the stability properties depend upon the height of the power spectral density curve at zero frequency and not upon the variance. Based on the results of Khas'miniskii (1967), Mitchell and Kozin established a relationship between the stability conditions for the cases of Gaussian white noise coefficient and of wide band physical noise. Ly (1974) extended the Infante theorem for the case of simultaneous parametric excitations in both damping and stiffness terms. Ly employed the theory of envelopes when the excitation probability densities are given.

Podvintsev (1971) extended the analysis of the frequency domain due to Fedosov, et al., (1968) to determine the stability regions of dynamic systems under random parametric excitation with a physical spectrum. He defined the intersection of the mean and variance stability regions as the "absolute stability" of the system. Although Podvintsev did not correlate his results with the Infante theorem, a simple inspection revealed that the two boundaries are identical.

The concept of global stability in the Liapunov sense was adopted by Soeda and Umeda (1966,1970) for studying the behavior of control systems with random coefficients. Levit and Yakubovich (1972) developed an algebraic criterion for exponential stability in the mean square. Chow and Chiou (1981) combined the analyses of Lyubarskii and Robotnikov (1963) and Kozin (1965b) into one scheme to derive conditions for the mean square and almost sure asymptotic stability in terms of the size and the correlation length of a bounded random parametric excitation. Geman (1982) developed a mathematical treatment based on the law of large numbers to derive the almost sure stability condition of a set of non-linear differential equations coupled through random coefficients. Ahmadi and Glockner (1982) obtained approximate conditions for almost sure stability of the equilibrium state of linear systems. The so called "Liapunov index" was introduced by Auslender and Mil'shtein (1982) who examined the stochastic stability of linear systems of Itô and Stratonovich types.

The technique of Kronecker product established originally by Bellman (1954) was employed by Bergen (1960) and Bharucha (1961). Bergen showed that the system is stable with probability one if all the eigenvalues of a specified matrix lie within a unit circle. Bharucha considered the case of piecewise constant coefficients of linear systems. He assumed two stochastic structures for the coefficient process: a sequence of independent identically distributed random matrices and a finite Markov chain of random matrices. Darkhovskii and Leibovich (1971) determined the conditions of stochastic stability and estimated the response mathematical expectations for automatic control systems described by a multidimensional state vector with piecewise linear coefficient matrices and white noise external excitation.

Sunahara, et al., (1977) combined the stochastic averaging method developed by Khas'miniskii (1966) and the Liapunov direct method into a new approach. They established a number of theorems dealing with sufficient conditions for asymptotic stability of certain types of non-linear systems.

7.1.3 PERTURBATION TECHNIQUES

Perturbation techniques were used to investigate the stochastic stability of linear and non-linear stochastic systems. Ivovich (1969) examined the stability of a plane rope grid subjected to random parametric excitation. Cubic elastic non-linearity was found to have no effect on the mean square stability. Katsnel'son, et al., (1971) developed an approximate method based on the perturbation technique to investigate the stochastic stability of linear systems under stationary narrow band excitation. The stability was established according to a monotone decrease of the mean square response. Averaging was performed over a set of initial conditions with respect to the realizations of the parametric excitations. Secular terms which emerged from the integration process were found to occur at zero and twice the natural frequency of the system. The procedure led to a stability criterion identical to the Weidenhammer (1964) condition. Kolovskii and Troitskaya (1972) obtained a system of recurrence equations for which the mean square value of their solution was examined for stability. They considered three examples including a two degree-of-freedom system with random coefficients. Wedig (1972a,b,c) examined the stability of the moments of a linear system excited by a low-pass process obtained by passing Gaussian white noise through a low-pass linear filter. A set of moment equations which are weakly coupled with higher order moments were truncated by dropping the higher moment terms. The stability boundaries were then obtained by expanding the eigenvalues and the corresponding eigenvectors of the homogeneous moment equations in power series. The envelope of all critical spectral densities of the excitation defines the limit of an instability region in mean square. The lowest point of the region was located at twice the natural frequency of the system. Wedig further considered linear systems excited by bandpass filtered white noise. In the time domain he represented the filter by two uncoupled differential equations of first order. A stability map comprised of two threshold spectral density values of the excitation was obtained. The mean square stability of the equilibrium position is mainly governed by a damping factor of order ε ($\varepsilon>0$) in the case of narrow band excitation. It is of order ε^2 if the excitation is wide band. Wedig (1973) obtained the stability regions of a two degree-of-freedom system in the neighborhood of principal and combination parametric resonance conditions. The regions exhibited the typical 'V' shape of the deterministic counterpart. Mirkina (1977) employed a multi-scale expansion to determine the first two regions of stohastic stability.

The mean square stability of unimodal response of a two degree-of-freedom structure with autoparametric coupling was determined by J.W. Roberts (1980) by using a perturbation technique. The stability of the same system was previously investigated by Ibrahim and J. W. Roberts (1977) who used moment functions and a Gaussian closure scheme. The two approaches exhibited the 'V' shape with the bottom located at the condition of internal resonance $r = \omega_2/\omega_1 = 0.5$ (where ω_1 and ω_2 are the system normal mode frequencies). The validity of the two solutions was examined experimentally by J. W. Roberts (1980) who found that at exact internal resonance the experimental instability threshold agreed with the perturbation solution. However, the experimental boundary exhibited wider instability region than those predicted analytically.

The perturbation theory of semi-groups of operators was employed by Gopalsamy (1976) to obtain sufficient conditions for the mean square stability of a class of randomly excited systems governed by partial differential equations.

7.1.4 OTHER MATHEMATICAL METHODS

The mathematical theorems of Lie groups and Lie algebra have been used to develop a number of techniques for studying the behavior of dynamic systems with state-dependent noise. Willems (1975a,b) and Brockett (1976) employed the Lie algebra to investigate the asymptotic properties of stochastic linear differential equations of the Itô type. For the case of linear systems with coefficients represented by colored multiplicative noise, the "Lie-theoretic" methods were employed by Blankenship (1975) and Willsky, et al.,

(1975). Blankenship (1977) adopted the Coppel (1975) inequality (used originally for linear deterministic differential equations) to derive stability criteria for linear stochastic systems. Willems and Aeyels (1976) used the Lie algebra to determine the moment stability of linear systems with colored noise coefficients. They found that the stability conditions depend only on the values of the excitation spectral densities at zero frequency. Blankenship and Papanicolaou (1978) analyzed the stability properties of control systems subjected to external noise disturbances. It may be noticed that these theorems have mainly been used in automatic control problems, but have not yet been implemented for random parametric vibration problems.

The stability of higher order moments was investigated by Moskvin and Smirnov (1975) and Willems (1977). They showed that the stability of higher order moments becomes more restrictive as the moment order increases.

A new measure of stochastic stability in terms of entropy has recently been introduced. With reference to dynamic systems, the entropy is defined as a measure of the degree of randomness in the system. A highly organized situation possesses little entropy, whereas a disorganized one possesses more entropy. The concept of entropy is a general statistical characteristic of a system. Phillis (1982) obtained conditions that ensure monotonic reduction of the entropy of a dynamic system described by stochastic differential equations of the Itô type.

7.1.5 APPLICATIONS

The stochastic stability of several dynamic systems such as elastic structures, liquid free surface in moving containers, aeroelastic structures, ships, and satellites has been examined by several investigators who used various stochastic stability techniques. The dynamic stability of an elastic column subjected to random axial load was determined by Ariaratnam (1967), Mitchell (1968b), Lepore and Shah (1968), Infante and Plaut (1969), Katsnel'son, et al., (1971), Wu (1971), Parthasarathy (1972), and Parthasarathy and Evan-Iwanowski (1978). Plaut and Infante (1970) established a general procedure to determine the almost sure asymptotic stability "in the large" for continuous elastic systems subjected to random parametric excitation. Wu (1971) examined the stochastic stability of columns with different boundary conditions. Lepore and Shah (1970) and Lepore and Stoltz (1971b) obtained mean square stability criteria for circular and rectangular plates. Wu (1971) determined the almost sure asymptotic stability boundaries of simply supported flat rectangular plates. Tylikowski (1978,1979) examined the effects of non-linearities on the dynamic stability of rectangular plates and cylindrical shells. He concluded that stability conditions for the linearized plates imply the stability of non-linear plates described by symmetric Kármán equations.

The dynamic stability of thin shells was studied by Lepore and Stoltz (1971a,1972,1974) and Kul'terbaev (1979). Kul'terbaev found two possible types of responses: oscillations about the snap-through and motion about a non-snap-through state. The transition from one stable stationary response to another one takes place for suitably slow changes of a wide band parametric loading. Lepore and Stoltz (1973) considered the effects of non-linearities on the dynamic stability of cylindrical shells. They showed that the stability regions may be increased by including the system non-linearities. Kurnik and Tylikowski (1983) and Tylikowski (1984) obtained sufficient stability conditions for asymptotic stability, almost sure asymptotic stability, and uniform stochastic stability for cylindrical shells in terms of the intensities of the radial and axial loads and the damping coefficient. A region of uncertainty was obtained numerically and was attributed to be a result of the Liapunov functional method. Ahmadi (1977a,b,1979) examined the mean square stability of a number of aeroelastic structural elements including plates, thin shells and airplane wings in turbulent flow.

The stability boundaries of an axially moving thin elastic strip subjected to random parametric excitation were analyzed by Milstead (1975) and Kozin and Milstead (1979). The stability regions were given in terms of the standard deviation of the random loading for the first mode response. It was found that the stability region is reduced as the mode order increases until it approaches asymptotically a critical region determined solely by the stiffness of the strip. Fontenot, et al., (1965) and Mitchell (1968a) gave independent treatments for the liquid free stability under random vertical acceleration. Price (1975) and Muhuri (1980) examined the asymptotic stability of the mean and mean square of the roll motion of a ship in a random seaway. Lin and Ariaratnam (1980) analyzed the uncoupled torsional behavior of a suspension bridge in turbulent wind based on a linear analytical model. The stability boundaries of the first and second statistical moments were obtained in terms of the mean wind velocity and the spectral level of the turbulence component. It was indicated that the first moment stability assures that the mean response diminishes with time while excursion from the mean can be unbounded as revealed from the stability boundary of the second moment.

7.2 BASIC DEFINITIONS

The study of stochastic stability requires a number of definitions evolved originally from the deterministic stability theory. Frequently these definitions are given in terms of the norm of the state vector. A brief description of the common forms of the norm will be given in this section. In addition, the basic definitions of deterministic and stochastic stability modes will be discussed.

7.2.1 FORMS OF THE NORM

i. The Euclidean (Frobenius) Norm of the state vector $\underset{\sim}{X}$ is given by the expression

$$||\underset{\sim}{X}|| = (\sum_{i=1}^{n} X_i^2)^{1/2} \tag{7.1}$$

ii. The Simple Absolute (Taxi-Cab) Norm is defined by the sum of the absolute values of the elements of the state vector $\underset{\sim}{X}$

$$||\underset{\sim}{X}|| = \sum_{i=1}^{n} |X_i| \tag{7.2}$$

iii. The General Quadratic Form is expressed as a positive definite quadratic function $Q(\underset{\sim}{X})$

$$Q(\underset{\sim}{X}) = \sum_{i=1}^{n} \sum_{j=1}^{n} p_{ij}X_iX_j = \underset{\sim}{X}^T \underline{P}\, \underset{\sim}{X} \tag{7.3}$$

where p_{ij} are the elements of the matrix $\underline{P}$ which can be regarded as a "weighing" matrix used in determining the norm

$$||\underset{\sim}{X}|| = (\underset{\sim}{X}^T \underline{P}\, \underset{\sim}{X})^{1/2} = (\sum_{i=1}^{n} \sum_{j=1}^{n} p_{ij}X_iX_j) \tag{7.4}$$

$Q(\underset{\sim}{X})$ is said to be positive definite if $Q(\underset{\sim}{X})>0$ for all values of the vector $\underset{\sim}{X}$ and $Q(\underset{\sim}{X}=\underset{\sim}{0})=0$.

$Q(\underset{\sim}{X})$ is said to be positive semi-definite if $Q(\underset{\sim}{X})\geq 0$ for all values of the vector $\underset{\sim}{X}$ but there exists at least one $\underset{\sim}{X}$ other than zero such that $Q(\underset{\sim}{X})=0$.

The conditions under which a specific matrix $\underline{P}$ will lead to a positive definite quadratic function depend upon the nature of the sub-determinants of $\underline{P}$. That is, given a non-zero $n\times n$ matrix $\underline{P}$, the quadratic form (7.3) will be positive definite if, and only if,

$$\Delta_k = \det[p_{ij}] > 0 \qquad \begin{matrix} k=1,2,\ldots,n \\ i,j=1,2,\ldots,k \end{matrix} \tag{7.5}$$

For linear systems the form (7.3) is the most common norm used as a Liapunov function.

7.2.2 DETERMINISTIC CONCEPT OF LIAPUNOV STABILITY

Consider the non-autonomous set of differential equations

$$\dot{\underset{\sim}{X}} = \underset{\sim}{f}(\underset{\sim}{X},t), \qquad \underset{\sim}{X} \in R \tag{7.6}$$

where the functions $\underset{\sim}{f}$ satisfy the Lipschitz condition of bounded increments, i.e.,

$$\left| f_i(\underset{\sim}{X},t) - f_i(\underset{\sim}{Y},t) \right| \leq K \left| \underset{\sim}{X} - \underset{\sim}{Y} \right| \qquad \text{for all } t \in [t_0,T] \tag{7.7}$$

in which K is a constant.

The stability of the equilibrium solution $\underset{\sim}{X}(t;X_0,t_0)=\underset{\sim}{0}$, with the initial conditions $\underset{\sim}{X}(t_0)=\underset{\sim}{X}_0$, can be expressed in terms of the Liapunov definitions of stability:

<u>Definition I (Liapunov Stability)</u>

The equilibrium solution $\underset{\sim}{X}(t;\underset{\sim}{X}_0,t_0)=\underset{\sim}{0}$ is said to be uniformly stable if for a given arbitrary $\varepsilon > 0$, there exists a $\delta(\varepsilon,t_0)>0$ such that for $\left\| \underset{\sim}{X}_0 \right\| < \delta$, it follows that

$$\sup_{t \geq t_o} \left\| \underset{\sim}{X}(t;\underset{\sim}{X}_0,t_0) \right\| < \varepsilon \tag{7.8}$$

<u>Definition II (Asymptotic Liapunov Stability)</u>

The equilibrium solution $\underset{\sim}{X}(t;\underset{\sim}{X}_0,t_0)=\underset{\sim}{0}$ is said to be asymptotically stable if it satisfies (7.8) and if there exists a $\delta > 0$ such that for $\left\| \underset{\sim}{X}_0 \right\| < \delta$, it follows that

$$\lim_{t \to 0} \left\| \underset{\sim}{X}(t;\underset{\sim}{X}_0,t_0) \right\| = 0 \tag{7.9}$$

A dynamic system is said to be asymptotically stable "in the large" if condition (7.9) holds for all possible initial conditions including the entire space of $\underset{\sim}{X}$. Cesari (1963) showed that if the functions $f_i(\underset{\sim}{X},t)$ are linear in $\underset{\sim}{X}$, then condition (7.9) implies Liapunov stability and is therefore a necessary and sufficient condition for asymptotic Liapunov stability.

The Liapunov function $V(\underset{\sim}{X},t)$ is a scalar energy-like function specified in a closed bounded region S in the $\underset{\sim}{X}$ space containing the origin. $V(\underset{\sim}{X},t)$ is said to be continuous positive semi-definite in the domain S if $V(\underset{\sim}{0},t)=0$, $(t \geq t_0)$, $V(\underset{\sim}{X},t) \geq 0$, and $V(\underset{\sim}{X},t)$ has continuous partial derivatives with respect to $\underset{\sim}{X}$ and t. On the other hand, $V(\underset{\sim}{X},t)$ is said to be continuous positive definite in S if $V(\underset{\sim}{0},t)=0$, $V(\underset{\sim}{X},t)>0$, and $V(\underset{\sim}{X},t)$ has continuous partial derivatives with respect to $\underset{\sim}{X}$ and t.

The time derivative of $V(\underset{\sim}{X},t)$ may be written in the form

$$\frac{d}{dt} V(\underset{\sim}{X},t) = \lim_{h \to 0} \frac{1}{h} \left\{ V(\underset{\sim}{X},t+h) - V(\underset{\sim}{X},t) \right\}$$

$$= \frac{\partial}{\partial t} V(\underset{\sim}{X},t) + \left(\text{grad } V(\underset{\sim}{X},t) \right) \cdot f(\underset{\sim}{X},t) \tag{7.10}$$

In terms of the Liapunov function $V(\underset{\sim}{X},t)$ the following two theorems establish the stability of the system (Hahn, 1963):

THEOREM 7.1. If in the neighborhood of the origin of the state space there exists a positive definite function $V(\underset{\sim}{X},t)$ such that its time derivative along any trajectory of (7.6) is negative semi-definite in S, then the equilibrium solution $\underset{\sim}{X}(0)=\underset{\sim}{0}$ is stable.

THEOREM 7.2. If in the neighborhood of the origin of the state space there exists a positive definite function $V(\underset{\sim}{X},t)$ such that its time derivative along the trajectory of (7.6) is negative definite in S, then the origin is said to be uniformly asymptotically stable.

7.2.3 MODES OF STOCHASTIC STABILITY

Any stability property considered for deterministic dynamic systems can be extended for stochastic systems in different ways depending on the mode of stochastic convergence described in chapter 3. The stochastic stability of the equilibrium solution $\underset{\sim}{X}(t)=\underset{\sim}{0}$ of dynamic systems with random coefficients $\underset{\sim}{\xi}(t)$ described by the set of stochastic differential equations

$$\dot{\underset{\sim}{X}}(t) = \underset{\sim}{f}(\underset{\sim}{X},t) + \underline{G}(\underset{\sim}{X},t)\underset{\sim}{\xi}(t) \qquad (7.11)$$

will be the subject of this chapter.

The Liapunov function of system (7.11) is usually considered for the truncated deterministic system (7.6). The stability of the equilibrium position of (7.11) may be examined in terms of one of the definitions listed in table (7.1). The first six definitions deal with the response sample behavior on the infinite half-line (t_0,∞). Because the analysis of the statistical behavior is easier than the sample stability treatment, most studies have been directed to derive stability conditions for various moments. The next three definitions pertain to the stability of moments. The last two definitions are based on the entropy stability concept. These two definitions have not yet been applied to practical dynamic problems.

It is clear that condition (7.16) implies inequality (7.14). With the aid of Chebyshev inequality it can be shown that the stability in mean square (7.12) for K=2 implies stability in probability. Thus stability with probability one and stability in mean square are stronger than stability in probability.

Lin and Prussing (1982) provided an excellent physical interpretation of three modes of stochastic stability. These are stability in probability (7.14), stability with probability one (7.16), and stability in the K-th moment (7.12). Definition (7.14) is concerned with the distribution of the values of $|X_j(t)|$ over the entire ensemble, and it means that for any small departure δ from the equilibrium state, the samples that have a magnitude equal to or greater than ε' at any time $t>t_0$ will constitute only a small fraction of the entire

TABLE 7.1 Modes of Stochastic Stability

Stability Mode	Stability Condition ($\varepsilon,\varepsilon' > 0$)	
i. Stability in the K-th moment	$E[\sup_{t \geq t_o} \| \underset{\sim}{X}(t;\underset{\sim}{X}_0,t) \|^K] < \varepsilon$ $\| X \|^K = \sum_i^n \| X_i \|^K$	(7.12)
ii. Asymptotic stability in the K-th moment	if (7.12) holds, and $\lim_{T\to\infty} E[\sup_{t \geq T} \| \underset{\sim}{X}(t;\underset{\sim}{X}_0,t_0) \|^K] = 0$	(7.13)
iii. Stability in probability	$P\{\sup_{t \geq t_o} \| \underset{\sim}{X}(t;\underset{\sim}{X}_0,t_o) \| > \varepsilon'\} < \varepsilon$ such that $\| \underset{\sim}{X}_0 \| < \delta(\varepsilon,\varepsilon',t_o)$	(7.14)
iv. Asymptotic stability in probability	if (7.14) holds, and $\lim_{T\to\infty} P\{\sup_{t \geq T} \| \underset{\sim}{X}(t;\underset{\sim}{X}_0,t_0) \| \geq \varepsilon\} = 0$	(7.15)
v. Almost sure stability	$P\{\lim_{\|\underset{\sim}{X}_0\| \to 0} \sup_{t > t_o} \| \underset{\sim}{X}(t;\underset{\sim}{X}_0,t_0) \| = 0\} = 1$	(7.16)
vi. Almost sure asymptotic stability	if (7.16) holds, and $\lim_{T\to\infty} P\{\sup_{t \geq T} \| X(t;X_0,t_0) \| = 0\} = 1$	(7.17)
vii. Liapunov stability in the mean	$\lim_{\|\underset{\sim}{X}_0\| \to 0} E[\| \underset{\sim}{X}(t;\underset{\sim}{X}_0,t_0) \|] = 0, \quad t \geq t_0$	(7.18)
viii. Liapunov asymptotic stability of the K-th moment	if (7.18) holds, and $\lim_{t\to\infty} E[\| \underset{\sim}{X}(t;\underset{\sim}{X}_0,t_0 \|^K] = 0$	(7.19)
ix. Exponential stability of the K-th moment	for α, β, and $\delta > 0$, such that $\| \underset{\sim}{X}_0 \|^K < \delta$, $E[\| \underset{\sim}{X}(t;\underset{\sim}{X}_0,t_0) \|^K] < \beta \| X_0 \|^K \exp\{-\alpha(t-t_0)\}$ $t > t_0$	(7.20)
x. Monotonic entropy stability	$dH(t)/dt < 0, \quad t \geq 0$	(7.21)
xi. Asymptotic entropy stability	$\lim_{t\to\infty} H(t) = -\infty$ where the total entropy H(t) is defined by relation (2.12)	(7.22)

ensemble. Condition (7.16) requires that each sample record is first examined for all $t>t_o$ to determine its maximum absolute value. The maximum absolute values and their times of occurrence are different from one sample record to another, but stability with probability one requires that those samples which have maximum absolute values equal to or exceeding a very small specified value form only a small fraction of the entire ensemble such that their probability of occurrence is negligible. Stability in the K-th moment is measured by estimating the average over the entire ensemble of the K-th power of $X_i(t)$ at a given $t_j>t_o$. Experimentally substantial difficulties are encountered in establishing stability regions. The difficulties mainly arise in the specification of the infinitesimal character of the parameters involved in the stability definitions and in the measurement of deviations from unperturbed trajectory, starting from sufficiently small values.

7.3 STABILITY OF THE MOMENTS

7.3.1 ANALYTICAL TREATMENT

The stability analysis of moments is identical to the treatment of deterministic stability if the differential equations of the response moments form a closed set. A system is said to be stable if its response moments of all orders are stable. However, stability of moments may not always provide a satisfactory measure for the true stability characteristics of the system response. This view was originally raised by Kozin (1972) and can be demonstrated by considering the example of a dynamic system described by the first order linear differential equation

$$dX = aX\,dt + \sigma X\,dB(t) \tag{7.23}$$

For this system the K-th moment can be obtained by direct application of the FPK equation

$$E[X^K(t)] = X_o^K \exp\{(a - \sigma^2/2)\,Kt + \sigma^2K^2t/2\} \tag{7.24}$$

It is clear that the system is stable in the K-th moment if, and only if,

$$a < \sigma^2(1-K)/2 \tag{7.25}$$

Thus for $a<0$ the first order moment is exponentially stable, but higher order moments are not. For $a<-\sigma^2/2$, the first and second moments are exponentially stable, but higher moments are not. In other words, one can show that the first K moments are stable and all higher moments are unstable.

Applying the Itô calculus the solution of (7.23) may be written in the form

$$X(t) = X_o\exp\{(a - \sigma^2/2)t + \sigma B(t)\} \tag{7.26}$$

According to the law of iterated logarithm, it is known that the sample function of the Brownian motion approaches the limiting value $\sqrt{2t\log(\log t)}$ as $t\to\infty$ with probability one. Therefore, the stability of the sample solution (7.26) is determined by the algebraic sign of $(a-\sigma^2/2)$ only. That is, $a<\sigma^2/2$ is a necessary and sufficient condition for the equilibrium solution to be asymptotically stable with probability one. It may be noticed that this condition does not correspond to any moment stability condition.

Stability of first order moments is a rather weak criterion since it yields no information about sample stability. However, by using Schwarz's inequality, the asymptotic stability of the first moments is found to be a necessary condition for asymptotic stability of second moments. Accordingly, some information may be given about the location of a stable second moment region. In some cases, e.g., those considered by Mitchell and Kozin (1974), it was found that the general size and location of the first moment stability region is more superior than the second moment stability region in that it is closer to the almost sure sample stability region.

In most cases the stability of the mean and mean square are of main concern to the dynamicist. For linear systems which have closed moment equations, the stability of moments of any order can be determined by using the Routh-Hurwitz criteria.

THEOREM 7.3. Exponential Stability of the K-th Moment (Nevel'son and Khas'miniskii (1966a))

Consider the system of the Itô equations

$$d\underset{\sim}{X}(t) = \underset{\sim}{f}(\underset{\sim}{X},t)\,dt + \underline{G}(\underset{\sim}{X},t)\,d\underset{\sim}{B}(t) \tag{7.27}$$

where $\underset{\sim}{B}(t)$ possesses the following statistical properties

$$E[B_i(t)] = 0, \quad E[B_i^2(t)] = 2t \tag{7.28}$$

The following operator is related to (7.27)

$$L(\cdot) = \frac{\partial(\cdot)}{\partial t} + \sum_{i=1}^{n} f_i(\underset{\sim}{X},t)\frac{\partial(\cdot)}{\partial X_i} + \sum_{i=1}^{n}\sum_{j=1}^{n} a_{ij}(\underset{\sim}{X},t)\frac{\partial^2(\cdot)}{\partial X_i \partial X_j} \tag{7.29}$$

where $[a_{ij}] = \underline{G}\underline{G}^*$, and $\underline{G}^*$ is the adjoint matrix of $\underline{G}$.

The operator L plays the same role as the Liapunov operator in deterministic stability analysis.

Statement of Theorem 7.3. The equilibrium solution $\underset{\sim}{X}(\underset{\sim}{X}_0,t)=0$ of system (7.27) is said to be exponentially K-stable for $t\geq t_0$ if there exists a function $V(\underset{\sim}{X},t)$ satisfying the following conditions:

$$C_1 \mid \underset{\sim}{X} \mid^K < V(\underset{\sim}{X},t) < C_2 \mid \underset{\sim}{X} \mid^K \tag{7.30}$$

$$LV(\underset{\sim}{X},t) < - C_3 \mid \underset{\sim}{X} \mid^K \tag{7.31}$$

$$\left| \frac{\partial V}{\partial X_i} \right| < C_4 \mid \underset{\sim}{X} \mid^{K-1} \qquad i=1,2,\ldots,n \tag{7.32}$$

where C_i are positive constants.

For the special case of linear systems

$$dX_i = \sum_{j=1}^{n} A_{ij}(t)\, X_j dt + \sum_{k=1}^{n} \sum_{j=1}^{n} G_{ij}(t)\, X_k dB_j(t) \tag{7.33}$$

the following theorems were established by Nevel'son and Khas'miniskii (1966a):

THEOREM 7.4. The equilibrium solution of system (7.33) is said to be exponentially K-stable if there exists a homogeneous function $V(\underset{\sim}{X},t)$ of degree K and it satisfies conditions (7.30) through (7.32).

THEOREM 7.5. The equilibrium solution of (7.33) is said to be asymptotically K-stable if for some positive definite homogeneous function $W(\underset{\sim}{X})$ of order K, there is a positive definite homogeneous function $V(\underset{\sim}{X})$ such that

$$LV(\underset{\sim}{X}) = - W(\underset{\sim}{X}) \tag{7.34}$$

Example 7.1: Determine the moment stability conditions of the system (7.23). The generating operator of (7.23) is

$$L(\cdot) = aX \frac{\partial(\cdot)}{\partial X} + \sigma^2 X^2 \frac{\partial^2(\cdot)}{\partial X^2} \tag{i}$$

Let $V(\underset{\sim}{X}) = \mid X \mid^K$, and applying (i) to $V(\underset{\sim}{X})$ gives

$$LV(\underset{\sim}{X}) = K \mid X \mid^K [a + \sigma^2 (K - 1)] \tag{ii}$$

From theorem 7.5 it follows that for asymptotic K-stability of the solution $X(t;X_0,t_0)=0$ as $t\to\infty$, it is necessary and sufficient that

$$a < \sigma^2 (1 - K) \tag{iii}$$

This condition and condition (7.25) are identical for K=1. For $K \geq 2$ the two conditions are not the same.

The first and second moment equations of system (7.33) are

$$\dot{\underset{\sim}{m}}_1(t) = \underline{A}(t)\, \underset{\sim}{m}_1(t) \tag{7.35}$$

$$\dot{\underline{m}}_2(t) = \underline{A}(t)\, \underline{m}_2(t) + \underline{m}_2(t)\, \underline{A}^T(t) + \sum_{j=1}^{n} \underline{G}_j(t)\underline{m}_2(t)\, \underline{G}_j^T(t) \tag{7.36}$$

where $\underset{\sim}{m}_1(t) = E[\underset{\sim}{X}(t)]$, and $\underline{m}_2(t) = E[\underset{\sim}{X}(t)\underset{\sim}{X}^T(t)]$

These equations indicate that the stability of the first and second moments are equivalent to the stability of the undisturbed deterministic systems (7.35) and (7.36), respectively. It is convenient to handle (7.36) in a vector form. Because $\underline{m}_2(t)$ is symmetric $E[X_iX_j] = E[X_jX_i]$, one can expand $\underline{m}_2(t)$ into a system of $n(n+1)/2$ linear differential equations, where n is the number of state vector equations. Let the elements of $n(n+1)/2$ of $\underline{m}_2(t)$ be grouped to form a vector $\underset{\sim}{m}_2$. The new form of (7.36) is

$$\dot{\underset{\sim}{m}}_2 = \bar{\underline{A}}\, \underset{\sim}{m}_2 \tag{7.37}$$

where $\bar{\underline{A}}$ is the new coefficient matrix which results from grouping the right-hand terms of (7.36). System (7.37) can be examined for stability by one of the standard deterministic techniques. The solution of (7.37) may be written in the form

$$\underset{\sim}{m}_2 = \underset{\sim}{m}_0 \exp(\lambda t) \tag{7.38}$$

Substituting (7.38) into (7.37) yields the characteristic equation

$$a_n\lambda^n + a_{n-1}\lambda^{n-1} + \dots + a_1\lambda + a_0 = 0 \tag{7.39}$$

A necessary and sufficient condition that (7.39) has only roots with negative real parts is that the coefficients a_i and the values of the Hurwitz determinants H_j all be positive, i.e.,

$$H_1 = a_1 > 0,\quad H_2 = \begin{vmatrix} a_1 & a_3 \\ a_0 & a_2 \end{vmatrix} > 0,\qquad H_3 = \begin{vmatrix} a_1 & a_3 & a_5 \\ a_0 & a_2 & a_4 \\ 0 & a_1 & a_3 \end{vmatrix} > 0$$

and

$$H_n = \begin{vmatrix} a_1 & a_3 & a_5 & a_7 & \cdots \\ a_0 & a_2 & a_4 & a_6 & \cdots \\ 0 & a_1 & a_3 & a_5 & \cdots \\ 0 & a_0 & a_2 & a_4 & \cdots \\ \cdots & \cdots & \cdots & \cdots & a_n \end{vmatrix} > 0,\qquad H_n = a_nH_{n-1} \tag{7.40}$$

Example 7.2: Determine the stability boundaries of the first and second moments of the column response of example 5.1.

The first and second moment equations of the column are:

$$\begin{Bmatrix} \dot{m}_{10} \\ \dot{m}_{20} \end{Bmatrix} = \begin{bmatrix} 0 & 1 \\ -\omega_n^2 & -2\zeta\omega_n \end{bmatrix} \begin{Bmatrix} m_{10} \\ m_{01} \end{Bmatrix} \qquad \text{(i)}$$

$$\begin{Bmatrix} \dot{m}_{20} \\ \dot{m}_{11} \\ \dot{m}_{02} \end{Bmatrix} = \begin{bmatrix} 0 & 2 & 0 \\ -\omega_n^2 & -2\zeta\omega_n & 1 \\ \sigma^2 & -2\omega_n^2 & -4\zeta\omega_n \end{bmatrix} \begin{Bmatrix} m_{20} \\ m_{11} \\ m_{02} \end{Bmatrix} \qquad \text{(ii)}$$

It is clear that the first moments are stable if both ω_n^2 and $2\zeta\omega_n$ are positive. To examine the stability of the second moments we determine the eigenvalues of the coefficient matrix of (ii) from the characteristic equation

$$\lambda^3 + 6\zeta\omega_n\lambda^2 + (8\zeta^2\omega_n^2 + 4\omega_n^2) + 2(4\zeta\omega_n^3 - \sigma^2) = 0 \qquad \text{(iii)}$$

Thus the column will be asymptotically stable in the second moment if

$$\zeta > \sigma^2/4\omega_n^3 \qquad \text{(iv)}$$

7.3.2 STABILITY OF N-TH ORDER LINEAR SYSTEMS

The stochastic differential equation of n-th order may be written in the form

$$Y^n + \big(a_1 + W_1(t)\big)\, Y^{n-1} + \dots + \big(a_n + W_n(t)\big)\, Y = 0 \qquad (7.41)$$

where $W_i(t)$ are, in general, correlated white noise processes with the statistical properties

$$E[W_i(t)] = 0, \qquad E[W_i(t)W_j(s)] = \sigma_{ij}\delta(t-s) \qquad (7.42)$$

It is possible to write (7.41) as a set of stochastic first order differential equations of the n-th dimensional process

$$\{X_1, X_2, \dots, X_n\} = \{Y, \dot{Y}, \dots, Y^{n-1}\} \qquad (7.43)$$

This transformation results in the set of the Itô equations

$$dX_i = X_{i+1}dt$$

$$dX_n = -\sum_{i=1}^{n} a_i X_{n-i+1}dt - \sum_{i=1}^{n}\sum_{j=1}^{n} G_{ij}X_{n-i+1}dB_i(t) \qquad (7.44)$$

where $\underline{\underline{G}}\underline{\underline{G}}^T = \underline{\sigma}$

THEOREM 7.6 (Nevel'son and Khas'miniskii, 1966b). For the equilibrium position of system (7.44) to be asymptotically stable in the second moments, it is necessary and sufficient that the Routh-Hurwitz conditions be satisfied, that is

$$H_i > 0, \quad i=1,2,\ldots,n \tag{7.45}$$

and $\quad H_n > H/2 \qquad (7.46)$

where

$$H = \begin{vmatrix} q_{nn}^{(0)} & q_{nn}^{(1)} & q_{nn}^{(2)} \cdots & q_{nn}^{(n-1)} \\ 1 & a_2 & a_4 \quad \cdots & 0 \\ 0 & a_1 & a_3 \quad \cdots & 0 \\ \cdots\cdots & \cdots\cdots & \cdots\cdots & \cdots \\ \cdots\cdots & \cdots\cdots & \cdots\cdots & a_n \end{vmatrix} \tag{7.47}$$

The determinant H is obtained by replacing the elements of the first row by the terms $q_{nn}^{(i)}$, $i=0,1,\ldots,n-1$. The $q_{nn}^{(i)}$ are related to the power and cross-power spectral densities of $W_i(t)$ through the relationship

$$q_{nn}^{(n-k-1)} = \sum_{i+j=2(n-k)} (-1)^{j+1} \sigma_{ij}, \quad k=0,1,\ldots,(n-1) \tag{7.48}$$

Example 7.3: Determine the mean square stability conditions of the system

$$\ddot{Y} + \{2\zeta\omega + \xi_1(t)\}\ \dot{Y} + \{\omega^2 + \xi_2(t)\}\ Y = 0 \tag{i}$$

where $\xi_1(t)$ and $\xi_2(t)$ are sample functions from Gaussian stochastic processes. Introducing the Wong-Zakai correction terms, equation (i) can be written in terms of the uncorrelated white noise processes $W_1(t)$ and $W_2(t)$

$$\ddot{Y} + \{2\zeta\omega - \sigma_{11}/2 + W_1(t)\}\ \dot{Y} + \{\omega^2 + W_2(t)\}\ Y = 0 \tag{ii}$$

Here we define σ_{11} and σ_{22} as the power spectral densities of $W_1(t)$ and $W_2(t)$, respectively.

Introducing the transformation $X_1=Y$, and $X_2=\dot{Y}$, equation (ii) can be written in the state vector form

$$\begin{Bmatrix} dX_1 \\ dX_2 \end{Bmatrix} = \begin{bmatrix} 0 & 1 \\ -\omega^2 & 2\zeta\omega - \sigma_{11}/2 \end{bmatrix} \begin{Bmatrix} X_1 \\ X_2 \end{Bmatrix} dt + \begin{bmatrix} 0 & 0 \\ -X_2 & X_1 \end{bmatrix} \begin{Bmatrix} dB_1(t) \\ dB_2(t) \end{Bmatrix} \tag{iii}$$

From the second equation of (iii) the matrix $\underline{\sigma}$ is constructed

$$\underline{\sigma} = \begin{bmatrix} \sigma_{11} & 0 \\ 0 & \sigma_{22} \end{bmatrix} \tag{iv}$$

Applying theorem 7.6 for n=2, relation (7.48) gives

$$q_{22}^{(0)} = \sigma_{11}, \qquad q_{22}^{(1)} = -\sigma_{22} \tag{v}$$

with $a_1 = 2\zeta\omega - \sigma_{11}/2$, and $a_2 = \omega^2$

The second moment asymptotic stability conditions are

$$H_1 = a_1 = 2\zeta\omega - \sigma_{11}/2 > 0, \quad \text{or} \quad 4\zeta\omega > \sigma_{11} \tag{vi}$$

$$H_2 = a_1 a_2 - 0 = \omega^2(2\zeta\omega - \sigma_{11}/2) > 0,$$

$$H_2 - H/2 > 0, \text{ or } \omega^2(2\zeta\omega - \sigma_{11}/2) - \frac{1}{2}\begin{vmatrix} \sigma_{11} & -\sigma_{22} \\ 1 & \omega^2 \end{vmatrix} > 0$$

or

$$2\zeta\omega^3 - \omega^2\sigma_{11} - \frac{1}{2}\sigma_{22} > 0 \tag{vii}$$

Example 7.4: Torsional Stability of Suspension Bridges

The influence of wind loads on the dynamic behavior of suspension bridges has been the subject of several studies since the damage of the Tacoma Narrows bridge. Suspension bridges which have bluff deck structure, such as girder-stiffened bridges of H-section, are usually exposed to strong vortex wakes. These wakes can cause extreme torsional flutter stability (Simiu and Scanlan, 1978).

Wind loads acting on bridges can be classified into two categories: buffeting loads which appear as non-homogeneous terms in the equation of motion, and parametric excited loads (or self-excited loads). Beliveau, et al., (1977) gave an account for the effect of the induced aerodynamic forces on the static stability (divergence) and dynamic stability (flutter) on suspension bridges. The coupling between vertical and torsional motions induced through self-excited aerodynamic loading was known to be very strong for high wind speeds. When the wind speed reached near the critical region, the bridge motion was almost completely dominated by torsional motion.

Lin and Ariaratnam (1980) examined the effect of turbulent components in the wind velocity on the stability boundary of the torsional motion of suspension bridges. They assumed that the turbulent components are weakly stationary wide band random processes for which the bridge response can be approximated as a Markov process. It was postulated that the flow is spatially smooth as viewed by the structure, and the changes in the aerodynamic indicial functions due to velocity variation can be neglected.

Figure (7.1) shows a typical cross section of a suspension bridge free to move in the vertical and torsional freedoms. Since the motion of the bridge is dominated by the torsional motion in the critical region, the analysis will be confined to the stability of torsional response. The simplified torsional equation of motion is

$$I(\ddot{\alpha} + 2\zeta\omega\dot{\alpha} + \omega^2\alpha) = \frac{1}{2}\rho(2B^2)\,\partial C_M/\partial\alpha\{X_\alpha U^2(t)\dot{\alpha}(t) + U^2(t)\alpha + \int_{-\infty}^{t} U^2(t)[X_{m\alpha}(t-\tau) - 1]\frac{\partial\alpha(\tau)}{\partial\tau}\,d\tau\} \qquad (i)$$

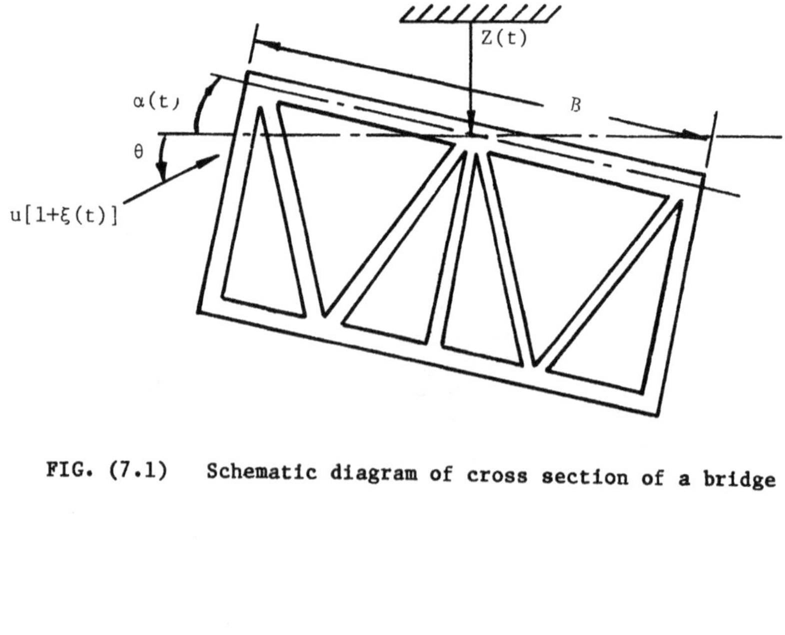

FIG. (7.1) Schematic diagram of cross section of a bridge

where α = the modal torsional angle,

I = torsional moment of inertia per unit span length,

ω = the natural frequency, ζ = the damping ratio,

ρ = air density, B = width of the bridge,

$U = u[1 + \xi(t)]$, u = the mean wind velocity,

$u\xi(t)$ = the wind turbulent fluctuation velocity,

$X_{m\alpha}$ = the aerodynamic indicial function,

X_α and $\partial C_M/\partial\alpha$ are experimentally determined aerodynamic constants.

For the full derivation of equation (i) the reader may refer to the papers by Scanlan, et al., (1974), Beliveau, et al., (1977), and Lin and Ariaratnam (1980).

For small random perturbation the approximation $U^2 \cong u^2[1 + 2\xi(t)]$ is introduced, and equation (i) can be written in the form

$$\ddot{\alpha} + 2\omega_o\zeta_o\dot{\alpha} + \omega_o^2\alpha = 2a_1\xi(t)\alpha + 2\omega_o a_2\xi(t)\dot{\alpha}$$
$$+ a_1 \int_{-\infty}^{t} [1 + 2\zeta(\tau)][C_1 \exp(-\gamma_1(t - \tau))$$
$$+ C_2\exp(-\gamma_2(t - \tau))]\dot{\alpha}(\tau)\, d\tau \qquad \text{(ii)}$$

where $a_1 = \rho u^2 \frac{B^2}{I} (\partial C_M/\partial\alpha)$, $\omega_o = (\omega^2 - a_1)^{1/2}$

$a_2 = a_1 X_\alpha/\omega_o$, $\zeta_o = (2\omega\zeta - a_1 X_\alpha)/2\omega_o$

and the indicial function $X_{m\alpha}$ has been replaced by the expression (Scanlan, et al., 1974)

$$X_{m\alpha}(t) = 1 + C_1 \exp(-\gamma_1 t) + C_2 \exp(-\gamma_2 t)$$

where C_1, C_2, γ_1, and γ_2 are constants.

Applying Leibnitz's rule to equation (ii), and letting $\alpha=Y_1$ and $\alpha=Y_2$, yields the state vector form

$$\dot{Y}_1 = Y_2$$
$$\dot{Y}_2 = -\omega_o^2 Y_1 - 2\omega_o\zeta_o Y_2 + a_1(Y_3 + Y_4) + 2a_1\xi(t)Y_1 + 2\omega_o a_2\xi(t)Y_2$$
$$\dot{Y}_3 = -\gamma_1 Y_3 + C_1[1 + 2\xi(t)]Y_2 \qquad \text{(iii)}$$
$$\dot{Y}_4 = -\gamma_2 Y_4 + C_2[1 + 2\xi(t)]Y_2$$

Now introducing the Wong-Zakai correction terms allows the physical wide band process to be replaced by a white noise process and equation (iii) can be written in the form

$$dY_i = f_i dt + \sum_{j=1}^{4} G_{ij} dB_j(t) \tag{iv}$$

where $B_j(t)$ are independent unit Brownian motion processes,

$$\begin{Bmatrix} f_1 \\ f_2 \\ f_3 \\ f_4 \end{Bmatrix} = \begin{bmatrix} 0 & 1 & 0 & 0 \\ -\omega_o(\omega_o - 2a_1 a_2 D_o) & -2\omega_o(\zeta_o - a_2^2\omega_o D_o) & a_1 & a_1 \\ 2C_1 a_1 D_o & C_1(1 + 2a_2\omega_o D_o) & -\gamma_1 & 0 \\ 2C_2 a_1 D_o & C_2(1 + 2a_2\omega_o D_o) & 0 & -\gamma_2 \end{bmatrix} \begin{Bmatrix} X_1 \\ X_2 \\ X_3 \\ X_4 \end{Bmatrix}$$

$$= \underline{F}\,\underset{\sim}{X} \tag{v}$$

$$\underline{G}\underline{G}^T = 4D_o \cdot$$

$$\begin{bmatrix} 0 & 0 & 0 & 0 \\ 0 & (a_1X_1 + \omega_o^2 a_2X_2)^2 & C_1X_2(a_1X_1 + \omega_o a_2X_2) & C_2X_2(a_1X_1 + \omega_o a_2X_2) \\ 0 & C_1X_2(a_1X_1 + \omega_o a_2X_2) & (C_1X_2)^2 & C_1C_2X_2^2 \\ 0 & C_2X_2(a_1X_1 + \omega_o a_2X_2) & C_1C_2X_2^2 & C_2X_2^2 \end{bmatrix} \tag{vi}$$

and D_o is the spectral density of $\xi(t)$.

The first order moment equations are

$$\dot{\underset{\sim}{m}}_1 = \underline{F}\,\underset{\sim}{m}_1 \tag{vii}$$

where $\underset{\sim}{m}_1^T = \{m_{1000}, m_{0100}, m_{0010}, m_{0001}\}$

The second order moment equations are

$$\dot{\underset{\sim}{m}}_2 = \underline{A}\,\underset{\sim}{m}_2 \tag{viii}$$

where

$$\underset{\sim}{m}_2^T = \{m_{2000},\ m_{0200},\ m_{0020},\ m_{0002},\ m_{1100},\ m_{1010},\ m_{1001},\ m_{0110},$$
$$m_{0101},\ m_{0011}\}$$

and the elements of $\underline{A}$ are given by Lin and Ariaratnam (1980).

The stability of the first and second moments may be inspected by determining the eigenvalues of the matrices $\underline{F}$ and $\underline{A}$, respectively. It may be noticed that the stability boundaries of the first and second moments are influenced by the mean wind velocity u, the spectral density of its fluctuation D_0, and the bridge structure parameters. Figure (7.2) shows the stability boundaries for the first and second moments on the velocity-excitation level plane $(u/u_{cr}, D_0/\zeta)$. Here u_{cr} is the critical wind speed at which flutter occurs. These boundaries were determined for the following structure parameters

$$B = 16.4 \text{ m},\ \zeta = 0.01,\ \omega = 0.74 \quad \text{rad/sec}, \quad I = 8.98\text{x}10 \quad \text{kg-m}^2/\text{m}$$

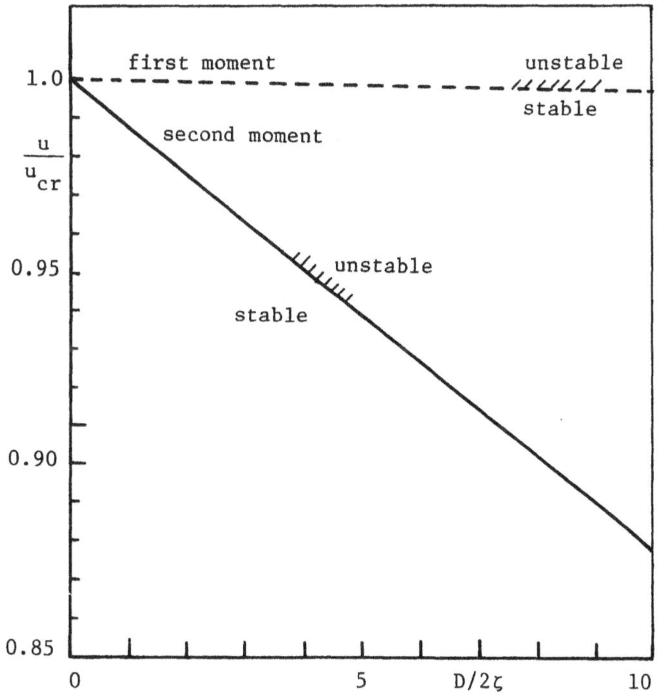

FIG. (7.2) Stability boundaries for bridge torsional motion (Lin and Ariaratnam, 1980)

and for the aerodynamic parameters

$$\rho = 1.226 \text{ kg/m}^3, \quad X_\alpha = 1.52, \; \partial C_M/\partial\alpha = 0.93, \quad C_1 = 1.64,$$

$$C_2 = -51.61, \; \gamma_1 = 0.38u/B, \quad \gamma_2 = 19.74u/B$$

The stability boundaries shown in fig. (7.2) indicate that if the second moment is stable, the first moment will also be stable. The results of Lin and Ariaratnam seem to be in contradiction with those observed by Irwin and Schuyler (1978) who found that wind turbulence can sometimes have a stabilizing effect. Lin and Ariaratnam attributed this discrepancy to certain factors ignored in their modeling such as the imperfect correlation of the turbulence field along the bridge span.

7.3.3 STABILITY OF INFINITE COUPLED MOMENTS

It has been indicated in chapter 5 that the differential equations for the response moments form an infinite coupled set for two classes of systems. These systems are either non-linear or linear with random coefficients generated from a shaping filter excited externally by a white noise process. The stability of the response moments cannot, therefore, be determined unless the differential equations of the moments are closed by a proper closure scheme. In this section the Gaussian closure schemes will be applied to linear and non-linear systems. However, the reader must bear in mind that these schemes may lead to stability boundaries which differ from those obtained by other methods such as the stochastic averaging methods.

Case 1: Linear Systems

Example 7.5: Determine the stability boundaries of the first and second order moments of the response of the system described in example 5.3.

From the general moment equation one can generate four equations for the first moments and ten equations for the second moments. Five of these moments are known from the properties of the filter equation. These are:

$$m_{0010} = m_{0001} = m_{0011} = 0$$

$$m_{0020} = \pi S_o/(2\zeta_z\omega_z^2) = C, \quad m_{0002} = \omega_z^2 C \tag{1}$$

Accordingly, the fourteen equations are reduced to nine equations. Inspection of the resulting equations shows that the second order moment equations contain terms of third order. Since we are interested in the stability of the equilibrium configuration, the third order moments will be replaced by lower order moments by using the Gaussian cumulant scheme which yields

$$m_{2010} = 2m_{1000}\, m_{1010},\; m_{1011} = m_{1000}\, m_{0011} = 0,$$

$$m_{1110} = m_{1000}\, m_{0110} + m_{0100}\, m_{1010},\; m_{1020} = C\, m_{1000} \qquad \text{(ii)}$$

The first three relations in (ii) result in non-linear terms, while the fourth relation is linear. The linearized moment differential equations are

$$\dot{\underset{\sim}{m}} = \underline{A}\, \underset{\sim}{m} \qquad \text{(iii)}$$

where the elements of matrix $\underline{A}$ are given by Bolotin (1972), and

$$\underset{\sim}{m}^T = \{m_{1000}, m_{0100}, m_{2000}, m_{0200}, m_{1100}, m_{1010}, m_{1001}, m_{0110}, m_{0101}\}$$

The stability boundaries of the first and second moments can be determined by using the Routh-Hurwitz criteria. For the system parameters $\zeta = 0.025$, $\omega_z\zeta_z/\omega_n = 0.02$, and $C = 1$, Bolotin (1972) and Bolotin and Moskvin (1972) obtained the stability boundaries shown in fig. (7.3). These boundaries show two critical regions: one at twice the natural frequency of the column and the other at its natural frequency. Bolotin and Moskvin (1973) and Bolotin (1984) extended the analysis of the stability of moments to include coupled two degree-of-freedom systems. The stability boundaries for canonical and non-canonical systems were obtained. For canonical coupled systems the stability boundaries exhibit instability in the neighborhood of combination resonances of the types $\omega_z = \omega_1 + \omega_2$ and $\omega_z = \frac{1}{2}(\omega_1+\omega_2)$. For non-canonical systems, Bolotin obtained instability region close to combination resonance of the difference type $\omega_z = |\omega_1-\omega_2|$.

<u>Case 2: Systems with Autoparametric Coupling</u>

Example 7.6: Ibrahim and J.W. Roberts (1977) and J.W. Roberts (1980) examined the moment stability of the equilibrium position of the coupled cantilever beam of the autoparametric vibration absorber shown in fig. (1.5b). For this system there are two possible responses. The first is the uncoupled motion in which the main system vibrates with identically zero motion of the cantilever beam when the system conditions are remote from the internal resonance. The second is the coupled motion which takes place when the cantilever transverse natural frequency is arranged to be in the vicinity of one-half the main system frequency. The boundaries between the two regions may be obtained by the method of moment functions in conjunction with the cumulant truncation scheme.

The primary system is represented by the discrete elements K_1, C_2, and M_1 and is excited by the random force $F(t)$. The cantilever motion

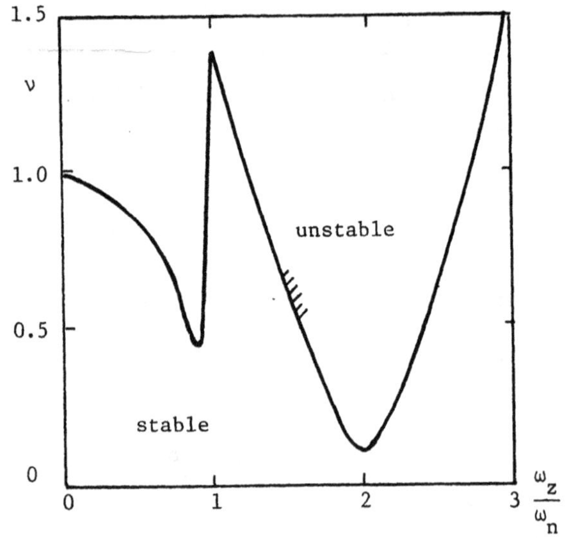

FIG. (7.3) Moment stability boundaries of the equilibrium position of an elastic column under filtered white noise (Bolotin, 1972)

will be represented by its fundamental mode. The system equations of motion are:

for the main system

$$(M + m)\ddot{x} + C_1\dot{x} + K_1 x - \frac{6}{5\ell} m(\dot{y}^2 + y\ddot{y}) = F(t) \qquad \text{(i)}$$

and for the coupled cantilever

$$m\ddot{y} + C_2\dot{y} + (K_2 - \frac{6}{5\ell} m\ddot{x})y + \frac{36}{25\ell^2} my(y\ddot{y} + \dot{y}^2) = 0 \qquad \text{(ii)}$$

It is convenient to write these equations in a dimensionless form. The following transformation is introduced

$$X = x/x_0, \quad Y = y/x_0, \quad \tau = \omega_1 t, \quad R = m/(M + m), \quad r = \omega_2/\omega_1,$$

$$\varepsilon = 6x_0/5\ell, \quad \zeta_1 = C_1/2(M + m)\omega_1, \quad \zeta_2 = C_2/2m\omega_2, \quad \mu = 1 - R,$$

$$\omega_1^2 = K_1/(M + m), \qquad \omega_2^2 = K_2/m \qquad \text{(iii)}$$

where x_0 is some convenient reference displacement of the main mass.

The equations of motion become

$$X'' + 2\zeta_1 X' + X - \varepsilon R(Y'^2 + YY'') = W(\tau)$$
$$Y'' + 2\zeta_2 rY' + (r^2 - \varepsilon X'')\, Y + \varepsilon^2(Y'^2 + YY'')\, Y = 0 \qquad \text{(iv)}$$

where prime denotes differentiation with respect to τ. The excitation function $W(\tau)$ is

$$W(\tau) = F(\tau/\omega_1)/[(M + m)\omega_1^2 x_0] \qquad \text{(v)}$$

It is assumed that the real excitation $F(t)$ is a zero mean, stationary Gaussian process with a smooth spectral density S_0 extending to some frequency well above ω_1, where

$$S_0 = \frac{1}{2\pi}\int_{-\infty}^{\infty} E[F(\tau)\, F(\tau + \bar{\tau})]\, \exp(-i\bar{\omega}\tau)\, d\bar{\tau} \qquad \text{(vi)}$$

In this case $W(\tau)$ may be regarded as an ideal white noise with a smooth spectral density $2D$, where D is given by the expression (after using (v) and relation (4.12))

$$D = \pi S_0/[(M + m)^2\omega_1^3 x_0^2] \qquad \text{(vii)}$$

The reference displacement x_0 may now be chosen to be the root mean square displacement of the primary system in the absence of the coupled cantilever motion. With $y(t)=0$ the first equation of (i) gives

$$x_0 = \{E[x^2(t)]\}^{1/2} = \frac{1}{(M+m)}\left(\pi S_0/2\zeta_1\omega_1^3\right)^{1/2} \qquad \text{(viii)}$$

and the coupling parameter takes the form

$$\varepsilon = \frac{6x_0}{5\ell} = \frac{6}{5\ell\,(M+m)}\left(\frac{\pi S_0}{2\zeta_1\omega_1^3}\right)^{1/2} \qquad \text{(ix)}$$

The non-linear acceleration terms in equations (iv) will be removed by successive elimination and terms up to order ε^2 will be retained. The system equations of motion can be written in terms of stochastic integrals. Introducing the coordinate transformation $X_1 = X$, $X_2 = Y$, $X_3 = X'$, and $X_4 = Y'$, equations (iv) become

$$X_1' = X_3, \qquad X_2' = X_4,$$

$$X_3' = -2\zeta_1 X_3 - X_1 + W(\tau) + \varepsilon R[X_4^2 - 2r\zeta_2 X_2 X_4 - r^2 X_2] + \varepsilon^2 R[W(\tau)X_2^2$$
$$- 2\zeta_1 X_2^2 X_3 - X_1 X_2^2]$$

$$X_4' = -2\zeta_2 rX_4 - r^2X_2 + \varepsilon[W(\tau)X_2 - X_1X_2 - 2\zeta_1X_2X_3] + \varepsilon^2\mu[2\zeta_2 rX_2^2X_4$$

$$+ r^2X_2^3 - X_2X_4^2] \qquad \text{(x)}$$

In these equations $W(\tau)$ has been replaced by the formal derivative of the Brownian motion $B(\tau)$. A general differential equation of the system response moments can be obtained by using one of the techniques described in chapter 5, with the result

$$m_{ijk\ell}' = im_{i-1,j,k+1,\ell} + jm_{i,j-1,k,\ell+1} - 2\zeta_1 km_{ijk\ell} - km_{i+1,j,k-1,\ell}$$

$$- 2\zeta_2 r\ell m_{ijk\ell} - r^2\ell m_{i,j+1,k,\ell-1} + Dk(k-1)m_{ij,k-2,\ell}$$

$$+ \varepsilon\{Rk[m_{ij,k-1,\ell+2} - 2\zeta_2 rm_{i,j+1,k-1,\ell+1} - r^2m_{i,j+2,k-1,\ell}]$$

$$+ \ell[-2\zeta_1 m_{i,j+1,k+1,\ell-1} - m_{i+1,j+1,k,\ell-1}] + 2Dk\ell m_{i,j+1,k-1,\ell-1}\}$$

$$+ \varepsilon^2\{RK[-2\zeta_1 m_{i,j+2,k\ell} - m_{i+1,j+2,k-1,\ell}] + \mu\ell[2\zeta_2 rm_{i,j+2,k\ell}$$

$$+ r^2m_{i,j+3,k,\ell-1} - m_{i,j+1,k,\ell+1}] + D\ell(\ell-1)m_{i,j+2,k,\ell-2}\}$$

$$+ 2RDk(k-1)m_{i,j+2,k-2,\ell}. \qquad \text{(xi)}$$

where the following notation applies

$$m_{ijk\ell} = E[X_1^i\, X_2^j\, X_3^k\, X_4^\ell]$$

Inspection of equation (xi) shows that for non-zero coupling parameter ε, the moment equations of any specific order are coupled with higher order moments, so (xi) represents an infinite hierarchy set. Further inspection reveals that the first and second order moment equations are satisfied by the unimodal stationary solution

$$m_{2000} = E[X^2] = D/2\zeta_1 = 1,$$

$$m_{0020} = E[X'^2] = D/2\zeta_1 = 1, \quad \text{and all other moments zero.} \qquad \text{(xii)}$$

It will be assumed that the response processes are nearly Gaussian and the cumulant truncation scheme may be applied. The mean square stability of solution (xii) may be determined by introducing the following perturbations to the unimodal stationary solution

$$\underset{\sim}{m}_{ijk\ell} = \underset{\sim}{m}_{ijk\ell}^{o} + \underline{\delta}_{ijk\ell}\exp(\lambda\tau) \qquad \text{(xiii)}$$

where $\underset{\sim}{m}^{o}$ is the unimodal solution (xii) and the vector $\underset{\sim}{\delta}$ represents a set of small perturbations. Neglecting products of δ's yields the following linearized equations

$$[\underline{A} - \lambda \underline{I}] = \underset{\sim}{0} \qquad \text{(xiv)}$$

where $\underline{A}$ is a 14×14 matrix which is given by Ibrahim and J. W. Roberts (1977).

The stability boundaries of the first and second moments were determined numerically for a range of system parameters. The effect of the excitation parameter ε on the stability boundaries for different damping parameters is shown in figs. (7.4a,b). These curves reveal the significance of the internal resonance in determining the behavior of the system. The stability regions are centered upon a value of internal resonance detuning parameter $r = 0.5$, showing that unstable random motions of the coupled system are possible when the cantilever natural frequency is arranged to be in the vicinity of half the main system natural frequency. In fig. (7.4a) it is seen that beyond a certain value of ε, any increase in ζ_1 leads to an increased width of the unstable region. This feature suggests that the damping of the primary system has a destabilizing effect over a certain excitation level. Such destabilizing effect of the damping is a common result in deterministic non-Hamiltonian systems. In the present case the possibility exists that the effect is due to the change of critical internal phase relationships with increasing damping. However, it is necessary to consider that the coupling parameter ε also contains $\zeta_1^{-1/2}$. This implies that in a given system, any increase of ζ_1 causes a reduction in ε and the cumulative effect is almost certainly a stabilizing one. The destabilizing influence of ζ_1 shown in fig. (7.4a) must therefore be interpreted as existing if, for any increase of ζ_1, adjustment of the excitation spectral density is made to maintain the value of the coupling parameter ε.

An interesting feature of the results which is not shown in figs. (7.4a,b) is that the mass ratio R has no effect on the stability boundaries. J.W. Roberts (1980) analyzed the structure of the matrix $\underline{A}$ and showed that one of the submatrices, which governs the stability of the system, is free of the mass ratio R and ε^2 and, therefore, cannot influence the stability. As a result, it appears that the stability boundaries are unaffected by the presence of non-linear reaction of the coupled system in the equation of motion of the main system.

7.4 ALMOST SURE STABILITY THEOREMS

This section deals with various theorems pertaining to the almost sure stability (with probability one) of linear systems described by the set of differential equations

$$\dot{\underset{\sim}{X}} = [\underline{A} + \underset{\sim}{\xi}(t)]\, \underset{\sim}{X} \qquad (7.49)$$

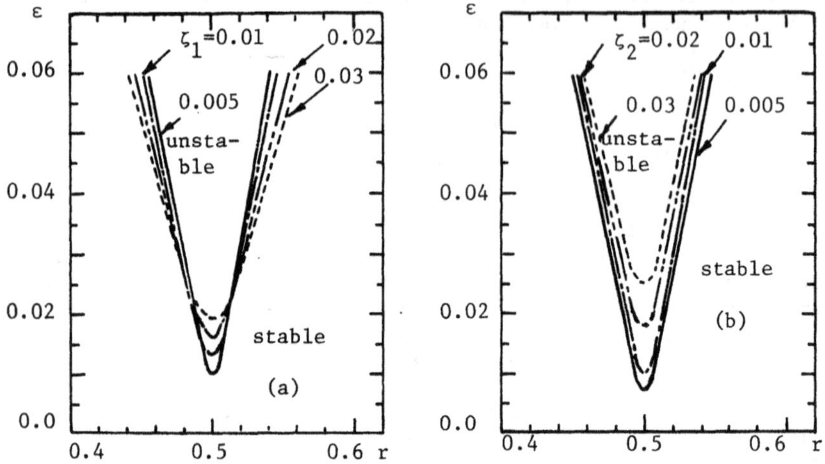

FIG. (7.4) Effect of excitation level on the stability boundaries for (a) different values of main mass damping ratio ζ_1 and (b) different values of cantilever damping ratio ζ_2

where $\underline{A}$ is a constant n×n matrix and $\underline{\xi}(t)$ is a n×n matrix. The non-zero elements of $\underline{\xi}(t)$ are stochastic processes $\{\xi_{ij}(t);t\in[0,\infty)\}$. It will be assumed that the elements $\xi_{ij}(t)$

i. are continuous on the interval $0<t<\infty$ with probability one.

ii. are strictly stationary. (7.50)

iii. satisfy an ergodic property, guaranteeing the equality of time averages and ensemble averages with probability one.

Based on these assumptions Kozin (1963), Caughey and Gray (1965), Infante (1968) and others developed the following theorems.

7.4.1 ALMOST SURE STABILITY (KOZIN, 1963)

Kozin developed a theorem based on the Gronwall-Bellman lemma and the strong law of large numbers for strictly stationary processes. According to Bellman (1953), the lemma assumes two positive functions, u(t) and v(t), such that u(t) is continuous with probability one and v(t) is Lebesgue integrable in every finite interval. Let C be a positive constant such that

$$u(t) \leq C + \int_0^t u(s)v(s)ds \tag{7.51}$$

then for $t \geq 0$

$$u(t) \leq C \exp\left(\int_0^t v(s)ds\right) \tag{7.52}$$

Now consider the unperturbed system

$$\dot{\tilde{X}} = \underline{A}\,\tilde{X} \tag{7.53}$$

Cesari (1963) showed that if λ_u is an upper-bound of the real parts of the eigenvalues of the stability matrix $\underline{A}$, then there exists a number $b>0$ such that

$$||\tilde{X}|| < ||b|| \cdot ||\tilde{X}_0|| \exp(\lambda_u t) \tag{7.54}$$

THEOREM 7.7. If the solution of the deterministic system (7.53) is asymptotically stable in the large, and if the non-zero elements of the random excitation matrix $\underline{\xi}(t)$ in (7.49) satisfy conditions (7.50), then there is a constant $a = \lambda_u/b > 0$ which depends on the matrix $\underline{A}$ such that

$$E[\,||\underline{\xi}(t)||\,] < a \tag{7.55}$$

which implies that the equilibrium solution of the system (7.49) is almost surely asymptotically stable in the large.

7.4.2 ALMOST SURE STABILITY (CAUGHEY AND GRAY, 1965)

Caughey and Gray considered linear systems described by (7.49) subject to conditions (7.50). In addition, the unperturbed system (7.53) is assumed to be asymptotically stable in the large. This is guaranteed if $\underline{A}$ has eigenvalues with negative real parts. If $\underline{A}$ is a stability matrix, there exists a symmetric, real, strictly positive definite matrix $\underline{P}$ (known as the Liapunov matrix) such that (Bellman, 1960)

$$\underline{A}^T\underline{P} + \underline{P}\underline{A} = -\underline{I} \tag{7.56}$$

Accordingly, there exists an orthogonal transformation matrix $\underline{q}$, such that

$$\underline{q}^T\underline{q} = \underline{I}, \quad \underline{q}^T\underline{P}\underline{q} = \lceil \lambda \rfloor \tag{7.57}$$

where $\lceil \lambda \rfloor$ is a diagonal matrix with positive elements. The matrix $\underline{P}$ possesses a unique square root given by

$$\underline{P}^{1/2} = \underline{q}[\lambda^{1/2}]\,\underline{q}^T, \quad \text{also} \quad \underline{P}^{-1/2} = \underline{q}[\lambda^{-1/2}]\,\underline{q}^T \tag{7.58}$$

With these definitions, two theorems given by Caughey and Gray will be stated:

THEOREM 7.8. The system (7.49) is almost surely stable if

$$\left.\begin{array}{rl} \text{i.} & \text{the matrix } \underline{A} \text{ is a stability matrix,} \\ \text{ii.} & \text{the non-zero elements of the matrix } \underline{\xi}(t) \text{ satisfy} \\ & \text{conditions (7.50), and} \\ \text{iii.} & E[\,\|\underline{P}^{-1/2}\,\underline{\xi}^T(t)\,\underline{P}^{1/2} + \underline{P}^{1/2}\,\underline{\xi}(t)\,\underline{P}^{-1/2}\|\,] < 1/\lambda_{max} \end{array}\right\} \quad (7.59)$$

THEOREM. 7.9. Let the excitation matrix $\underline{\xi}(t)$ be written in the form

$$\underline{\xi}(t) = \sum_{i=1}^{R} \xi_i(t)\,\underline{G}_i \tag{7.60}$$

where $\xi_i(t)$ are scalar functions of time, $\underline{G}_i$ are constant matrices, $R<n^2$, and n is the order of the state coordinates vector $\underset{\sim}{X}$. System (7.49) is almost surely asymptotically stable in the large if

$$\left.\begin{array}{rl} \text{i.} & \underline{A} \text{ is a stability matrix,} \\ \text{ii.} & \text{all } \xi_i(t) \text{ in (7.60) satisfy properties (7.50),} \\ \text{iii.} & \sum_{i=1}^{R} |\mu^i|_{max}\, E[\,|\xi_i(t)|\,] < 1/\lambda_{max} \end{array}\right\} \quad (7.61)$$

where $|\mu^i|_{max}$ represents the largest eigenvalue of the matrix

$$\underline{B}_i = \underline{P}^{-1/2}\,\underline{G}_i^T\underline{P}^{1/2} + \underline{P}^{1/2}\,\underline{G}_i\underline{P}^{-1/2} \tag{7.62}$$

7.4.3 ALMOST SURE STABILITY (INFANTE, 1968)

This theorem is based on the properties of pencils of quadratic forms (Gantmacher, 1959a,b). In linear algebra, a regular pencil P of quadratic form is given in the form

$$P = \underset{\sim}{X}^T\,\underline{D}\,\underset{\sim}{X} - \underset{\sim}{X}^T\,\underline{B}\,\underset{\sim}{X} \tag{7.63}$$

where $\underline{D}$ and $\underline{B}$ are n×n real symmetric matrices with $\underline{B}$ positive definite. The characteristic equation of the pencil (7.63) is given by the determinant

$$\det|\underline{D} - \lambda\underline{B}| = 0 \tag{7.64}$$

where λ_i are the eigenvalues of the pencil.

Let λ_{max} and λ_{min} be the maximum and minimum eigenvalues of the pencil. Since the matrix $\underline{DB}^{-1}$ has the same eigenvalues as the pencil, we can write

$$\lambda_{1,\text{ or } n}\,[\underline{DB}^{-1}] = \text{min, or max}\ \frac{\underset{\sim}{X}^T\,\underline{D}\,\underset{\sim}{X}}{\underset{\sim}{X}^T\,\underline{B}\,\underset{\sim}{X}} \tag{7.65}$$

where the eigenvalues of the regular pencil are labeled in the non-descending order

$$\lambda_1[\underline{D}\underline{B}^{-1}] \leq \lambda_2[\underline{D}\underline{B}^{-1}] <\ldots< \lambda_n[\underline{D}\underline{B}^{-1}] \tag{7.66}$$

Consider the quadratic Liapunov function

$$V(\tilde{X}) = \tilde{X}^T \underline{B} \tilde{X} \tag{7.67}$$

Evaluating the time derivative of $V(\tilde{X})$ along the trajectories of (7.49) gives

$$\frac{d}{dt} V(\tilde{X}) = \tilde{X}^T\{(\underline{A} + \xi(t))^T\underline{B} + \underline{B}(\underline{A} + \xi(t))\}\tilde{X} \tag{7.68}$$

then, along the trajectories of (7.49) we define

$$\lambda(t) = \dot{V}(\tilde{X})/V(\tilde{X}) = [\tilde{X}^T\{(\underline{A} + \xi(t))^T\underline{B} + \underline{B}(\underline{A} + \xi(t))\}\tilde{X}]/\tilde{X}^T\underline{B}\,\tilde{X} \tag{7.69}$$

Comparing relations (7.65) and (7.69) shows that

$$[(\underline{A} + \xi(t))^T\underline{B} + \underline{B}(\underline{A} + \xi(t))] = \underline{D}$$

From the extremal properties of pencils, the following inequality is obtained

$$\lambda_{min}\,[(\underline{A} + \xi(t))^T + \underline{B}(\underline{A} + \xi(t))\underline{B}^{-1}] \leq \lambda(t)$$

$$< \lambda_{max}[(\underline{A} + \xi(t))^T + \underline{B}(\underline{A} + \xi(t))\underline{B}^{-1}] \tag{7.70}$$

Integrating the first order differential equation (7.69) gives

$$\begin{aligned} V(\tilde{X}(t)) &= V(\tilde{X}(t_0))\exp\{\int_{t_0}^{t} \lambda(s)ds\} \\ &= V(\tilde{X}(t_0))\exp\{[\frac{1}{t-t_0}\int_{t_0}^{t} \lambda(s)\,ds](t - t_o)\} \end{aligned} \tag{7.71}$$

Introducing the ergodic assumption

$$\lim_{t\to\infty} \frac{1}{t-t_0}\int_{t_0}^{t} \lambda(s)\,ds = E[\lambda(t)] \tag{7.72}$$

and taking limits of both sides of (7.71) gives

$$\lim_{t\to\infty} V(\tilde{X}(t)) = V(\tilde{X}(t_0))\exp\{\lim_{t\to\infty} (E[\lambda(t)](t - t_0))\} \tag{7.73}$$

Equation (7.73) reveals that if $E[\lambda(t)]\leq-\varepsilon$ for some $\varepsilon > 0$, then $V(\tilde{X}(t))$ is bounded and converges to zero as $t\to\infty$. Since $V(\tilde{X}) = \tilde{X}^T\underline{B}\tilde{X}$

and $\underline{B}$ is positive definite, then $V(\underset{\sim}{X})>0$ for all $\underset{\sim}{X} \neq 0$ and $V(\underset{\sim}{X})=0$ for $\underset{\sim}{X}=\underset{\sim}{0}$.

THEOREM 7.10. The system (7.49) is said to be almost surely stable if the following condition holds

$$E[\lambda_{max}(\underline{D}\underline{B}^{-1})] = E[\lambda_{max}(\underline{A}^T + \underline{\xi}^T(t) + \underline{B}(\underline{A} + \underline{\xi}(t))\underline{B}^{-1})] < -\varepsilon \tag{7.74}$$

This condition constitutes the main result of Infante's theorem. In order to satisfy condition (7.74), it is necessary that the eigenvalues of the matrix $\underline{A}$ have negative real parts which, in most cases, require the inclusion of positive damping in (7.49).

Corollary 7.1. If, for some positive definite matrix $\underline{B}$ and some $\varepsilon>0$,

$$E[\lambda_{max}(\underline{\xi}^T(t) + \underline{B}\underline{\xi}(t)\underline{B}^{-1})] \leq -\lambda_{max}(\underline{A}^T + \underline{B}\underline{A}\underline{B}^{-1}) - \varepsilon \tag{7.75}$$

then system (7.49) is almost surely asymptotically stable in the large.

Corollary 7.2. Let the system (7.49) be written in the form

$$\dot{\underset{\sim}{X}} = \underline{A}\,\underset{\sim}{X} + \left(\sum_{i=1}^{\ell} \underset{\sim}{f}_i(t)\underline{C}_i\right)\underset{\sim}{X} \tag{7.76}$$

where $\ell \leq n^2$ and $E[\underset{\sim}{f}_i(t)] = \underset{\sim}{0}$.

If, for some positive definite matrix $\underline{B}$ and some $\varepsilon>0$,

$$\sum_{i-1}^{\ell} \frac{1}{2} E[\,|f_i(t)|\,](\lambda_{max}[\underline{C}_i^T + \underline{B}\underline{C}_i\underline{B}^{-1}] - \lambda_{min}[\underline{C}_i^T + \underline{B}\underline{C}_i\underline{B}^{-1}])$$

$$\leq -\lambda_{max}[\underline{A}^T + \underline{B}\underline{A}\underline{B}^{-1}] - \varepsilon \tag{7.77}$$

then system (7.76) is almost surely asymptotically stable in the large.

It can be shown (Infante, 1968) that if the matrix $\underline{B}$ is obtained by using relation (7.56), the theorem and the two corollaries will lead to the same theorem of Caughey and Gray. In order to obtain sharper results, an optimal matrix $\underline{B}$ must be used. However, there is no standard procedure to construct $\underline{B}$ for every problem. For certain problems which involve one random parametric coefficient, the optimal matrix $\underline{B}$ can be expressed in terms of two parameters. These parameters can be determined by maximizing $E[\xi^2(\tau)]$.

7.4.4 ALMOST SURE STABILITY (KOZIN AND WU, 1973)

Kozin and Wu considered a linear system described by equation (7.49) such that $\underline{A}$ and $\underline{\xi}(t)$ are given in the form

$$\underline{A} = \begin{bmatrix} 0 & 1 \\ -1 & -2\zeta \end{bmatrix}, \qquad \underline{\xi}(t) = \begin{bmatrix} 0 & 0 \\ -\xi(t) & 0 \end{bmatrix} \tag{7.78}$$

THEOREM 7.11. If for $\ell > 0$, $C = (\ell + 1 - \zeta^2)^{-1/2}$ such that

$$\{(1 - 2P)\ell + 2PE\}\, C < 2\zeta \tag{7.79}$$

$$\text{where } P = P\{\xi(t) \geq \ell\} = \int_{\ell}^{\infty} p(\xi(t))\, d\xi,$$

$$E = E[\xi(t) \mid \xi(t) \geq \ell] = \{\int_{\ell}^{\infty} \xi(t)\, dP(\xi, t)\}/P$$

the equilibrium position of (7.49) is almost surely asymptotically stable. The value of ℓ is restricted to the limiting value

$$2\zeta(\zeta - 1) < \ell < 2(\zeta + 1) \tag{7.80}$$

This theorem establishes the stability boundary by utilizing an optimization technique for the undetermined parameter ℓ, under the constraint condition (7.80) and for a fixed value of the damping ratio ζ. A numerical algorithm is used by changing ℓ and the spectral density of the excitation until the maximum stability region is obtained. In addition, the probability distribution of the random excitation $\underline{\xi}(t)$ must be specified.

7.4.5 ALMOST SURE STABILITY (KOZIN AND MILSTEAD, 1979)

Kozin and Milstead developed a theorem to determine the almost sure stability of higher order systems. They expressed the Liapunov function of (7.49) in the quadratic form

$$V(\underset{\sim}{X}) = \underset{\sim}{X}^T \underline{P}\, \underset{\sim}{X} \tag{7.81}$$

where $\underline{P}$ is a symmetric positive-definite matrix. The total time derivative of $V(\underset{\sim}{X})$ along the trajectory of (7.49) is

$$\begin{aligned} \dot{V}(\underset{\sim}{X}) &= \dot{\underset{\sim}{X}}^T \underline{P}\, \underset{\sim}{X} + \underset{\sim}{X}^T \underline{P}\, \dot{\underset{\sim}{X}} \\ &= \underset{\sim}{X}^T\{(\underline{A} + \underline{\xi}(t))^T \underline{P} + \underline{P}(\underline{A} + \underline{\xi}(t))\}\, \underset{\sim}{X} \\ &= \underset{\sim}{X}^T(\underline{A}^T\underline{P} + \underline{P}\underline{A})\underset{\sim}{X} + \underset{\sim}{X}^T(\underline{\xi}^T(t)\underline{P} + \underline{P}\underline{\xi}(t))\underset{\sim}{X} \end{aligned} \tag{7.82}$$

According to a lemma given by Milstead (1975), the second expression in (7.82) can be written in the form

$$\underline{\xi}^T(t)\underline{P} + \underline{P}\underline{\xi}(t) = \sum_{m=1}^{n} \xi_{mm}(t)\underline{P} + \left(\hat{\underline{\xi}}^T(t)\underline{P} + \underline{P}\hat{\underline{\xi}}(t)\right) \tag{7.83}$$

Introducing (7.83) into (7.82) yields

$$\dot{V}(\underset{\sim}{X}) = \sum_{m=1}^{n} \xi_{mm}(t)V(\underset{\sim}{X}) + \underset{\sim}{X}^T(\underline{A}^T\underline{P}+\underline{P}\underline{A})\underset{\sim}{X} + \underset{\sim}{X}^T\left(\hat{\underline{\xi}}^T(t)\underline{P} + \underline{P}\hat{\underline{\xi}}(t)\right)\underset{\sim}{X} \tag{7.84}$$

where $1 \le n_1 \le n_2 < n$, $n_2 - n_1$ is the number of non-zero terms on the diagonal of $\underline{\xi}(t)$, and $\hat{\underline{\xi}}(t)$ is constructed such that its elements have one of the forms

$$\hat{\xi}_{ij}(t) = \begin{cases} \xi_{ij}(t) & i \neq j \\ \xi_{ij}(t) - \dfrac{1}{2}\displaystyle\sum_{m=1}^{n} \xi_{mm}(t) & i = j \end{cases} \tag{7.85}$$

Introducing the real unknown constant λ in (7.84) gives

$$\dot{V}(\underset{\sim}{X}) = \left\{-\lambda + \sum_{m=1}^{n} \xi_{mm}(t)\right\} V(\underset{\sim}{X}) + \underset{\sim}{X}^T(\lambda\underline{P} + \underline{A}^T\underline{P} + \underline{P}\underline{A})\,\underset{\sim}{X}$$

$$+\, \underset{\sim}{X}^T\{\hat{\underline{\xi}}^T(t)\,\underline{P} + \underline{P}\hat{\underline{\xi}}(t)\}\,\underset{\sim}{X} \tag{7.86}$$

Define a constant matrix $\underline{L} = [\ell_{ij}]$ which has the same form of the excitation matrix such that

$$\text{for } i \neq j: \begin{cases} \ell_{ij} = 0 & \text{if} \quad \xi_{ij} = 0 \\ \ell_{ij} = \hat{\ell}_{ij} \neq 0 & \text{if} \quad \xi_{ij} \neq 0 \end{cases} \tag{7.87}$$

$$\text{for } i=j, \begin{cases} \xi_{ij} = 0 \text{ then} & \hat{\ell}_{ii} = -\dfrac{1}{2}\displaystyle\sum_{m=1}^{n} \ell_{mm} \\ \xi_{ii} \neq 0 \text{ then} & \hat{\ell}_{ii} = \ell_{ii} - \dfrac{1}{2}\displaystyle\sum_{m=1}^{n} \ell_{mm} \end{cases} \tag{7.88}$$

The unknown matrix $\underline{L}$ is constructed such that it satisfies the relation

$$\lambda\underline{P} + \underline{A}^T\underline{P} + \underline{P}\underline{A} = \hat{\underline{L}}^T\underline{P} + \underline{P}\hat{\underline{L}} \tag{7.89a}$$

or

$$\lambda \underline{P} = (-\underline{A} + \hat{\underline{L}})^T \underline{P} + \underline{P}(-\underline{A} + \hat{\underline{L}}) \tag{7.89b}$$

Relation (7.89b) represents n equations in $(n^2 + 1)$ unknowns. These are the λ and ℓ_{ij}. It is possible, however, to develop n^2 equations from (7.89b) by using the Kronecker product (Bellman, 1960).

Kronecker Product: Let $\underline{A}$ be an n-dimensional matrix and $\underline{P}$ be another n-dimensional matrix. The n-dimensional matrix defined by

$$\underline{T} = [a_{ij}\underline{P}] \tag{7.90}$$

is called the Kronecker product of $\underline{A}$ and $\underline{P}$ and the following notation has been adopted in control theory literature

$$\underline{T} = \underline{A} \otimes \underline{P} = [a_{ij}\underline{P}] \tag{7.91}$$

where $\otimes$ denotes Kronecker product and it implies that every element of $\underline{A}$ is multiplied by the matrix $\underline{P}$ resulting in n^2-dimensional matrix $\underline{T}$. According to this definition, the Liapunov equation

$$\underline{A}^T \underline{P} + \underline{P}\underline{A} = -\underline{C} \tag{7.92}$$

may be written in the form

$$[\underline{A}^T \otimes \underline{I} + \underline{I} \otimes \underline{A}^T] \, \underset{\sim}{p} = \underset{\sim}{c} \tag{7.93}$$

where $\underset{\sim}{p}$ is a column vector of dimension n^2 formed by expanding the elements of the $n \times n$ matrix $\underline{P}$ such that the first n elements will be the first row, the second n elements will be the second row, and so on, i.e.,

$$\underset{\sim}{p}^T = \{p_{11}, p_{12}, \ldots, p_{1n}, p_{21}, p_{22}, \ldots, p_{2n}, \ldots, p_{nn}\} \tag{7.94}$$

and $\underset{\sim}{c}$ is a column vector of dimension n^2 formed by expanding the elements of the $n \times n$ matrix $\underline{C}$. $\underline{I}$ is the $n \times n$ identity matrix.

Equation (7.89b) may be written in the Kronecker product form

$$\lambda \underset{\sim}{p} = [(-\underline{A} + \hat{\underline{L}}) \otimes \underline{I} + \underline{I} \otimes (-\underline{A} + \hat{\underline{L}})^T] \, \underset{\sim}{p} \tag{7.95}$$

This equation is a characteristic equation with eigenvalues $\underset{\sim}{\lambda}$ and eigenvectors $\underset{\sim}{p}$. For each $(-\underline{A}+\underline{L})$ there are $\underline{n}$ eigenvalues λ_i and equation (7.95) gives n^2 numbers $(\lambda_i + \lambda_j)$. Since λ is restricted to be real, the real values obtained from the sums $\lambda_i + \lambda_j$ will be considered. The values of λ are determined in terms of ℓ_{ij} and the system matrix $\underline{A}$. Substituting (7.89a) into (7.86) gives

$$\dot{V}(\underset{\sim}{X}) = \left\{-\lambda + \sum_{m=1}^{n} \xi_{mm}(t)\right\} V(\underset{\sim}{X}) + \underset{\sim}{X}^T \left\{\left(\hat{\underline{L}} + \hat{\underline{\xi}}(t)\right)^T \underline{P} + \underline{P}\left(\hat{\underline{L}} + \hat{\underline{\xi}}(t)\right)\right\} \underset{\sim}{X} \tag{7.96}$$

It can be shown by direct substitution that

$$\underset{\sim}{X}^T\{(\hat{\underline{L}} + \hat{\underset{\sim}{\xi}}(t))^T\underline{P} + \underline{P}(\hat{\underline{L}} + \hat{\underset{\sim}{\xi}}(t))\}\ \underset{\sim}{X} = \sum_{k=1}^{n} \underset{\sim}{y}_k^T \underset{\sim}{z}_k \tag{7.97}$$

where $y_k^T = \{(\ell_{k1} + \xi_{k1}(t)), \ldots, (\ell_{km} + \xi_{km}(t))\}$

and $$\underset{\sim}{Z}^T = \{2\sum_{i=1}^{n} p_{ik}X_1X_i,\ 2\sum_{i=1}^{n} p_{ik}X_2X_i,\ \ldots,\ 2\sum_{i=1}^{n} p_{ik}X_{k-1}X_i,$$

$$p_{kk}X_k^2 - \sum_{\substack{i=1\\ i\neq j}}^{n} \sum_{j=1}^{n} p_{ij}X_iX_j,\ \ldots,\ 2\sum_{i=1}^{n} p_{ik}X_nX_i\}$$

According to a theorem by Milstead (1975), it is possible to show that

$$(\underset{\sim}{y},\underset{\sim}{z}) \leq (p_{kk}/\Delta)^{1/2}\Big[\sum_{i,j=1}^{n} \underline{P}_{ij}(\ell_{kj} + \xi_{kj}(t))(\ell_{ki} + \xi_{ki}(t))\Big]^{1/2} V(\underset{\sim}{X}) \tag{7.98}$$

where $\underline{P}_{ij}$ is the ij cofactor of the matrix $\underline{P}$ and Δ is the determinant of $\underline{P}$.

Applying (7.98) to equation (7.97) yields

$$\dot{V}(\underset{\sim}{X}) \leq \{-\lambda + \sum_{k=1}^{n} \xi_{kk}(t) + (p_{kk}/\Delta)^{1/2}\ \cdot$$

$$\Big[\sum_{i,j=1}^{n} \underline{P}_{ij}(\ell_{kj} + \xi_{kj}(t))(\ell_{ki} + \xi_{ki}(t))\Big]^{1/2}\} V(\underset{\sim}{X}) \tag{7.99a}$$

or

$$\dot{V}(\underset{\sim}{X}) < G(t)V(\underset{\sim}{X}) \tag{7.99b}$$

where G(t) is the expression between the braces in (7.99a).

Integrating inequality (7.99b) gives

$$V(\underset{\sim}{X}) < V(\underset{\sim}{X}_0)\exp\int_0^t G(\tau)d\tau, \qquad \underset{\sim}{X}(t{=}0) = \underset{\sim}{X}_0 \tag{7.100}$$

THEOREM 7.12. The equilibrium solution of (7.49) is almost surely asymptotically stable if

$$\lim_{t\to\infty} \frac{1}{t}\int_0^t G(\tau)\, d\tau = E[G(\tau)] \leq -\varepsilon \tag{7.101}$$

If the elements of $\underline{\xi}$ have zero mean, condition (7.101) can be written in the form

$$-\lambda + \sum_{k=1}^{n} (p_{kk}/\Delta)^{1/2} E[(\sum_{i,j=1}^{n} \underline{p}_{ij}(\ell_{kj} + \xi_{kj}(t))(\ell_{ki} + \xi_{ki}(t)))^{1/2}] < -\varepsilon \tag{7.102}$$

This condition is evaluated by using the probability distribution of the excitation to determine the expectation while optimizing the unknown constants $\{\ell_{ij}\}$ to generate the maximum stability region. In this case it is convenient to develop a numerical optimization algorithm which results in a systematic maximization of the stability region over the space of the $\{\ell_{ij}\}$ parameters. In each instance, the value of λ and the elements p_{ij}, calculated from (7.95) for a given set of $\{\ell_{ij}\}$, must be admissible in that the resulting Liapunov function must be positive definite. The results of this theorem are applied for general systems with more than one stochastic coefficient. However, for higher order systems the required computational time will increase significantly, partially because of the complicated form of the second term in (7.102).

7.4.6 COMPARISON OF ALMOST SURE STABILITY THEOREMS

Case 1:

The almost sure stability of the second order system

$$\ddot{Y} + 2\zeta \dot{Y} + \{1 + \xi(\tau)\} Y = 0 \tag{7.103}$$

was examined by several authors who applied various theorems and obtained different results. According to theorem 7.5, Kozin obtained the three stability boundaries

$$E[\xi^2(\tau)] < \frac{\pi}{32} \zeta^2(1 - \zeta^2) \qquad \zeta < 1 \tag{7.104a}$$

$$E[\xi^2(\tau)] < \frac{\pi(\zeta^2-1)}{32(1+\zeta)} (\zeta - \zeta^2 - 1) \qquad \zeta > 1 \tag{7.104b}$$

$$E[\xi^2(\tau)] < \frac{\pi}{72} (1 - \frac{1}{e})^2 \qquad \zeta = 1 \tag{7.104c}$$

where e is the Napierian base.

These boundaries are shown in fig. (7.5) by curves K. It is seen that as the damping ratio approaches the critical value $\zeta=1$, the excitation variance diminishes to zero which is unrealistic. This peculiar situation is a direct consequence of the inclusion of the upper bound of the eigenvalue. In an attempt to refine these results Ariaratnam (1967) used the unit impulse response of the deterministic system and obtained the following stability conditions

$$E[\xi^2(\tau)] < \pi\zeta^2(1 - \zeta^2)/2 \qquad 0 < \zeta < 1 \tag{7.105a}$$

$$E[\xi^2(\tau)] < \pi(\zeta - 1/e)/2 \qquad \zeta \approx 1 \tag{7.105b}$$

$$E[\xi^2(\tau)] < \frac{\pi}{2}(\zeta - \zeta^2 - 1 - 1/e)^2 \qquad \zeta > 1 \tag{7.105c}$$

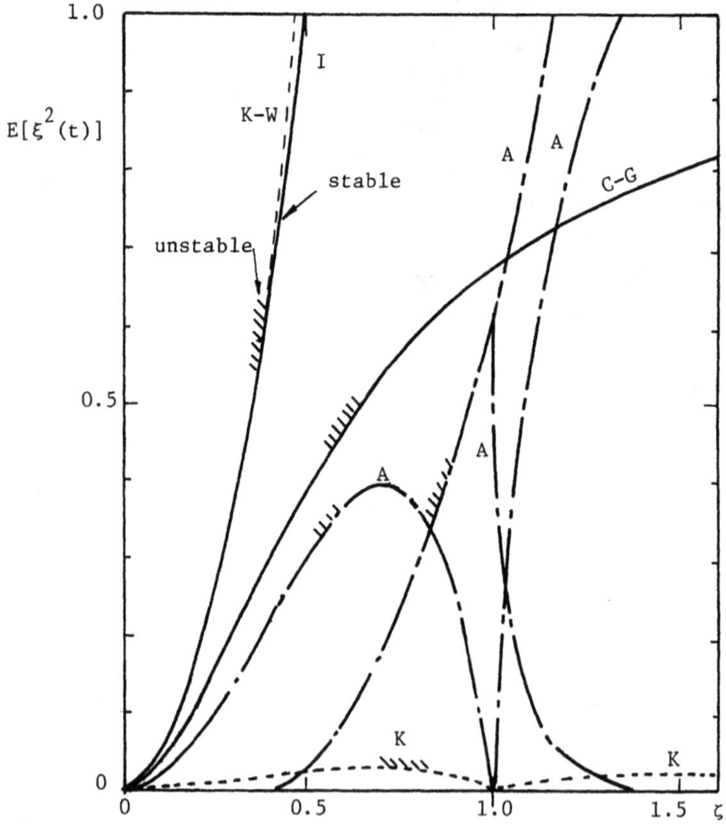

FIG. (7.5) Almost sure stability boundaries of stochastic damped Mathieu equation as obtained by various theorems

These boundaries are indicated in fig. (7.5) by curves A.

Caughey and Gray (1965) employed their theorem and derived the stability boundary for all ranges of the damping factor ζ

$$E[\xi^2(\tau)] < 4\zeta^2/(\zeta + 1 + \zeta^2)^2 \tag{7.106}$$

This condition is shown in fig. (7.5) by the curve C-G. However, this condition was found less sharp than the one predicted by Infante (1968), indicated by curve I,

$$E[\xi^2(\tau)] < 4\zeta^2 \tag{7.107}$$

The numerical algorithm of Kozin and Wu (1973) gives a rather significant improvement in the stability boundary shown by the K-W curve.

Case 2:

Infante (1968) and Kozin and Wu (1973) examined the almost sure stability of a second order system with damping coefficient which includes a random component. The system is described by the differential equation

$$\ddot{Y} + \{2\zeta + \xi(\tau)\}\dot{Y} + Y = 0 \tag{7.108}$$

The equilibrium position of this system is almost surely stable if

$$E[\xi^2(\tau)] < 4\zeta^2/(1 + \zeta^2) \tag{7.109}$$

This condition was obtained by Infante and is compared with the numerical solution of Kozin and Wu (for Gaussian noise) in fig. (7.6). It is seen from condition (7.109) that $E[\xi^2(\tau)] \to 4$ as the damping factor approaches infinity.

Khas'miniskii (1980, p. 234) indicated that when $\xi(\tau)$ is a white noise process, then the low frequency oscillations of the system become stable under a very strong perturbation of the damping coefficient.

Example 7.7 (Milstead, 1975): Determine the almost sure asymptotic stability of a second order system under simultaneous parametric excitations in damping and stiffness

$$\ddot{Y} + \{2\zeta + \xi_1(t)\}\dot{Y} + \{1 + \xi_2(t)\}Y = 0 \tag{i}$$

where $\xi_1(t)$ and $\xi_2(t)$ are independent stationary ergodic random processes with zero means.

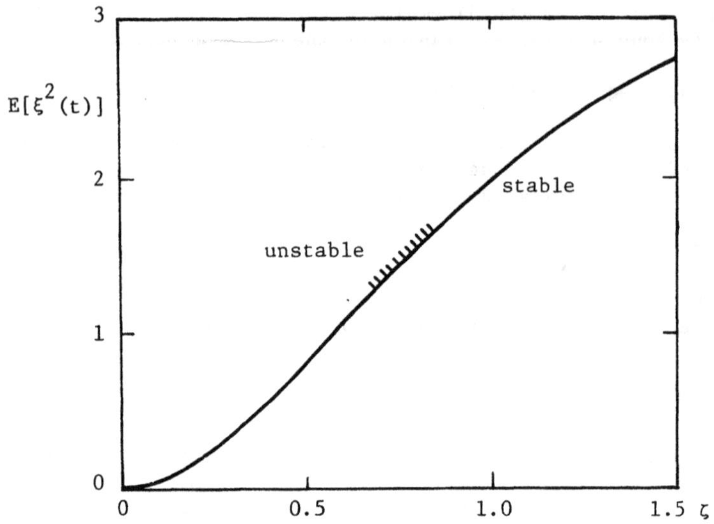

FIG. (7.6) Almost sure stability boundary of dynamic systems described by equation (7.108) (Infante, 1968)

Introducing the transformation $Y=X_1$, and $X_2=\dot{X}_1$, equation (i) can be written in the matrix form

$$\begin{Bmatrix} \dot{X}_1 \\ \dot{X}_2 \end{Bmatrix} = \begin{bmatrix} 0 & 1 \\ -1 & -2\zeta \end{bmatrix} \begin{Bmatrix} X_1 \\ X_2 \end{Bmatrix} + \begin{bmatrix} 0 & 0 \\ -\xi_2(t) & -\xi_1(t) \end{bmatrix} \begin{Bmatrix} X_1 \\ X_2 \end{Bmatrix} \tag{ii}$$

According to the almost sure stability of Kozin and Milstead outlined in section 7.4.5, the Liapunov matrix $\underline{P}$ may be expressed in the form

$$P = \begin{bmatrix} p_{11} & p_{12} \\ p_{12} & p_{22} \end{bmatrix}, \quad p_{11} > 0 \quad , p_{11}\, p_{22} - p_{12}^2 > 0 \tag{iii}$$

From equation (ii) it is seen that $-\xi_1(t)$ is the only non-zero diagonal element in the excitation matrix, and the time derivative of the Liapunov function becomes

$$\dot{V}(\underset{\sim}{X}) = -\xi_1(t)V(\underset{\sim}{X}) + \underset{\sim}{X}^T[(\underline{A}^T + \hat{\underline{\xi}}^T(t))\underline{P} + \underline{P}(\underline{A} + \hat{\underline{\xi}}(t))]\,\underset{\sim}{X} \tag{iv}$$

where

$$\hat{\xi}(t) = \begin{bmatrix} \xi_1(t)/2 & 0 \\ -\xi_2(t) & -\xi_1(t)/2 \end{bmatrix} \tag{v}$$

From (v) the matrix $\hat{\underline{L}}$ may be constructed in the form

$$\hat{L} = \begin{bmatrix} -\ell_2/2 & 0 \\ \ell_1 & \ell_2/2 \end{bmatrix} \tag{vi}$$

The eigenvalues of the matrix $[-\underline{A} + \hat{\underline{L}}]$ are

$$\lambda_{1,2} = \zeta \pm \{(\zeta + \ell_2/2)^2 - (1 + \ell_1)\}^{1/2} \tag{vii}$$

The roots $\lambda_{1,2}$ may represent a conjugate pair of complex roots and their sum gives the real value 2ζ. Substituting this value for λ in (7.89a) and solving for the elements of the matrix $\underline{P}$ gives

$$\begin{aligned} 2\zeta p_{11} &= -\ell_2 p_{11} + 2(1 + \ell_1)\, p_{22} \\ 2\zeta p_{12} &= -\ p_{11} + 2\zeta p_{12} + (1 + \ell_1)\, p_{22} \\ 2\zeta p_{22} &= -\ 2p_{12} + (4\zeta + \ell_2)\, p_{22} \end{aligned} \tag{viii}$$

Setting $p_{22} = 1$, and solving (viii) for the remaining values gives

$$p_{11} = 1 + \ell_1, \quad p_{12} = \zeta + \ell_2/2, \quad p_{22} = 1 \tag{ix}$$

Conditions (iii) require that

$$1 + \ell_1 > 0, \ \Delta = p_{11}p_{22} - p_{12}^2 = 1 + \ell_1 - (\zeta + \ell_2/2)^2 > 0 \tag{x}$$

The second condition, if it holds, leads to a negative value of the expression under the radical sign in (vii), and thus we set $\lambda = 2\zeta$.

Introducing p_{ij} into condition (7.102), it is possible to show that the system will be asymptotically stable with probability one if

$$-2\zeta + \Delta^{-1/2}E[\{(\ell_1 - \xi_2(t))^2 - 2(\zeta + \ell_2/2)(\ell_1 - \xi_2(t))(\ell_2 - \xi_1(t)) + (1 + \ell_1)(\ell_2 - \xi_1(t))^2\}^{1/2}] < -\varepsilon \tag{xi}$$

This expression may be simplified if $\underline{P}$ is positive definite, i.e.,

$$p_{11}p_{22} > p_{12}^2 \qquad \text{or} \qquad (p_{11}p_{22})^{1/2} > -\ p_{12} \tag{xii}$$

Thus the expression inside the expectation brackets in (xi) can be reduced into the form

$$\underline{p}_{11}(\ell_1 - \xi_2(t))^2 - 2\,\underline{p}_{12}(\ell_2 - \xi_2(t))(\ell_2 - \xi_1(t)) + \underline{p}_{22}(\ell_2 - \xi_1(t))^2$$
$$\leq \underline{p}_{11}(\ell_1 - \xi_2(t))^2 + 2\sqrt{p_{11}p_{22}}\,|\ell_1 - \xi_2(t)|\cdot|\ell_2 - \xi_1(t)|$$
$$+ p_{22}(\ell_2 - \xi_1(t))^2 \qquad \text{(xiii)}$$

Introducing (xiii) into condition (xi) gives

$$-2\zeta + \Delta^{-1/2}\{E[\,|\ell_1 - \xi_2(t)|\,] + (1+\ell_1)^{1/2}\,E[\,|\ell_2 - \xi_1(t)|\,]\}$$
$$= -2\zeta + \Delta^{-1/2}\{(\ell_1 + (1+\ell_1)^{1/2}) + 2(w_1(E_1 - \ell_1)$$
$$+ (1+\ell_1)^{1/2}w_2(E_2 - \ell_2))\} < -\varepsilon \qquad \text{(xiv)}$$

where $w_1 = \text{Prob}\{\xi_2(t) \geq \ell_1\}$, $\quad w_2 = \text{Prob}\{\xi_1(t) \geq \ell_2\}$

$E_1 = E[\xi_2(t)\,|\,\xi_2(t) \geq \ell_1]$, $\quad E_2 = E[\xi_1(t)\,|\,\xi_1(t) \geq \ell_2]$

Figure (7.7) shows a comparison between the boundaries given by (xi) and (xiv) for the case of $\zeta=1$. It is seen that condition (xiv) provides a somewhat weaker condition as would be expected.

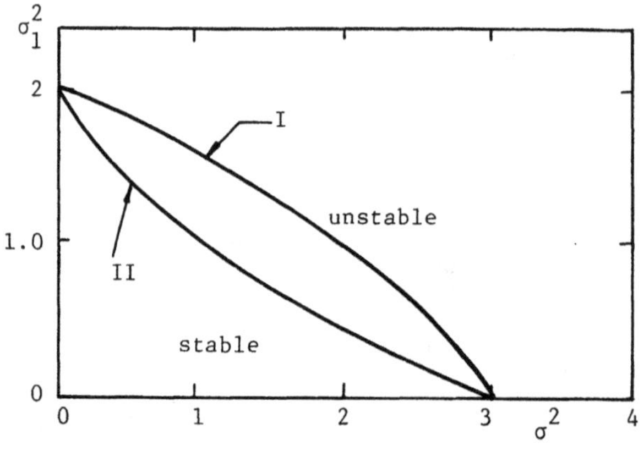

FIG. (7.7) Almost sure stability boundaries of example (7.7) according to I equation (xi) and II equation (xiv) (Kozin and Milstead, 1979)

CHAPTER 8

Parametric Random Response

8.1 INTRODUCTION AND HISTORICAL REVIEW

It has been shown in chapter 7 that the modes of stochastic stability of the equilibrium position can usually be determined on the basis of a linearized approximation to the system equations of motion. If the equilibrium is unstable, the linearized equations do not provide a unique bounded solution. On the other hand, if the inherent non-linearities are included in the mathematical modeling, the solution trajectories which emanate from an unstable equilibrium often end up in bounded limit cycles. Moreover, the non-linear modeling also allows the dynamicist to predict a wide range of complex response characteristics such as multiple solutions, jump phenomenon, internal resonance, and the like. Bolotin (1964) and Barr (1980) indicated that the non-linearity can enter the structural model through the elastic restoring forces, the inertial forces, and other internal or external dissipating forces.

The general theory of random vibration of linear time-invariant systems is well documented by Crandall and Mark (1963), Robson (1963), Lin (1967), Newland (1975), and Nigam (1983). For Gaussian excitations the theory allows direct evaluation of the response statistics. However, if the system is non-linear or involves random time coefficients, the classical methods fail to determine the response which, in fact, is not Gaussian even if the excitation is Gaussian. During the past twenty years a number of techniques have been developed to investigate the response properties of non-linear systems. These techniques include:

(i) Markov methods based on the Fokker-Planck-Kolmogorov equation or the Itô stochastic calculus;

(ii) Gaussian and non-Gaussian closure schemes;

(iii) stochastic averaging methods which have been described in chapter 6;

(iv) equivalent linearization methods developed originally by Caughey (1963b) and by others (see recent surveys by Spanos (1981) and J. B. Roberts (1981));

(v) perturbation techniques (Crandall, 1963);

(vi) functional series representation (J. B. Roberts, 1981); and

(vii) simulation methods (digital and analog).

These approaches have been applied to systems with various forms of non-linearities. However, there can be no general rule about the suitability of any method for a particular non-linear system. Furthermore, the application of various techniques to the same system may lead to different results. The first three methods have been extensively used in random parametric problems and systems involving autoparametric coupling. Recently, Dimentberg (1980b) and Bolotin (1984) have addressed a limited number of problems involving random parametric coefficients.

Bounded responses of linear time-variant damped systems may exist if the system is subjected to external excitation. In this case the response is governed by the parametric and external random excitations, while the stochastic stability depends only on the parametric excitation. The existence and uniqueness of solutions of stochastic differential equations have been examined by Caughey (1971). Brissaud and Frisch (1974) outlined some mathematical approaches to analyze the response of linear systems subjected to parametric and external excitations. They introduced the concept of Green's function which satisfies the homogeneous part of the system equations of motion.

The probabilistic description of the response of dynamic linear systems involving time variation in the inertia coefficient (mass or mass moment of inertia) and subjected to external random excitations was examined by Szopa (1976,1977,1979,1981,1982) and Szopa and Wojtylak (1979). These studies have been carried out through the application of stochastic Volterra integral equations of the second kind (Bharucha-Reid, 1972).

One class of problems which has attracted the attention of mathematicians and dynamicists is the random response and stability of helicopter rotor blades in atmospheric turbulent flow. The dynamic behavior of such systems is usually described by linear time-variant differential equations with non-homogeneous random terms. Various aspects of random vibrations of rotor blades have recently been reviewed by Gaonkar (1980,1981). For those who are not familiar with helicopter rotor dynamics, they may refer to the recent monograph by Johnson (1980). A brief description of the basic blade dynamics will be given in section (8.4). At high advance ratio severe vibrations can occur in blade flapping or flap bending modes. Gaonkar and Hohenemser (1969) described three flight regimes of lifting rotor craft where the rotor generates high intensity turbulence in the rotor disk leading to substantial random blade loads. The first is the transition from hovering to forward flight, and vice versa, where at advance ratios between 0.05 and 0.1 dynamic blade loads and vibrations are often severe because of strongly non-uniform inflow and turbulent wake recirculation. The second is the vertical descent in the vortex state which also causes large random blade loads. The third is a condition with partial blade stall where unsteady stall phenomena in conjunction with air turbulence can produce sizable random loads.

The random nature of air flow was examined experimentally by Grant (1966) who also measured the rotor blade damping by adding a random velocity component to the airstream. Gaonkar (1971b,c,d) developed a numerical algorithm to determine the aerodynamic damping parameter or the inertia number of lifting rotor blades in forward flight from the measured response variance. The numerical algorithm consisted of computing the state transition matrix and the one-dimensional integration to obtain the blade response variance matrix. Hohenemser and Crews (1973) conducted a series of model tests to examine the unsteady wake effects. They found that the effects are remarkable at zero advance ratio and low collective pitch, but remain significant at high advance ratios and collective pitch settings.

The blade flapping response to atmospheric turbulence at small advance ratio (less than one) was determined by Gaonkar and Hohenemser (1969). In another paper they (1971) investigated the stochastic response of rigid flapping blades with elastic flapping restrained at high advance ratio. They considered vertical turbulence components and large ratio of turbulence scale length over the rotor radius length. The response was obtained by employing a mixed time-frequency domain analysis suggested by Sveshnikov (1966). A frequency decomposition of the quasi-stationary excitation was performed and the response to individual frequencies was then determined by integrating the relevant differential equations numerically. The non-analytic coefficients of the differential equation were replaced by 16-term Fourier series. Finally, the responses to individual frequencies were superposed to obtain statistical information such as the variance of the flapping angle. On the basis of computed threshold crossing expectation, the blade response was found to behave as a quasi-coherent narrow band random process. Gaonkar and Hohenemser (1972) employed this procedure to determine the effect of a non-stationary inflow ratio.

Gaonkar (1972) introduced shaping filters describing Gaussian random excitations to lifting rotors with feedback controls. A set of first order differential equations for the response variance were generated and solved numerically. The effectiveness of two types of control mechanisms (coning and tilt integral feedbacks) on the time history of the root mean square flapping response revealed that the peak is reduced by almost 25%.

For a general linear time-variant system under external random forcing the steady state response is, in general, a non-stationary process even if the external excitation is stationary. This fact motivated Wan and Lakshmikantham (1973) to use a direct time domain approach based on generating a set of deterministic differential equations for the covariance and correlation functions of the response. They obtained the flapping response of a single rotor blade subjected to a stationary, exponentially correlated excitation modulated by an envelope function. For the case of high advance ratio they confirmed the results of Gaonkar and Hohenemser (1971). Wan (1973a) developed approximate solutions for the covariance and correlation equations based on a two-variable expansion.

Gaonkar (1974b) extended the analysis of blade flapping response to include the expectation of total number of peak rates and the probability density functions conditional on the occurrence of a peak over the rotor disk. The average number of peaks above arbitrary threshold level and the effects of non-uniform vertical turbulence over the rotor disk were determined by Gaonkar and Subramanian (1977) and Gaonkar (1977). The analysis identified threshold ranges above which peak distribution functions over one rotor revolution could be approximated by the statistics of threshold upcrossings.

Recently, Prussing and Lin (1982,1983) have extended the analysis of deterministic stability of Ormiston and Hodges (1972) and Peters (1975a,b) to include the influence of longitudinal, lateral, and vertical turbulent velocity components to determine the stability of response moments of the coupled flap-lag motion. The first moment stability condition has been found to be strongly affected by the spectral density of the vertical random velocity component, which appears in the expression of the smaller eigenvalue of the coupled system. The form of this eigenvalue indicated that the lead-lag damping decreases quadratically with the blade angle of attack and, hence, with the blade lift. The critical value of the lift was found to be higher when the vertical turbulence component is taken into account. In other words, the presence of vertical turbulence was found to enhance the stability of the coupled motion. The same feature was also found in the mean square stability boundary. For the case of flap-torsion coupled motion the situation was reversed and the vertical turbulence provided a destabilizing effect as verified by Fujimori (1978) and Fujimori, et al., (1979).

The response statistics of coupled torsion-flapping rotor blade to atmospheric turbulence were determined numerically by Gaonkar, et al., (1972), Gaonkar (1974a) and Fuh, et al., (1983). They showed that the flapping maximum standard deviation could be reduced by introducing flapping or coning angle feedbacks. However, the torsion maximum standard deviation was not appreciably affected by any type of feedback. In addition, the blade torsional response to atmospheric turbulence at high rotor advance ratio was found to be detrimental unless the torsional blade stiffness was several times greater than that for the static torsional divergence limit in the region of maximum reversed flow. On the other hand, the flapping-torsion coupling was found to have little effect on flapping motion. The results of Gaonkar, et al., (1972) and Gaonkar (1974a) did not provide the full effect of atmospheric turbulence since only the vertical component was included in the equations of motion. This led Fuh, et al., (1983) to determine the blade response to both random parametric and external excitations. In this case the turbulent velocity components in the blade rotational plane appeared in the coefficients of the coupled equations of motion. The non-homogeneous terms included these components as well as the velocity component perpendicular to the rotational plane. The statistical average of the blade response was nearly the same as the deterministic response which depends only on the selected trim condition (a condition for which the pitching moments vanish.) The mean square response was found to be strongly affected by the level of turbulence.

These studies, however, did not exhibit the absorber effect observed in linear coupled deterministic systems. Furthermore, they did not take into account the effects of parametric combination resonance. Dimentberg and Isikov (1983) studied the response of two degree-of-freedom systems under conditions of sum and difference combination resonances. The existence of combination resonance was identified by the presence of non-zero correlation coefficient for the centered response amplitude. Thus in order to identify the combination resonance, the centered components of the squares of the response amplitudes have to be isolated and their correlation functions are then estimated.

More refined analysis will indicate that the rotor generates turbulence which is very likely to vary along the blade span with a correlation length less than or equal to the blade length. In this case the excitation must be described by a random field (Vanmarcke, 1983) in which the time and space are the independent parameters. Wan (1972,1973b) developed a method based on the formulation of spatial correlation functions of the response process. The method was applied to the problem of purely rigid blade (Wan, 1974a,1980) and flexible blade with zero bending stiffness (Wan and Lakshmikantham, 1974). It was shown that these extreme cases may result in delimiting the range of the response statistics for flexible blades with non-vanishing bending stiffness.

When the parametric excitation is a white noise process, the response of the system constitutes a Markov process and the response transition probability density function can be determined by solving the Fokker-Planck equation of the system. The solutions of the Fokker-Planck equation have been obtained for a certain class of first order dynamic systems. Caughey and Dienes (1961) and Atkinson and Caughey (1968) derived the probability density for a class of piecewise linear systems subjected to simultaneous parametric and external random white processes. The conditions for exact solutions of single and coupled non-linear systems were established by Caughey and Payne (1967) and Caughey and Ma (1982). Dimentberg (1980c,

1982) obtained the probability density of a special class of second order linear and non-linear systems involving white noise parametric and external excitations. The solutions were valid under certain criteria such as the almost sure instability conditions of the corresponding parametric systems.

If the excitations were not white noise processes, the response can be approximated by a Markov vector by introducing shaping linear filters between a white noise input and the system itself. The filters, which are directly excited by white noises, serve as generators of physical random excitations and the response of the extended system is then described by a Markov process. However, this approach may result in a rather complicated system for which the corresponding Fokker-Planck equation may not be solved analytically. Alternatively, one can generate a set of differential equations for various statistical response functions such as moments and semi-invariants. For linear time-invariant systems these differential equations are consistent to any order. However, for non-linear systems and for linear parametric systems coupled with linear shaping filters, the resulting equations form a system of infinite hierarchy. The infinite coupled equations must then be truncated by using a proper closure scheme.

Closure schemes are classified into two classes: Gaussian and non-Gaussian. The first class is based on the assumption that the response is Gaussian distributed, thus they are referred to as Gaussian closures. These include the cumulant (semi-invariant) scheme, central moment method, and the mean square technique. These methods were developed and used by Richardson (1964), Bellman and Richardson (1968), Newland (1965), Haines (1967), Sancho (1968,1969,1970a,b,c), Adomian (1971), Adomian and Malakian (1979), Iyenger and Dash (1976,1978), and Dash and Iyenger (1982). The second class includes those methods which take into account the deviation of the response from normality; hence, they are known as non-Gaussian closure schemes. The probability density function of the response may be expressed in terms of one of the asymptotic expansions such as the Edgeworth or Gram-Charlier series (Cramèr, 1946). These approaches were developed and applied for general dynamic systems by Kuznetsov, et al., (1960), Dashevskii (1967), Dashevskii and Liptser (1967), Beran (1968), Nakamizo (1970), Assaf and Zirkle (1976), Bover (1978a,b), Crandall (1980, 1981), and Beaman and Hedrick (1981a,b). Recently, Wu and Lin (1984), Ibrahim, et al., (1985), Ibrahim and Heo (1985), and Ibrahim and Soundararajan (1985) have employed the cumulant-neglect closure to various types of non-dynamic systems with random coefficients.

Gaussian closure schemes were used to determine the response of non-linear single degree-of-freedom systems with white noise coefficients. Ibrahim (1978c,1982), Ariaratnam (1980) and Ibrahim and Soundararajan (1983) determined the stationary mean square response of dynamic systems involving various types of non-linearities. The results were compared with those obtained by the stochastic averaging method. The comparison revealed that the Gaussian closure led to results consistent with the mean square instability condition $D/2\zeta>1$ while the averaging solution is compatible with the almost sure instability condition $D/2\zeta>2$, (here 2D is the excitation spectral density, and ζ is the system damping factor.)

The random response of two degree-of-freedom systems with autoparametric coupling was examined by Ibrahim and Roberts (1976) and Schmidt (1977a,b). Ibrahim and Roberts used the cumulant Gaussian closure scheme to determine the mean square response in the neighborhood of the internal resonance condition. The results indicated that when the system is close to the internal resonance, large random motions of the coupled system (which is not directly excited) take place and give rise to a suppression effect on the main system. The main features of these results were confirmed by Schmidt (1977a) who used the stochastic averaging technique.

The effects of non-linearities on the system response to narrow band random excitation were determined by Model (1978). Model considered the behavior of the system under combination parametric resonance and excluded the possibility of autoparametric instability. He applied the method of integro-differential equations (Schmidt, 1975) and showed that the probability of one amplitude response corresponds to the minimum probability of the other mode amplitude.

In addition to the stochastic averaging techniques presented in chapter 6, this chapter will address various other methods to determine the response of dynamic linear and non-linear systems to random parametric excitations. One remarkable feature of this chapter is that it includes the application of the non-Gaussian closure schemes to non-linear systems with random coefficients, and resolves some discrepancies observed in the literature.

8.2 LINEAR TIME-VARIANT SYSTEMS UNDER RANDOM EXTERNAL EXCITATION

The random response of linear time-variant systems subjected to external random excitation can be determined by solving the dynamic moment equations. This section outlines another two additional methods for analysis of the response statistics of linear systems. The first is referred to as the direct method and deals with systems involving non-homogeneous random terms. The second is known as the spatial correlation method and is suitable for linear systems subjected to random field external excitation.

8.2.1 THE DIRECT METHOD

The dynamic behavior of linear time-variant systems subjected to external random excitations may be described by the state vector differential equation

$$\dot{\underset{\sim}{X}}(t) = \underline{A}(t)\underset{\sim}{X}(t) + \underline{G}(t)\underset{\sim}{W}(t), \qquad \underset{\sim}{X}(t=t_0) = \underset{\sim}{X}_0 \tag{8.1}$$

The elements of the vector $\underset{\sim}{W}(t)$ are white noise processes with zero means and have autocorrelation matrix $\underline{D}$ which is positive definite.

Consider the homogeneous part of system (8.1)

$$\dot{\underset{\sim}{X}}(t) = \underline{A}(t)\underset{\sim}{X}(t) \tag{8.2}$$

The solution of (8.2) may be written in the form

$$\underset{\sim}{X}(t) = \exp\{\underline{A}(t)(t - t_0)\}\ \underset{\sim}{X}(t_0) \tag{8.3}$$

with $\Delta t = t - t_0$, the exponential part can be represented by the expansion

$$\exp(\underline{A}\Delta t) = \underline{I} + \underline{A}\Delta t + \frac{1}{2!}\underline{A}^2(\Delta t)^2 + \ldots \tag{8.4}$$

where $\underline{I}$ is the identity matrix.

Introducing the fundamental matrix which is known as the state transition matrix

$$\underline{\Phi}(t,t_0) = \exp\{\underline{A}(t)(t-t_0)\} \tag{8.5}$$

for time-varying systems the transition matrix $\underline{\Phi}(t,t_0)$ is a function of two variables rather than the difference Δt. From (8.3) and (8.2) the state transition matrix must satisfy the differential equation

$$\dot{\underline{\Phi}}(t,t_0) = \underline{A}(t)\ \underline{\Phi}(t,t_0), \quad \underline{\Phi}(t_0,t_0) = \underline{I} \tag{8.6}$$

The homogeneous solution given by (8.3) can be written in terms of $\underline{\Phi}(t,t_0)$

$$\underset{\sim}{X}(t) = \underline{\Phi}(t,t_0)\ \underset{\sim}{X}(t_0) \tag{8.7}$$

In most cases it is not possible to derive an analytical solution for system (8.6). Instead the response of (8.2) is determined by numerical integration. Two main properties of the solution are outlined by Van Trees (1968). These are:

$$\underline{\Phi}(t_2,t_0) = \underline{\Phi}(t_2,t_1)\ \underline{\Phi}(t_1,t_0) \tag{8.8}$$

$$\underline{\Phi}^{-1}(t_1,t_0) = \underline{\Phi}(t_0,t_1) \tag{8.9}$$

The solution of the non-homogeneous system described by equation (8.1) can be written in the form

$$\underset{\sim}{X}(t) = \underline{\Phi}(t,t_0)\underset{\sim}{X}(t_0) + \int_{t_0}^{t} \underline{\Phi}(t,s)\underline{G}(s)\underline{W}(s)\ ds \tag{8.10}$$

Alternatively, the solution of (8.1) may be expressed in terms of the impulse response $\underline{h}(t,s)$

$$\underset{\sim}{X}(t) = \int_{-\infty}^{\infty} \underline{h}(t,s)\underset{\sim}{W}(s)\, ds \tag{8.11}$$

The effect of the initial condition ($t=-\infty$) can be omitted from (8.10)

$$\underset{\sim}{X}(t) = \int_{-\infty}^{t} \underline{\Phi}(t,s)\ \underline{G}(s)\underset{\sim}{W}(s)\, ds \tag{8.12}$$

Comparing (8.11) and (8.12) yields

$$\underline{h}(t,s) = \underline{\Phi}(t,s)\ \underline{G}(s) \qquad\qquad t \geq s \tag{8.13}$$

It is evident that both $\underline{\Phi}$ and $\underline{G}$ will depend on the state representation of the system such that the matrix response is unique.

Now it is possible to evaluate the differential equation of the response covariance matrix $\underline{C}_x(t)$ along the trajectory of system (8.1)

$$\dot{\underline{C}}_x(t) = \frac{d}{dt} E[\underset{\sim}{X}(t)\underset{\sim}{X}^T(t)]$$

$$= \underline{A}(t)\underline{C}_x(t) + \underline{C}_x(t)\underline{A}^T(t) + E[\underset{\sim}{X}(t)\underset{\sim}{W}^T(t)]\underline{G}^T(t)$$

$$+ \underline{G}(t)\ E[\underset{\sim}{W}(t)\ \underset{\sim}{X}^T(t)] \tag{8.14}$$

The cross-correlation between the state vector $\underset{\sim}{X}(t)$ and the white noise excitation vector $W(\tau)$ is

$$\underline{R}_{xw}(t,\tau) = E[\underset{\sim}{X}(t)\underset{\sim}{W}^T(\tau)] \tag{8.15}$$

Substituting for $\underset{\sim}{X}(t)$ from (8.10) gives

$$\underline{R}_{xw}(t,\tau) = E[\{\underline{\Phi}(t,t_0)\underset{\sim}{X}(t_0) + \int_{t_0}^{t} \underline{\Phi}(t,s)\underline{G}(s)\underset{\sim}{W}(s)ds\}\underset{\sim}{W}^T(\tau)] \tag{8.16}$$

Since the initial state $\underset{\sim}{X}(t_0)$ is independent of $\underset{\sim}{W}(\tau)$ for $\tau > t_0$, relation (8.16) can be written in the form

$$\underline{R}_{xw}(t,\tau) = \int_{t_0}^{t} \underline{\Phi}(t,s)\underline{G}(s)E[\underset{\sim}{W}(s)\underset{\sim}{W}^T(\tau)]\, ds$$

$$= \int_{t_0}^{t} \underline{\Phi}(t,s)\underline{G}(s)\underline{D}\delta(s-\tau)\, ds \tag{8.17}$$

If $\tau > t$, then $\underline{R}_{xw}(t,\tau)$ is zero. If $\tau = t$, the delta function is assumed to be symmetric since it is the limit of a covariance function and only one-half of the area at the right end point is considered. Introducing the property $\underline{\Phi}(t,t) = \underline{I}$ into (8.17) gives

$$\underline{R}_{xw}(t,t) = E[\underset{\sim}{X}(t)\underset{\sim}{W}^T(t)] = \frac{1}{2}\,\underline{G}(t)\,\underline{D} \tag{8.18}$$

Similarly,

$$\underline{R}_{wx}(t,t) = E[\underset{\sim}{W}(t)\underset{\sim}{X}^T(t)] = \frac{1}{2}\,\underline{D}\underline{G}(t) \tag{8.19}$$

Substituting (8.18,19) into (8.14) gives

$$\dot{\underline{C}}_x(t) = \underline{A}(t)\underline{C}_x(t) + \underline{C}_x(t)\underline{A}^T(t) + \underline{G}(t)\underline{D}\underline{G}^T(t), \quad \underline{C}_x(t=0) = \underline{0} \tag{8.20}$$

Equation (8.20), along with the associated initial state, determines the covariance matrix $\underline{C}_x$ completely. The solution (8.20) can then be used to calculate the correlation matrix of the response. For all $t > \tau$, the following set of differential equations for the response correlation function is obtained

$$\frac{d}{dt}\,E[\underset{\sim}{X}(t)\underset{\sim}{X}^T(\tau)] = \underline{A}(t)E[\underset{\sim}{X}(t)\underset{\sim}{X}^T(\tau)] + \underline{G}(t)E[\underset{\sim}{W}(t)\underset{\sim}{X}^T(\tau)] \tag{8.21}$$

Since $\underline{R}_{wx}(t,\tau) = \underline{0}$ for $t > \tau$, (8.20) becomes

$$\dot{\underline{R}}_{xx}(t,\tau) = \underline{A}(t)\,\underline{R}_{xx}(t,\tau), \quad \text{with } \underline{R}_{xx}(t_0,t_0) = \underline{C}_x(t_0) \tag{8.22}$$

Wan and Lakshmikantham (1973) provided an alternative form

$$\frac{d}{ds}\,\underline{R}_{xx}(t,t+s) = \underline{A}(t+s)\underline{R}_{xx}(t,t+s), \quad \text{with } \underline{R}_{xx}(t,t+0) = \underline{C}_x(t) \tag{8.23}$$

In principle the response covariance matrix $\underline{C}_x(t)$ has to be determined first by solving (8.20). The solution can be used as initial conditions for the initial-value problem (8.22).

8.2.2 THE SPATIAL CORRELATION METHOD

The direct method treats the excitations as random processes which depend only on time. When the random loading depends on time and the spatial coordinates, the random process is referred to as a random field. In this case the dynamic behavior of structural systems subjected to random field excitations $\underset{\sim}{F}(X,t)$ is described by the initial-boundary-value problem

$$\frac{\partial}{\partial t}\,\underset{\sim}{u}(x,t) + L\{\underset{\sim}{u}(x,t)\} = \underset{\sim}{F}(x,t), \qquad t>0,\ x \in V \tag{8.24}$$

with the initial conditions

$$\underset{\sim}{u}(x,0) = \underset{\sim}{0}, \qquad x \in \bar{V} \tag{8.25}$$

and the boundary condition on the boundary surface S of volume V

$$B(\underset{\sim}{u}) = 0, \qquad t>0,\ x \in S \tag{8.26}$$

where L and B are linear partial differential operators in the spatial coordinate x, with coefficients dependent upon both x and t. $\bar{V}$ is the closure of volume V, and the random vector loading $\underset{\sim}{F}(x,t)$ may be expressed in terms of an envelope function matrix $\underline{M}(x,t)$ multiplied by a random process vector $\underset{\sim}{\xi}(t)$

$$\underset{\sim}{F}(x,t) = \underline{M}(x,t)\,\underset{\sim}{\xi}(x,t) \tag{8.27}$$

The elements of $\underset{\sim}{\xi}$ are assumed Gaussian with zero mean. The excitation $\underset{\sim}{\xi}(x,t)$ may also be assumed temporally uncorrelated with autocorrelation function

$$\underline{R}_{\xi}(x,t \mid y,\tau) = E[\underset{\sim}{\xi}(x,t)\underset{\sim}{\xi}^{T}(y,\tau)] = \underline{R}_{s}(x,y,\tau)\delta(t-\tau) \tag{8.28}$$

with $\underline{R}_{s}^{T}(x,y,\tau) = \underline{R}_{s}(y,x,t)$, and y refers to the value of x at τ.

The solution of system (8.24) can be derived in terms of the associated Green's function $G(x,t \mid x',t')$. The Green function $G(x,x')$ associated with the non-homogeneous equation $L(u)=F(x)$ satisfies the differential equation

$$LG(x,x') = \delta(x - x') \tag{8.29}$$

It is possible to represent the solution $Lu=F(x)$ in terms of $G(x,x')$ by the integral

$$u(x) = \int_{-\infty}^{\infty} F(x')G(x,x')\,dx' \tag{8.30}$$

and the solution of (8.24) can be written in the form

$$\underset{\sim}{u}(x,t) = \int_{V}\int_{0}^{t} \underline{G}(x,t \mid x',t')\,\underset{\sim}{F}(x',t')\,dt'\,dx' \tag{8.31}$$

The cross-spatial correlation between the response coordinate vector and the random loading vector is

$$E[\underset{\sim}{u}(x,t)\,\underset{\sim}{F}^{T}(y,t)] = \frac{1}{2}\int_{V} \underline{G}(x,t \mid x',t)\underline{M}(x',t)\underline{R}_{s}(x',y,t)\underline{M}^{T}(y,t)\,dx'$$

$$= \frac{1}{2}\,\underline{M}(x,t)\underline{R}_{s}(x,y,t)\underline{M}^{T}(y,t) \tag{8.32}$$

The Green function $\underline{G}(x,t \mid x',t)$ associated with system (8.24) can be shown to be spatial delta correlated (Wan, 1973b)

$$\underline{G}(x,t \mid x',t) = \underline{I}\delta(x - x') \tag{8.33}$$

From (8.32) it is possible to write

$$E[\underset{\sim}{F}(x,t)\underset{\sim}{u}^T(y,t)] = \frac{1}{2}\,\underline{M}(x,t)\underline{R}_s(x,y,t)\underline{M}^T(y,t) \tag{8.34}$$

The spatial correlation matrix $\underline{U}(x,y,t)$ of the response vector $\underset{\sim}{u}(x,t)$ is defined by the relation

$$\underline{U}(x,y,t) = E[\underset{\sim}{u}(x,t)\,\underset{\sim}{u}^T(y,t)] \tag{8.35}$$

This matrix can be determined by formulating an initial-boundary value problem for $\underline{U}$. This is carried out by taking the ensemble average of the time derivative of (8.35)

$$\frac{d}{dt}\,E[\underset{\sim}{u}(x,t)\underset{\sim}{u}^T(y,t)] = E[\underset{\sim}{\dot{u}}(x,t)\underset{\sim}{u}^T(y,t)] + E[\underset{\sim}{u}(x,t)\underset{\sim}{\dot{u}}^T(y,t)] \tag{8.36}$$

Using equation (8.24), eliminating $(\underset{\sim}{\dot{u}},\underset{\sim}{\dot{u}}^T)$, and introducing relations (8.32) and (8.34) gives

$$\frac{\partial}{\partial t}\,\underline{U} + \frac{\partial}{\partial x}\,L(\underline{U}) + \left\{\frac{\partial}{\partial y}\,L(\underline{U}^T)\right\}^T = \underline{M}(x,t)\,\underline{R}_s(x,y,t)\underline{M}^T(y,t) \tag{8.37}$$

This equation is a matrix partial differential equation in the time domain $0<t<\infty$, and the space domain (x,y) (V,V). Thus the initial-boundary value problem is completely specified by (8.37), where the initial condition is

$$\underline{U}(x,y,0) = \underline{0} \qquad (x,y) \quad (V,V) \tag{8.38}$$

and the boundary conditions are, for $t>0$

$$\frac{d}{dx}\,B[\underline{U}(x,y,t)] = 0 \qquad (x,y) \quad (S,V)$$

$$\frac{d}{dy}\,B[\underline{U}^T(x,y,t)] = 0 \qquad (x,y) \quad (V,S) \tag{8.39}$$

The solution of the matrix $\underline{U}$ can be obtained by numerical integration. If $\underline{M}(x,t)$ is continuous in x, and $\underline{R}_s(x,y,t)$ is continuous in both x and y, then the right-hand side expression of equation (8.37) will also be continuous in both x and y. The solution $\underline{U}$, therefore, will be continuous in x and y, in particular in the subspace $x=y$, and $\underline{U}(x,x,t)$ is therefore a well defined covariance matrix. Wan (1973b,1974a,b) discussed the significant reduction in computing time of the numerical integration as compared with the Green function method or its variants.

Special Case: Temporally-Correlated Excitation $\underset{\sim}{\xi}(x,t)$

In this case a shape filtered white noise is introduced and the system response becomes Markov process. The problem will be drastically simplified if the excitation is assumed to be uniformly distributed over the spatial coordinate, i.e.,

$$\underset{\sim}{\xi}(x,t) = \underset{\sim}{\xi}(t) \tag{8.40}$$

and $\underset{\approx}{R}_{\xi}(x,t;y,\tau) = \underset{\approx}{R}_{\xi}(t,\tau)$

where $\underset{\sim}{\xi}(t)$ is the response of the linear filter

$$\dot{\underset{\sim}{\xi}}(t) + \underline{\nu}(t)\, \underset{\sim}{\xi}(t) = \underline{F}'(t)\, \underset{\sim}{W}(t) \tag{8.41}$$

where $\underline{\nu}(t)$ and $\underline{F}'(t)$ are known matrix functions of time t, and $\underset{\sim}{W}(t)$ is a white noise vector.

The process $\underset{\sim}{\xi}(t)$ is treated as an unknown vector of n elements and the two equations (8.24,41) form a large system of dimension 2n.

8.3 THE MOMENT EQUATIONS METHOD

The random response of linear classical systems with random uncertainties in their coefficients and subjected to external excitations can be solved by generating the moment equations of the response. The method will be demonstrated by two examples.

Example 8.1: Consider the second order system

$$\ddot{Y} + \{2\zeta + \xi_1(\tau)\}\, \dot{Y} + \{1 + \xi_2(\tau)\}\, Y = \xi_3(\tau) \tag{i}$$

where dot denotes differentiation with respect to the dimensionless time $\tau = \omega t$, ω is the natural frequency of the system, and $\xi_i(\tau)$ are independent Gaussian random processes.

Introducing the coordinate transformation $X_1=Y$, and $X_2=\dot{Y}$, system (i) is equivalent to the two equations:

$$\begin{aligned} dX_1 &= X_2 d\tau \\ dX_2 &= \{-(2\zeta - D_1)X_2 - X_1\}\, d\tau - X_2 dB_1(\tau) - X_1 dB_2(\tau) + dB_3(\tau) \end{aligned} \tag{ii}$$

where $\xi_i(\tau) = dB_i(\tau)/d\tau$, and $E[dB_i^2(\tau)] = 2D_i d\tau$

The FPK equation of system (ii) is

$$\frac{\partial}{\partial\tau} p(\underset{\sim}{X},\tau) = \frac{\partial}{\partial X_1}\{X_2 p(\underset{\sim}{X},\tau)\} + \frac{\partial}{\partial X_2}\{[(2\zeta - D_1)X_2 + X_1]\, p(\underset{\sim}{X},\tau)\}$$

$$+ \frac{\partial^2}{\partial X_2^2}\{(D_1X_2^2 + D_2X_1^2 + D_3)\, p(\underset{\sim}{X},\tau)\} \tag{iii}$$

The differential equation of the joint moment $m_{ij}=E[X_1^i X_2^j]$ is

$$\dot{m}_{ij} = im_{i-1,j+1} - j(2\zeta - jD_1)m_{ij} - jm_{i+1,j-1} + j(j-1)D_2 m_{i+2,j-2} + j(j-1)D_3 m_{i,j-2} \qquad \text{(iv)}$$

Inspection of equation (iv) shows that the first order moments are dependent on the intensity of the damping coefficient uncertainty D_1. It is obvious that the effect of D_1 is to reduce the system damping and if it exceeds 2ζ, the system becomes unstable in the first moment.

The second order moment equations are:

$$\begin{aligned} \dot{m}_{20} &= 2m_{11} \\ \dot{m}_{02} &= -4(\zeta - D_1)m_{02} - 2m_{11} + 2D_2 m_{20} + 2D_3 \qquad \text{(v)} \\ \dot{m}_{11} &= m_{02} - (2\zeta - D_1)m_{11} - m_{20} \end{aligned}$$

The stationary solution of (v) is

$$m_{20} = m_{02} = D_3/\{2(\zeta - D_1) - D_2\}, \qquad m_{11} = 0 \qquad \text{(vi)}$$

It is seen that, in the absence of $\xi_1(\tau)$, the system is stable in the mean square if

$$D_2/2\zeta < 1 \qquad \text{(vii)}$$

which is independent of the external excitation $\xi_3(\tau)$.

It is possible to obtain a stationary solution for the system FPK equation (iii) if the random parametric coefficients are related such that

$$D_1 = D_2 \qquad \text{(viii)}$$

Under this condition the stationary probability density of the response is

$$p(X_1,X_2) = p_0\{\frac{D_3}{D_2} + X_1^2 + X_2^2\}^{-\delta} \qquad \text{(ix)}$$

where $\delta = \frac{\zeta}{D_2} + \frac{1}{2}$, and the constant p_0 is determined from the normalized condition

$$\int_{-\infty}^{\infty}\int p(X_1,X_2)\, dX_1 dX_2 = 1 \qquad \text{(x)}$$

According to formula 4 of section 3.241 of Gradshteyn and Ryzhik (1980) tables, the constant p_o is

$$p_o = \frac{1}{\pi}(\delta - 1)\{D_3/D_1\}^{\delta-1} \qquad \text{(xi)}$$

provided $D_3/D_2 > 0$ and $\delta > 1$ or

$$D_2/2\zeta < 1 \qquad \text{(vii)}$$

which coincides with the condition of the mean square stability of the homogeneous system.

If the external excitation $\xi_3(\tau)$ is removed, the probability density $p(X_1, X_2)$ will have a non-integrable singularity at $\underset{\sim}{X} = \underset{\sim}{0}$ and $p(X_1, X_2)$ will degenerate into a delta function, $\delta(0)$. Physically the system becomes dormant with zero oscillation at $\delta > 1$. Similar conclusions were obtained in chapter 6.

The stationary probability density of the response displacement is obtained by integrating (ix)

$$p(X_1) = \frac{1}{\sqrt{\pi}}(D_3/D_2)^{\delta-1}\Gamma(\delta - \tfrac{1}{2})(X_1 + D_3/D_1)^{-\delta+\frac{1}{2}}/\Gamma(\delta - 1) \qquad \text{(xii)}$$

The n-th order moment of the displacement is

$$E[X^n] = \begin{cases} \frac{1}{\sqrt{\pi}}(D_3/D_2)^{n/2}\Gamma(\frac{n+1}{2})\Gamma(\delta - \frac{n}{2} - 1)/\Gamma(\delta - 1) & n \text{ even} \\ 0 & n \text{ odd} \end{cases} \qquad \text{(xiii)}$$

provided $\delta > \frac{n}{2} + 1$.

Setting $n=2$, relation (xiii) gives the same mean square (xi) if $D_1 = D_2$. The inequality ($\delta > 1+n/2$) is identical to the stability condition of even order moments. It is interesting to note that within the range $1 < \delta < 2$, the response is unstable in the mean square although it is almost surely stable for $D_1 \neq 0$.

Example 8.2: Determine the stationary response of a two degree-of-freedom system described by the equations of motion

$$\begin{bmatrix} m_{11} & m_{12} \\ m_{12} & m_{22} \end{bmatrix} \begin{Bmatrix} \ddot{q}_1 \\ \ddot{q}_2 \end{Bmatrix} + \begin{bmatrix} k_{11} & 0 \\ 0 & k_{22} \end{bmatrix} \begin{Bmatrix} q_1 \\ q_2 \end{Bmatrix} + \ddot{\xi}(t) \begin{bmatrix} \mu_{11} & \mu_{12} \\ \mu_{22} & \mu_{22} \end{bmatrix} \begin{Bmatrix} q_1 \\ q_2 \end{Bmatrix}$$

$$= -\ddot{\xi}(t) \begin{Bmatrix} 1 + m_{22} \\ 0 \end{Bmatrix} \qquad \text{(i)}$$

The normal mode frequencies of system (i) are

$$\omega_{1,2}^2 = \frac{1}{2\mu}\{\omega_{11}^2 + \omega_{22}^2 \mp [(\omega_{11}^2 + \omega_{22}^2)^2 - 4\mu\omega_{11}^2\omega_{22}^2]^{1/2}\} \tag{ii}$$

where $\omega_{11}^2 = k_{11}/m_{11}$, $\omega_{22}^2 = k_{22}/m_{22}$, $\mu = 1 - (m_{12}/m_{11})(m_{12}/m_{22})$

Introducing the following transformation into principal coordinates y_1, y_2

$$\begin{Bmatrix} \bar{q}_1 \\ \bar{q}_2 \end{Bmatrix} = \begin{bmatrix} 1 & 1 \\ \phi_1 & \phi_2 \end{bmatrix} \begin{Bmatrix} y_1 \\ y_2 \end{Bmatrix} \tag{iii}$$

where the columns of the square matrix represent the eigenvectors corresponding to the eigenvalues ω_1 and ω_2, respectively.

The following non-dimensional parameters are introduced

$\tau = \omega_1 t$, $\quad y_1 = Y_1/q_1^o$, $\quad y_2 = Y_2/q_2^o$, $\quad \varepsilon = q_1^o/\ell_1$, $\quad r = \omega_2/\omega_1$, and

$\bar{q}_i = q_1/q_1^o$

where q_1^o is the root mean square response of the linear uncoupled system

$$\ddot{q}_1^o + 2\zeta_{11}\omega_{11}\dot{q}_1^o + \omega_{11}^2 q_1^o = -\ddot{\xi}(t), \qquad \omega_{11}^2 = k_{11}/m_{11}$$

It is not difficult to show that the mean square response of this system is

$$E[q_1^{o^2}] = D/(2\zeta_{11}\omega_{11}^3), \quad \text{thus } q_1^o = \{D/2\zeta_{11}\omega_{11}^3\}^{1/2} \tag{iv}$$

The equations of motion, in terms of the dimensionless principal coordinates y_i, are:

$$y'' + 2\zeta_1 y' + y_1 = \{A_1 + \varepsilon(A_2 y_1 + A_3 y_2)\} W(\tau)$$

$$y'' + 2\zeta_2 r y' + r^2 y_2 = \{B_1 + \varepsilon(B_2 y_1 + B_3 y_2)\} W(\tau) \tag{v}$$

where the random excitation $\ddot{\xi}(t)$ has been replaced by the white noise $W(t)$ of spectral density 2D.

Equations (v) are non-homogeneous coupled equations through the random parametric terms $y_2 W(\tau)$ and $y_1 W(\tau)$. The linear damping terms were introduced to account for energy dissipation for each normal mode. The coefficients A_i and B_i are functions of the system parameters.

The response Markov vector can be represented by introducing the coordinate transformation

$$\{y_1, y_2, \dot{y}_1, \dot{y}_2\} = \{X_1, X_2, X_3, X_4\} \tag{vi}$$

The equivalent Markov vector equations of equations (v) are:

$$\begin{aligned}
&\dot{X}_1 = X_3, \qquad \dot{X}_2 = X_4, \\
&\dot{X}_3 = -X_1 - 2\zeta_1 X_3 + (A_1 + \varepsilon A_2 X_1 + \varepsilon A_3 X_2) W(\tau) \\
&\dot{X}_4 = -r^2 X_2 - 2\zeta_2 r X_4 + (B_1 + \varepsilon B_2 X_1 + \varepsilon B_3 X_2) W(\tau)
\end{aligned} \tag{vii}$$

The general differential equation of the response moment of order $N=i+j+k+\ell$ is given in the form

$$\begin{aligned}
\dot{m}_{ijk\ell} = {} & i m_{i-1,j,k+1,\ell} + j m_{i,j-1,k,\ell+1} - k(m_{i+1,j,k-1,\ell} + 2\zeta_1 m_{ijk\ell}) \\
& - \ell(r^2 m_{i,j+1,k,\ell-1} + 2\zeta_2 r m_{ijk\ell}) + k(k-1)D(A_1^2 m_{ij,k-1,\ell} \\
& + 2\varepsilon A_1 A_2 m_{i+1,j,k-2,\ell} + 2\varepsilon A_1 A_3 m_{i,j+1,k-2,n} + \varepsilon^2 A_2^2 m_{i+2,j,k-2,\ell} \\
& + 2\varepsilon^2 A_2 A_3 m_{i+1,j+1,k-2,\ell} + \varepsilon^2 A_3^2 m_{i,j+2,k-2,\ell}) \\
& + 2k\ell D\{A_1 B_1 m_{ij,k-1,\ell-1} + \varepsilon(A_1 B_2 + A_2 B_1) m_{i+1,j,k-1,\ell-1} \\
& + \varepsilon(A_1 B_3 + A_3 B_1) m_{i,j+1,k-1,\ell-1} + \varepsilon^2 A_2 B_2 m_{i+2,j,k-1,\ell-1} \\
& + \varepsilon^2 (A_2 B_3 + A_3 B_2) m_{i+1,j+1,k-1,\ell-1} + \varepsilon^2 A_3 B_3 m_{i,j+2,k-1,\ell-1}\} \\
& + \ell(\ell-1)D(B_1^2 m_{ijk,\ell-2} + 2\varepsilon B_1 B_2 m_{i+1,jk,\ell-2} + 2\varepsilon B_1 B_3 m_{i,j+1,k,\ell-2} \\
& + \varepsilon^2 B_2^2 m_{i+2,jk,\ell-2} + 2\varepsilon^2 B_2 B_3 m_{i+1,j+1,k,\ell-2} + \varepsilon^2 B_3^2 m_{i,j+2,k,\ell-2})
\end{aligned} \tag{viii}$$

where $m_{ijk\ell} = E[X_1^i \, X_2^j \, X_3^k \, X_4^\ell]$

The stationary solution for the first and second order response moments are

$$m_{1000} = m_{0100} = m_{0010} = m_{0001} = m_{1010} = m_{0101} = 0$$

(ix-a)

$$m_{0002} = r^2 m_{0200} \, , \; m_{0110} = -m_{1001} \, , \; m_{1001} = \frac{1 - r^2}{2(\zeta_1 + \zeta_2 r)} m_{1100}$$

m_{2000}, m_{0200}, and m_{1100} are given by the solution of the three algebraic equations

$$\begin{bmatrix} a_1 & a_2 & a_3 \\ b_1 & b_2 & b_3 \\ c_1 & c_2 & c_3 \end{bmatrix} \begin{Bmatrix} m_{2000} \\ m_{0200} \\ m_{1100} \end{Bmatrix} = \begin{Bmatrix} a_4 \\ b_4 \\ c_4 \end{Bmatrix} \qquad \text{(ix-b)}$$

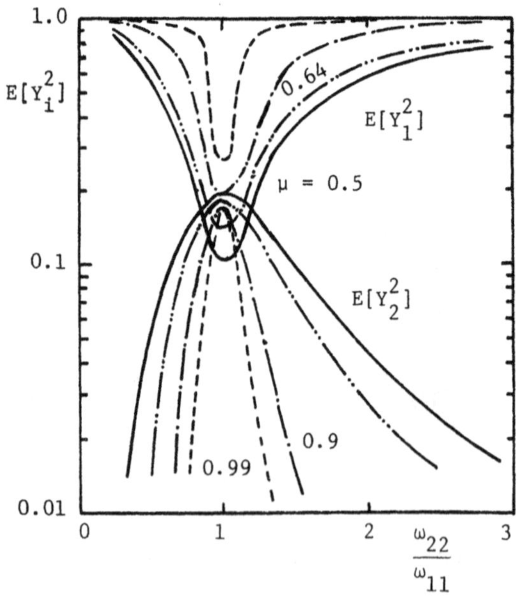

FIG. (8.1) Mean squares of linear coupled system for various values of mass ratio μ

where the constants a_i, b_i, and c_i are functions of the system parameters and the excitation spectral density D.

The solution of (ix-b) for the response mean squares m_{2000} and m_{0200} is shown in fig. (8.1) as function of the frequency ratio ω_{22}/ω_{11} for various values of the mass parameter μ. The influence of the parametric terms is shown in fig. (8.2). It is clear that these terms have very small contribution to the response mean squares.

It is possible to determine the mean squares of the response generalized displacements q_i by using transformation (iii). Figure (8.3) shows the response mean squares in terms of the dimensionless coordinates q_i for various values of the mass parameter μ.

It is seen that the random linear modal analysis results in the well known suppression effect in the neighborhood of frequency ratio $\omega_{22}/\omega_{11}=1$. The mean square responses exhibit strong interaction at frequency ratio $\omega_{22}/\omega_{11}=1$. The degree of interaction depends also upon the mass parameter μ which measures the degree of linear dynamic coupling. As $m_{12} \to 0$ the absorbing effect is decreasing.

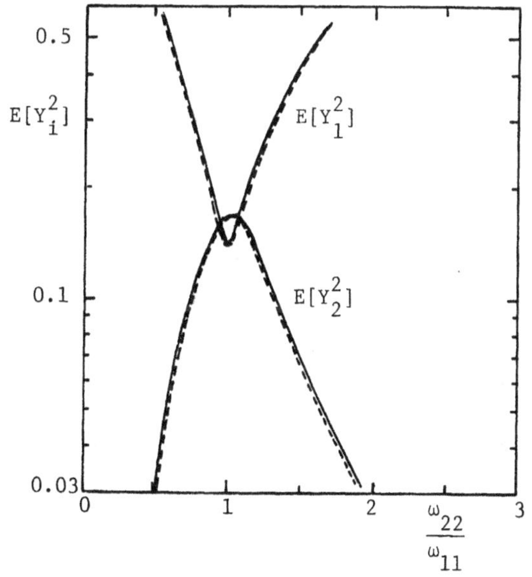

FIG. (8.2) Effect of random parametric excitation on the response for μ = 0.64
——— under external and parametric excitations
- - - - - - under external excitation

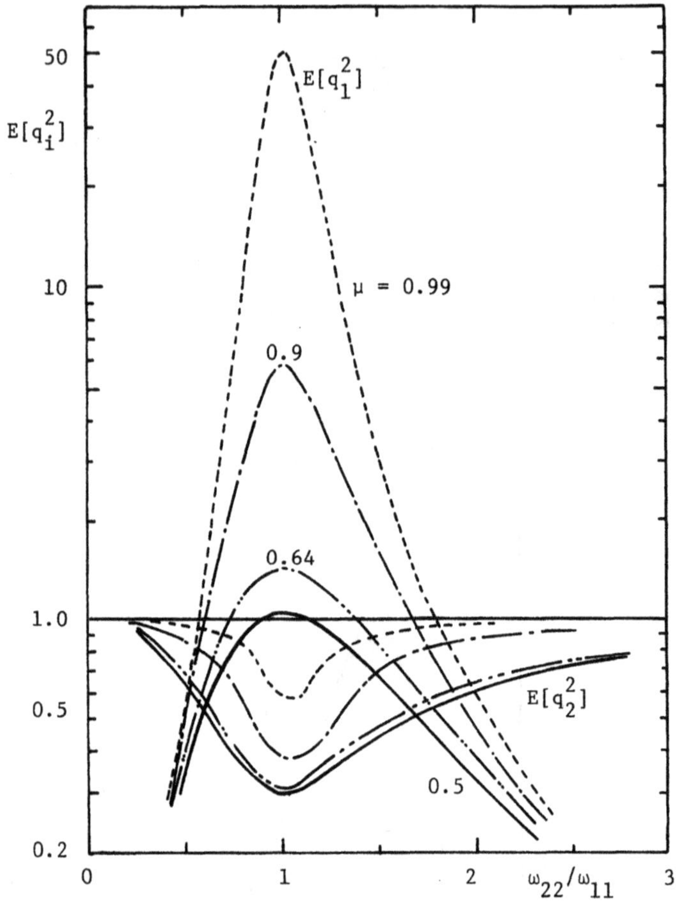

FIG. (8.3) Response mean squares in terms of generalized coordinates

Under random forced and parametric excitations the system response is mainly governed by the external excitation, and the response curves indicate that the influence of random parametric excitation is almost negligible. However, the role of the parametric coefficients is manifested in determining the stochastic stability of the moments.

If the mathematical model of the system response is refined to include the deflection non-linearity, the normal modes will exhibit autoparametric resonance (internal resonance.) However, the resulting moment equations will form an infinite hierarchy coupled set and Gaussian and non-Gaussian closure schemes may be used to solve for the system response statistics. This case will be considered in example 8.6.

8.4 RANDOM RESPONSE OF HELICOPTER ROTOR BLADES

8.4.1 BASIC MECHANICS OF HELICOPTER BLADES

The study of random parametric vibration of helicopter rotor blades in atmospheric turbulent flow is not a simple task, partly because the aerodynamic loading acting on each blade changes significantly in the course of each blade revolution. In addition, there is a possibility of structural dynamic interaction between the blade freedoms.

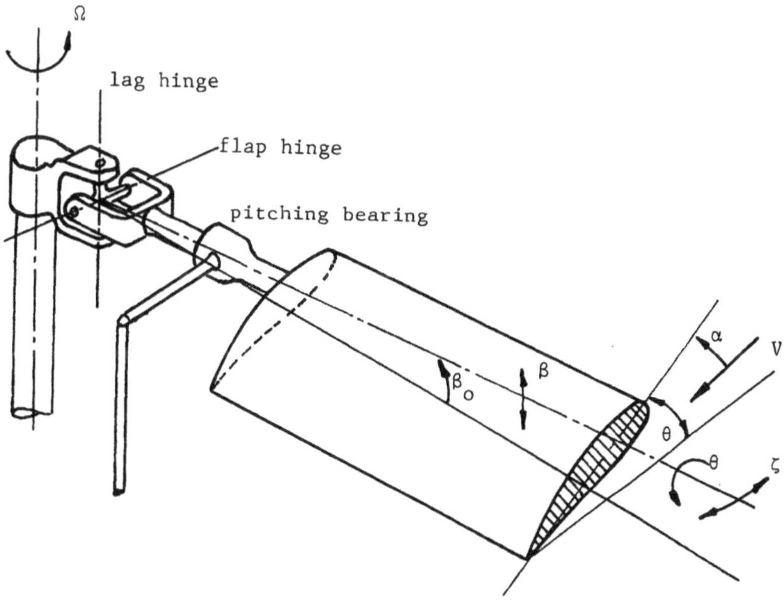

FIG. (8.4) Schematic diagram of an articulated rotor hub (Johnson, 1980)

A brief description of the blade mechanics and helicopter terminology will be given in this section as a guide to those who are not familiar with helicopter mechanics. The reader may consult Johnson (1980) for a comprehensive treatment of helicopter flight mechanics.

Figure (8.4) shows a rotor blade and its basic three motions: the flap angle β, the lag ζ, and the pitch θ (or feathering) motions. In the forward flight regime the blade on the advancing side [see fig. (8.5)] has greater velocity relative to the surrounding air than on the retreating side. If the angle of attack α is assumed constant, the rotor on the advancing side will have more aerodynamic lift than on the retreating side. This situation will result in a rolling moment which prevents steady level flight. The elimination of average rolling and pitching moments at the rotor head is via the cyclic pitch change. The variation of relative air speed across blade sections is balanced on average by a sinusoidal variation of blade pitch. The flapping response results from a combination of aerodynamic and inertia forces (and structural stiffness forces if the blade is "semi-rigid," i.e., not hinged) at any instant.

The ratio of the helicopter velocity component $V\cos\alpha$ to the rotor tip velocity ΩR is called the advance ratio $\mu = V\cos\alpha/\Omega R$.

The flow field relative to the blade consists of three velocity components: the tangential U_t, the radial U_r, and the perpendicular U_p. Let v, u, and w be the turbulent velocity components along the x, y, and z axes, respectively. The net expressions for the three velocity components are given by the expressions (Fujimori, 1978 and Johnson, 1980)

$$U_t = \Omega r + (V + r)\cos\alpha \sin\psi + u \cos\psi$$

$$U_r = (V + r)\cos\alpha \cos\psi - u \sin\psi \tag{8.42}$$

$$U_p = \lambda\Omega R + r\, d\beta/dt + \{(V + v)\cos\alpha \cos\psi - u \sin\psi\}\beta$$

where λ is the rotor inflow ratio given by the expression

$$\lambda = V\sin\alpha/\Omega R + \lambda_i \tag{8.43}$$

$\lambda_i = w/\Omega R$ is the induced inflow ratio, and $\psi = \Omega t$ is the azimuth angle.

It should be noticed that the advancing side contains a region of normal flow for which the air approaches the blade from the leading edge. On the retreating side the air velocity relative to the blade is directed from the trailing edge to the leading edge within a region, known as the reverse flow region, defined by the boundary

$$\Omega R(\mu \sin\psi - r/R) > 0$$

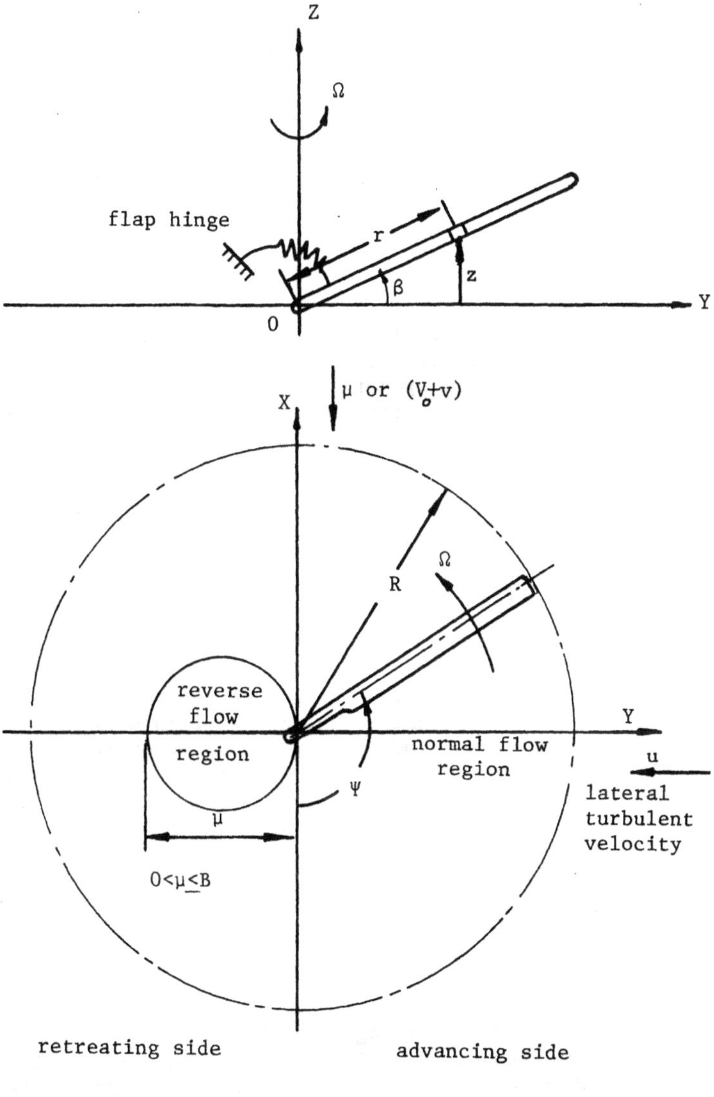

FIG. (8.5) Rotor blade in flapping motion under turbulent flow

This region is shown by a circle of diameter μ centered at $r/R=\mu/2$ on the left side of the y axis. For low advance ratio, the reverse flow region occupies a small region. It was shown by Johnson (1980) that the effects of the reverse flow region are negligible up to $\mu=0.5$. At higher advance ratios the aerodynamic forces of the reversed flow generate appreciable moment about the elastic axis. This moment results in a torsion of the blade which, in turn, interacts with the flapping motion.

For steady-state conditions, the behavior of the blade as it revolves must always be the same at a given azimuth ψ. In other words, the blade loads are periodic with a fundamental frequency Ω and period $T=2\pi/\Omega$. Thus the flap angle may be represented by the Fourier series

$$\beta(\psi) = \beta_0 + \beta_{1c} \cos\psi + \ldots + \beta_{1s} \sin\psi + \ldots \qquad (8.44)$$

where the mean value β_0 is called the coning angle shown in fig. (8.4). This angle results from the reaction to the mean blade lift. The first harmonic β_{1c} generates a variation of the flap angle given by $\beta_{1c} \cos\psi$ once per revolution. Thus the blade out-of-plane deflection is

$$Z_c = r\beta = r\beta_{1c} \cos\psi \qquad (8.45)$$

This deflection will cause the blade to describe a plane tilted forward about the y axis by an angle β_{1c}. Similarly, β_{1s} generates an out-of-plane deflection given by

$$Z_s = r\beta_{1s} \sin\psi \qquad (8.46)$$

which corresponds to a plane tilted to the left (toward the retreating side) about the x axis. The cyclic flap motions β_{1c} and β_{1s} represent the response to moments on the rotor disk. The combination of β_0, β_{1c}, and β_{1s} forms a cone that is tilted laterally and longitudinally. The higher harmonics $\beta_{nc} \cos n\psi$, $\beta_{ns} \sin n\psi$, are usually small ($n>2$), and the rotor flap motion is described primarily by β_0, β_{1c}, and β_{1s}.

The lag motion ζ may also be represented by the first three terms of the expansion

$$\zeta = \zeta_0 + \zeta_{1c} \cos\psi + \ldots + \zeta_{1s} \sin\psi + \ldots \qquad (8.47)$$

where ζ_0 is the mean lag angle of the blade relative to the rotor hub and shaft as shown in fig. (8.6). This angle is due to the reaction to the mean rotor torque. The first harmonic cyclic lag ζ_{1c} produces a lateral shift of the blades to the left. By simple analysis, it can be shown that ζ_{1c} will cause a lateral shift of the rotor center of mass. Similarly, ζ_{1s} produces a longitudinal shift of the blades in the plane of rotation and a longitudinal shift in the rotor center of mass.

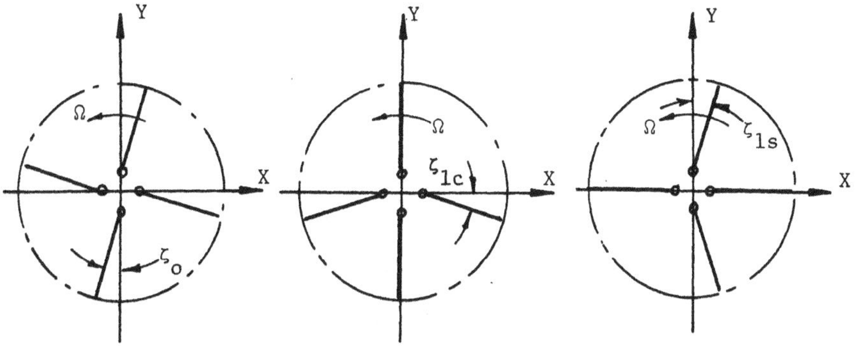

(a) mean lag ζ_o (b) lateral shift ζ_{1c} (c) longitudinal shift ζ_{1s}

FIG. (8.6) The first three blade lag harmonics

The third degree-of-freedom is the blade pitch angle and is represented by the Fourier series

$$\theta = \theta_0 + \theta_{1c} \cos\psi + \ldots + \theta_{1s} \sin\psi + \ldots \tag{8.48}$$

where the mean angle θ_0 is called the collective pitch. This angle controls the average blade force and, hence, the rotor thrust magnitude. The cyclic pitch angles θ_{1c} and θ_{1s} control the tip-path-plane tilt (i.e., 1/revolution flapping) and, hence, the thrust vector orientation. θ_{1c} controls the lateral orientation while θ_{1s} controls the longitudinal orientation. In studying the rotor flapping motion, the aerodynamic moments appear in the equation of motion in terms of a non-dimensional parameter known as the blade lock number γ. This parameter is defined as the ratio of the aerodynamic forces to the inertia forces

$$\gamma = \rho ac\, R^4/I_f$$

where I_f is the mass moment of inertia of the blade about the flap hinge, ρ is the air density, a is the blade lift curve slope, and c is the blade chord.

Wei (1978) derived expressions for θ_0, θ_{1c}, and θ_{1s} for advance ratio $\mu \leq 0.5$ to suppress the first harmonics in flapping for a trimmed motion condition.

For simplicity the aerodynamic forces are estimated from the linear, quasi-steady strip theory. However, at the blade tip the aerodynamic loading drops off and its effects may be estimated by assuming that the blade elements outboard of the radial station r=BR have profile drag but produce no lift. The parameter B is called the "tip loss factor" and is given by the Prandtl formula

$$B = 1 - \frac{1}{n}\sqrt{2C_T} \tag{8.49}$$

where n is the number of blades and C_T is the thrust coefficient

$$C_T = T/\rho A(\Omega R)^2$$

T is the rotor thrust, defined to be normal to the disk plane and positive when directed upward, and A is the rotor disk area.

The study of the random response and stability of helicopter blades in turbulent flow depends mainly on the mathematical modeling of the blade motion and the method of solving the equations of motion. The mathematical modeling may be formulated for one or more blade motions such as (i) the uncoupled flapping motion, (ii) the coupled flap-torsional motion, (iii) the coupled flap-lag motion, and (iv) the coupled flap-lag-torsional motion. The equations of motion are evolved from structural and aerodynamic models which are not considered in this section. The mathematical modeling of various blade motions was established by Sissingh (1968), Sissingh and Kuczynski (1970) and Peters (1975a,b).

8.4.2 UNCOUPLED FLAPPING MOTION

The random response of a rotor blade flapping motion in forward flight is governed by the linearized non-homogeneous differential equation (Sissingh, 1968)

$$\ddot{\beta} + \frac{\gamma}{2} C(\tau)\dot{\beta} + \{K_o^2 + \frac{\gamma}{2} K(\tau)\}\beta = -\{M_\lambda(\tau)\lambda(\tau) + M_\theta(\tau)\theta(\tau)\} \tag{8.50}$$

with the initial condition $\dot{\beta}(\tau=0)=0$. The blade is assumed rigid and centrally hinged with elastic restraint at its root as shown in fig. (8.5). The dot denotes differentiation with respect to the dimensionless time parameter $\tau=\psi=\Omega t$. The stiffness parameter K_o^2 is given by the relation

$$K_o^2 = 1 + (\omega_\beta/\Omega)^2 \tag{8.51}$$

where ω_β is the blade flapping natural frequency due to the elastic restraint. The time varying parameters $C(\tau)$ and $K(\tau)$ are the aerodynamic damping and stiffness coefficients, respectively. The expressions on the right-hand side of equation (8.50) represent the flap moment due to the effective angle of attack changes due to the blade inflow ratio $\lambda(\tau)$ and the collective pitch $\theta(\tau)$. Both $\lambda(\tau)$ and $\theta(\tau)$ are assumed to be stationary ergodic random processes. The functions $M_\lambda(\tau)$ and $M_\theta(\tau)$ are envelope or modulating functions, thus the right-hand side of (8.50) contains non-stationary random excitations. The time dependent coefficients $C(\tau)$, $K(\tau)$, $M_\lambda(\tau)$, and $M_\theta(\tau)$ are listed in table (8.1) for the three flow regions depicted in fig. (8.7).

TABLE (8.1) Coefficients of Equation (8.53), (Sissingh, 1968)

Coefficient	Region	Coefficient Expression
	I	$\frac{1}{4} B^4 + \frac{1}{3} \mu \sin\tau = C_I(\tau)$
$C(\tau)$	II	$C_I(\tau) + \frac{1}{48} \mu^4 (3 - 4\cos 2\tau + \cos 4\tau) = C_{II}(\tau)$
	III	$- C_I(\tau) = C_{III}(\tau)$
	I	$\frac{1}{3} B^3 \mu \cos\tau + \frac{1}{4} B^2 \mu^2 \sin 2\tau = K_I(\tau)$
$K(\tau)$	II	$K_I(\tau) - \frac{1}{24} \mu^4 (2\sin 2\tau - \sin 4\tau) = K_{II}(\tau)$
	III	$- K_1(\tau) = K_{III}(\tau)$
	I	$\frac{1}{3} B^3 + \frac{1}{2} B^2 \mu \sin\tau = M_{\lambda I}(\tau)$
$M_\lambda(\tau)$	II	$M_{\lambda I}(\tau) - \frac{1}{12} \mu^3 (3\sin\tau - \sin 3\tau) = M_{\lambda II}(\tau)$
	III	$- M_{\lambda I}(\tau) = M_{\lambda III}(\tau)$
	I	$\frac{1}{4} B^4 + \frac{2}{3} B^3 \mu \sin\tau + \frac{1}{4} B^2 \mu^2 (1 - \cos 2\tau) = M_{\theta I}(\tau)$
$M_\theta(\tau)$	II	$M_{\theta I}(\tau) - \frac{1}{48} \mu^3 (3 - 4\cos 2\tau + \cos 4\tau) = M_{\theta II}(\tau)$
	III	$- M_{\theta I}(\tau) = M_{\theta III}(\tau)$

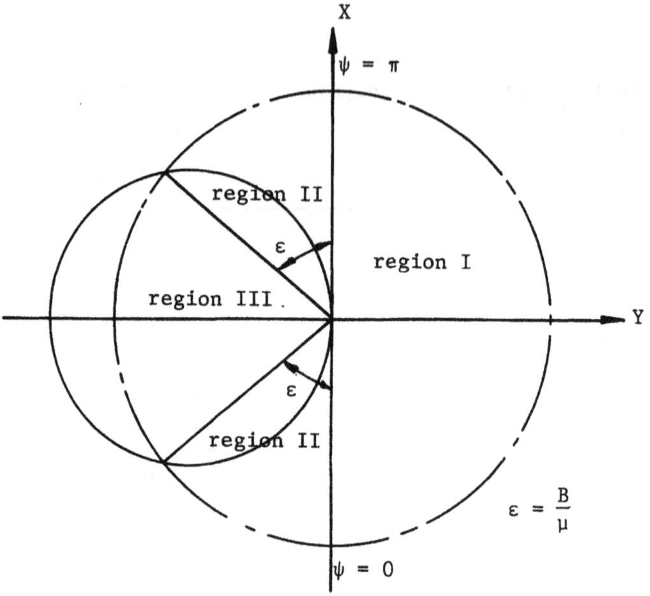

FIG. (8.7) Reversed flow region on the blade rotational plane (B<μ)

Gaonkar and Hohenemser (1971) studied the flapping motion due to the random inflow excitation $M_\lambda(\tau)\lambda(\tau)$. They (1969) and Wan and Lakshmikantham (1973) considered the effect of the change in the angle of attack through a pitch angle θ. In the present analysis the effect of $M_\theta(\tau)\theta(\tau)$ will be examined. The random process $\theta(\tau)$ is assumed to be generated from the white noise shaping filter

$$\dot{\theta}(\tau) + \nu\theta(\tau) = \sqrt{2\nu}\,\sigma W(\tau) \tag{8.52}$$

where $\nu = 2\mu/(L/R)$, L is the longitudinal turbulence scale, and $W(\tau)$ is a white noise with unit spectral density.

It is not difficult to show that $\theta(\tau)$ has zero mean with the steady state correlation function

$$R_{\theta\theta}(\tau+s,\tau) = \sigma^2\exp(-\nu\,|\,s\,|\,) \tag{8.53}$$

Introducing the coordinate transformation $\{X_1, X_2, X_3\} = \{\beta, \dot{\beta}, \theta\}$ equations (8.50) and (8.52) may be written in the state vector form

$$\begin{Bmatrix} \dot{X}_1 \\ \dot{X}_2 \\ \dot{X}_3 \end{Bmatrix} = \begin{bmatrix} 0 & 1 & 0 \\ -\left(K_o^2 + \frac{\gamma}{2} K(\tau)\right) & -\frac{\gamma}{2} C(\tau) & \frac{\gamma}{2} M_\theta(\tau) \\ 0 & 0 & -\nu \end{bmatrix} \begin{Bmatrix} X_1 \\ X_2 \\ X_3 \end{Bmatrix} + \begin{Bmatrix} 0 \\ 0 \\ \sqrt{2\nu}\,\sigma W(\tau) \end{Bmatrix} \tag{8.54}$$

Comparing (8.54) with the standard form (8.1), the matrix $\underline{G}(\tau)$ and the vector $\underset{\sim}{W}(\tau)$ are

$$\underline{G}(\tau) = \begin{bmatrix} 0 & 0 & 0 \\ 0 & 0 & 0 \\ 0 & 0 & \sqrt{2\nu}\,\sigma \end{bmatrix}, \qquad \underset{\sim}{W}(\tau) = \begin{Bmatrix} 0 \\ 0 \\ W(\tau) \end{Bmatrix} \tag{8.55}$$

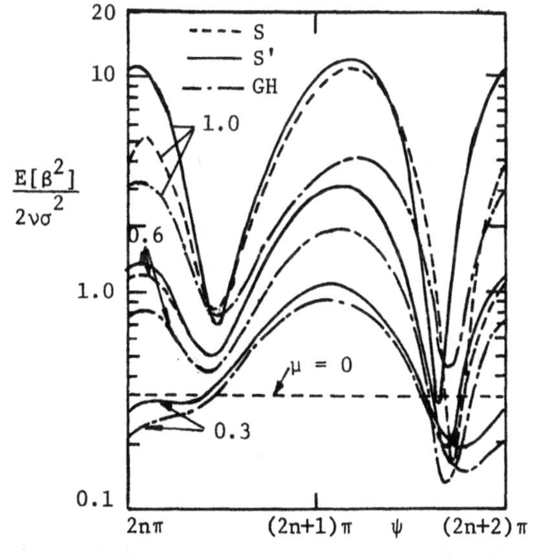

FIG. (8.8) Mean square flapping response for various blade models and μ values. γ = 4.0, ν = 0.5, K_o = 1.0, B = 1.0 (Wan and Lakshmikantham, 1973)

The spectral density matrix (8.2) becomes the identity matrix $\underline{D}=\underline{I}$. In view of the complexity of the coefficients $C(\tau)$, $K(\tau)$, and $\overline{M_{\theta}(\tau)}$, the covariance and autocorrelation functions of the response of system (8.54) may be determined by numerical integration of equations (8.20,22). Wan and Lakshmikantham (1973) obtained the numerical response statistics which are displayed in figs. (8.8-10). These figures show the non-stationary mean square of the flapping angle β and its time rate $\dot{\beta}$ for advance ratios $\mu=0$, 0.3, 0.6, and 1.0. These response curves were evaluated for fixed blade lock number $\gamma=4$, filter constant $\nu=0.5$, $K_o=1$ (hinged blade) and tip loss factor $B=1$. It is seen that the peak values of the response mean squares increase as the advance ratio increases. Figure (8.8) includes the mean square response according to three mathematical models of equation (8.50). These models are: the Sissingh model (denoted by S), the Sissingh model after omitting the effect of the reverse flow from table 8.1 (indicated by S'), and the Gaonkar and Hohenemser (GH) model which does not include the μ^2 terms and the effect of the

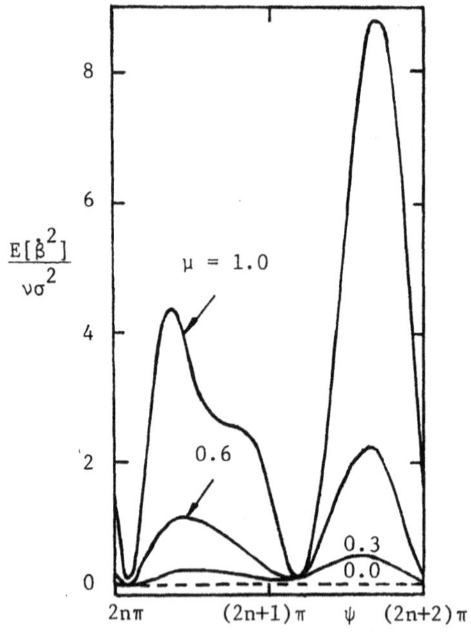

FIG. (8.9) Variance of $\dot{\beta}$ for various μ values, $\gamma = 4.0$, $B = 1.0$, $\nu = 0.5$, $K_o = 1.0$ (Wan and Lakshmikantham, 1973)

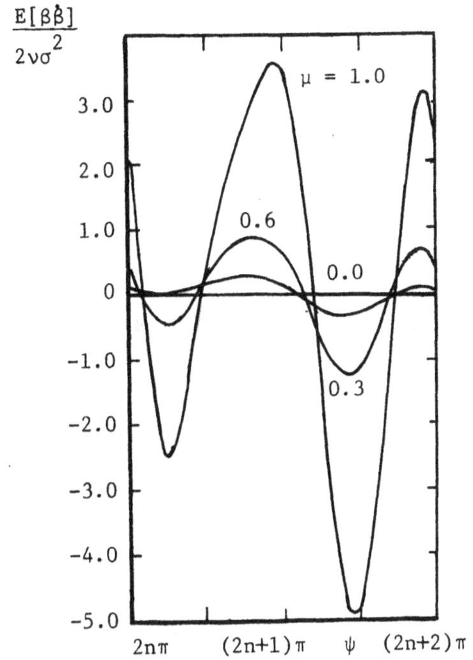

FIG. (8.10) Cross variance of $[\beta(\tau)\dot{\beta}(\tau)]$ for various μ values, $\nu = 0.5$, $\gamma = 4.0$, $B = 1.0$, $K_0 = 1.0$ (Wan and Lakshmikantham, 1973)

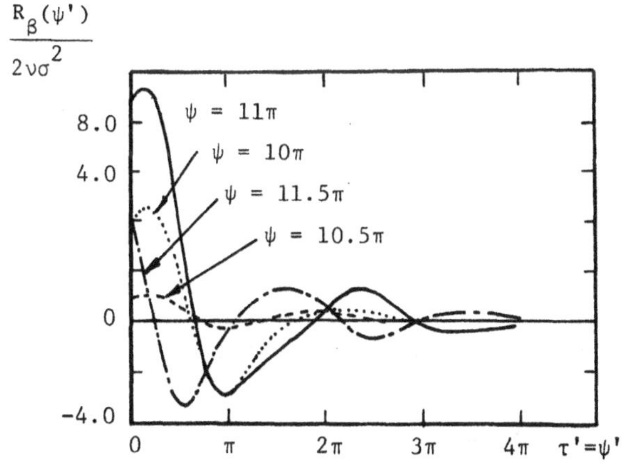

FIG. (8.11) Steady state autocorrelation of $\beta(\tau)$ for $\mu = 1.0$, $\gamma = 4$, $\nu = 0.5$, $B = 1.0$, $K = 1.0$ (Wan and Lakshmikantham, 1973)

reverse flow. It is evident that the higher order terms in μ^2 have a pronounced effect on the peak values. The reverse flow has less effect since it introduces flap motion of order μ^4. The mean square of the flapping velocity $E[\dot{\beta}^2(\tau)]$ and the joint moment $E[\beta(\tau)\dot{\beta}(\tau)]$, as evaluated according to the Sissingh model, are shown in figs. (8.9) and (8.10), respectively. Figure (8.11) demonstrates the variation of the response correlation function of the flapping angle for various values of the azimuth angle $\psi=\tau$. It is seen that the correlation becomes significantly weaker for correlation intervals that exceed one revolution.

8.4.3 COUPLED FLAPPING-TORSIONAL RANDOM RESPONSE

At relatively high advance ratio the aerodynamic forces of the reversed flow region generate appreciable moments about the blade elastic axis. These moments result in torsion of the blade which, in turn, interact with the flapping motion. The equations of motion of coupled flapping-torsional response were derived by Sissingh and Kuczynski (1970). The blade is assumed rigid in bending. Furthermore, the elastic axis, the center of mass, and the shear center of the blade section are assumed to coincide with the 25% chord line. It is also postulated that the torsion mode can be represented by a straight line through the rotor center ($\Delta\theta=r\delta/R$, where $\Delta\theta$ is the pitch angle change due to blade torsion, and δ is the torsional deflection of the blade tip.) Accordingly, the equations of motion are obtained in the form (Lin, et al., 1979)

$$\begin{Bmatrix} \ddot{\beta} \\ \ddot{\delta} \end{Bmatrix} + \frac{\gamma}{2}\begin{bmatrix} C(\tau) & 0 \\ 6Q\ell_{r\dot{\beta}}(\tau) & 6FC_{\delta}(\tau) \end{bmatrix}\begin{Bmatrix} \dot{\beta} \\ \dot{\delta} \end{Bmatrix} + \begin{bmatrix} K_o^2 + \frac{\gamma}{2}K(\tau) & -\frac{\gamma}{2}m_{\delta}(\tau) \\ 3\gamma Q\ell_{r\beta}(\tau) & \omega_{\delta}^2 + 3\gamma QK_{\delta}(\tau) \end{bmatrix}\begin{Bmatrix} B \\ \delta \end{Bmatrix}$$

$$= \begin{bmatrix} M_{\beta D} & M_{\beta u} & M_{\beta v} & M_{\beta\lambda} \\ M_{\delta D} & M_{\delta u} & M_{\delta v} & M_{\delta\lambda} \end{bmatrix}\begin{Bmatrix} 1 \\ \tilde{u}(\tau) \\ \tilde{v}(\tau) \\ \lambda(\tau) \end{Bmatrix} \tag{8.56}$$

where $\{\tilde{u},\tilde{v},\lambda\} = \frac{1}{\Omega R}\{u,v,w\}$ are the non-dimensional lateral, longitudinal, and induced vertical turbulence velocity components, respectively. The coefficients $C(\tau)$, $K(\tau)$, and $M_{\beta\lambda}$ are the same as those listed in table (8.1). The remaining coefficients are given by Fujimori, et al., (1979). $M_{\beta D}$ and $M_{\delta D}$ are the flapping and torsional deterministic forcing functions. ω_{δ} represents the natural torsional frequency per Ω. The constants F and Q are given by the expressions:

$$F = (I_p/I_f)(C/4R), \qquad Q = F(C/4R) \tag{8.57}$$

where I_p and I_f are the mass moment of inertia about pitching and flapping axes, respectively, and C is the blade chord.

$\tilde{u}(\tau)$, $\tilde{v}(\tau)$, and $\lambda(\tau)$ are assumed Gaussian random processes. If these terms are removed from equations (8.56), the resulting motion will represent either a trimmed (case of zero pitching moments) or untrimmed deterministic motion. On the other hand, if these terms are retained, they may be modeled as Gaussian white noise processes or can be generated from white noise filters. The response of the two cases will constitute a Markov vector process. With regard to the first modeling, Lin, et al., (1979) gave a qualitative analysis to justify the delta-correlation representation. When the rotor blade rotates at a speed much greater than the convection speed of turbulence, an observer moving with the blade will sense very rapid flow velocity even when the turbulence pattern itself is changing slowly. Fuh, et al., (1983) obtained the coupled flap-torsional response when $\tilde{u}(\tau)$, $\tilde{v}(\tau)$, and $\lambda(\tau)$ are represented by white noise processes. Gaonkar, et al., (1972) considered the case of turbulent velocity components generated from white noise linear filters. The blade responses of the two cases will be outlined.

Case i: Filtered White Noise Excitations

For simplicity the analysis will be restricted to a single random excitation belonging to the vertical turbulent velocity $\lambda(\tau)$ described by the linear filter

$$\dot{\lambda} + \nu\lambda = \sqrt{2\nu}\,\sigma_\lambda W(\tau) \tag{8.58}$$

Accordingly, the Markov vector is

$$\{X_1, X_2, X_3, X_4, X_5\} = \{\beta, \dot{\beta}, \delta, \dot{\delta}, \lambda\} \tag{8.59}$$

Equations (8.56) and (8.58) can be written in the vector form (8.1). The covariance matrix of the response vector $\underset{\sim}{X}$ is governed by the set of differential equations (8.20) which can be solved numerically. Gaonkar (1974a) determined the response statistics for the system constants μ=1.6, B=0.97, γ=4, K_0=1 and 1.3, and torsional natural frequencies ω_δ=8 and 10. The velocity of vertical turbulence is assumed to have a unit standard deviation and a typical longitudinal turbulence scale ratio (L/R)=12. By omitting the coupling terms in equations (8.56), the resulting responses β and δ are referred to as pure flapping and pure torsional motions, respectively. Figure (8.12) displays two response regimes for four combinations of K_0 and ω_δ. The first regime, given by solid curves, shows the effect of coupling on the blade response as compared by the second regime, indicated by dotted curves, which represents the uncoupled motions. It is seen that the coupling has negligible effect on the maximum standard deviation of the flapping motion. On the other hand, the coupling has a strong effect on the torsion response. It may be noticed that these results do not reveal the well known phenomenon of

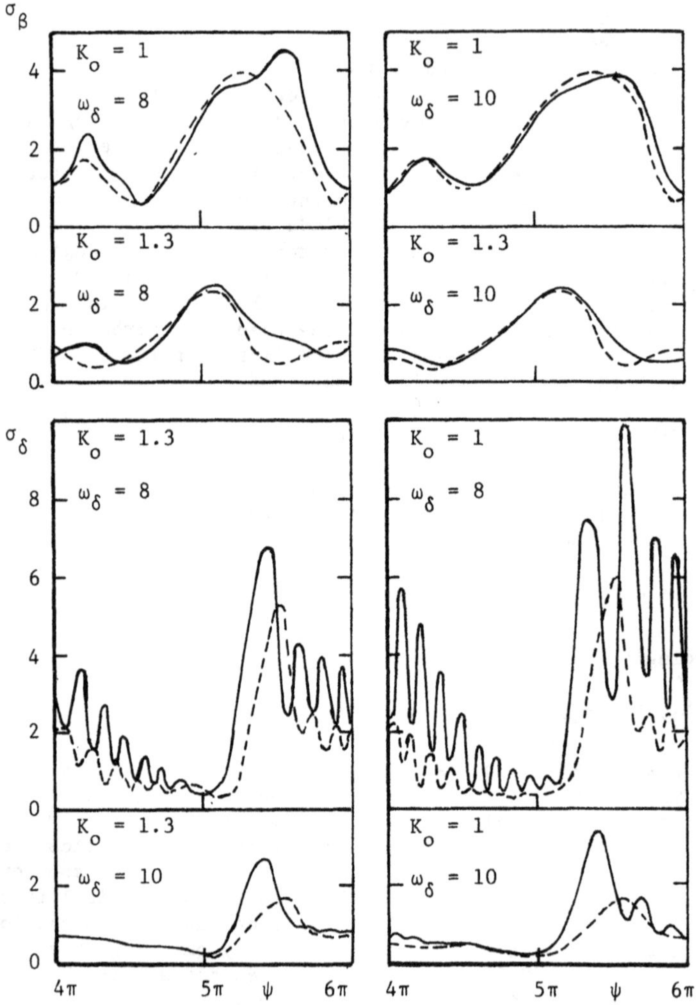

FIG. (8.12) **Standard deviations of flapping and torsional responses**
——— with flapping-torsion coupling
no coupling (Gaonkar, 1974)

vibration absorber in linear coupled systems as shown in example 8.2. The absence of absorber characteristics may be attributed to the fact that the natural frequencies of the system were remote from the conditions of linear vibration absorbers.

Case ii: White Noise Velocity Fluctuations

The actual elements of the damping and stiffness matrices given in equations (8.56) include components which are functions of the horizontal turbulence velocities $\tilde{u}(\tau)$ and $\tilde{v}(\tau)$. These random elements are given by Lin, et al., (1979) who showed that equations (8.56) involve periodic and random parametric coefficients in addition to the deterministic and random non-homogeneous terms. In this case the response coordinate vector is

$$\{X_1, X_2, X_3, X_4\} = \{\beta, \dot{\beta}, \delta, \dot{\delta}\}$$

The blade equations of motion (8.53) can be replaced by the first order differential equations

$$\dot{\underset{\sim}{X}} = \underline{A}(\tau)\underset{\sim}{X} + \underline{F}(\tau)\underset{\sim}{X} + \underset{\sim}{a}(\tau) + \underset{\sim}{f}(\tau) \tag{8.60}$$

where the matrix $\underline{A}(\tau)$ and the vector $\underset{\sim}{a}(\tau)$ include periodic elements, while $\underline{F}(\tau)$ and $\underset{\sim}{f}(\tau)$ contain random functions multiplied by modulated periodic functions. The matrices $\underline{A}(\tau)$ and $\underline{F}(\tau)$ represent deterministic and random parametric excitations, respectively. The elements of matrix $\underline{F}(\tau)$ are functions of the turbulent velocities $\tilde{u}(\tau)$ and $\tilde{v}(\tau)$. These parametric excitations govern the stability characteristics of the blade. The non-homogeneous excitations $\underset{\sim}{a}(\tau)$ and $\underset{\sim}{f}(\tau)$ do not affect the stability; however, they actually govern the level of the steady state response. The elements of the excitation vector $\underset{\sim}{f}(x)$ depend on the three turbulent flow velocities: $\tilde{u}(\tau)$, $\tilde{v}(\tau)$, and $\tilde{\lambda}(\tau)$

It is possible to write equation (8.60) in the form

$$\dot{\underset{\sim}{X}} = \underline{A}(\tau)\underset{\sim}{X} + \underline{G}(\underset{\sim}{X},\tau)\underset{\sim}{\xi}(\tau) + \underset{\sim}{a}(\tau) \tag{8.61}$$

where the vector $\underset{\sim}{\xi}(\tau)$ consists of the three turbulent flow velocities. The matrix $\underline{G}(X,\tau)$ is related to $\underline{F}(\tau)\underset{\sim}{X}$ and $\underset{\sim}{f}(\tau)$ through the following relations:

$$G_{i1} = F_{ik,\tilde{u}}\, X_k + f_{i\tilde{u}}\,, \qquad G_{i2} = F_{ik,\tilde{v}}\, X_k + f_{i\tilde{v}}$$

$$G_{i3} = f_{i\lambda}\,, \qquad G_{i4} = 0$$

The physical system (8.61) can be transformed into the Itô type equations (after including the Wong-Zakai correction terms)

$$dX_i = f_i(\underset{\sim}{X},\tau)\, d\tau + \sigma_{ij}(\underset{\sim}{X},\tau)\, dB_j(\tau) \tag{8.62}$$

where the drift and diffusion coefficients are given by the following expressions, respectively

$$f_i(\underset{\sim}{X},\tau) = A_{ij}(\tau)X_j + a_i(\tau) + D_{jk}[\frac{\partial}{\partial X_\ell} G_{ij}(\underset{\sim}{X},\tau)]G_{\ell k}(\underset{\sim}{X},\tau) \tag{8.63}$$

$$\sigma_{i\ell}\sigma_{j\ell} = 2D_{kr}G_{ik}(\underset{\sim}{X},\tau)G_{jr}(\underset{\sim}{X},\tau) \tag{8.64}$$

where $E[\xi_j(\tau)\xi_k(\tau+\tau')] = 2D_{jk}\delta(\tau')$

It is possible to show that the first order moment equations are given in the form

$$\dot{m}_i = E[f_i] = \bar{A}_{ij}E[X_j] + a_i + e_i \qquad i,j=1,2,3,4 \tag{8.65}$$

The elements of the stability matrix $\underline{A}$ are given by Fujimori, et al., (1979). The non-homogeneous terms a_i and e_i are due to the deterministic and stochastic excitations, respectively. The elements of e_i are obtained by the stochastic averaging and are given by the relation

$$e_i = F_{ij,r}\, f_{j,s}\, D_{rs} \tag{8.66}$$

Since the turbulent velocity components have zero means, equations (8.65) give the deterministic response which is governed by selected trim condition or the pitch control parameters.

The second order moment equations are, (k+r=2)

$$\dot{m}_{kr} = \bar{A}_{kj}m_{jr} + \bar{A}_{rj}m_{jk} + 2Q_{kjr\ell}m_{j\ell} + (e_k + f_k)m_r$$
$$+ (e_r + f_r)m_k + 2(Q_{kjr} + Q_{rjk})m_j + 2Q_{kr} \tag{8.67}$$

where $Q_{ijk\ell} = F_{ij,m}\, F_{k\ell,n}\, D_{mn} = Q_{k\ell ij}$, $\qquad Q_{ijk} = F_{ij,m}\, f_{k,n}\, D_{mn}$

$$Q_{ij} = f_{i,m}\, f_{j,m}\, D_{mn} = Q_{ji}$$

There are ten second order moment equations coupled with first order moments. It may be noticed that equations (8.65) are independent of second order moments and the two sets are closed. In this case the stability and the steady-state response of first and second order moments can be examined numerically. Within the stability domain of first and second order moments, the response moments converge to periodic solutions of the steady-state. The results of the numerical integration (reported by Fuh, et al., (1983)) showed that the turbulent flow does not change significantly the average response (first order moments). The mean square responses were found to be affected by the spectral density levels of the random flow

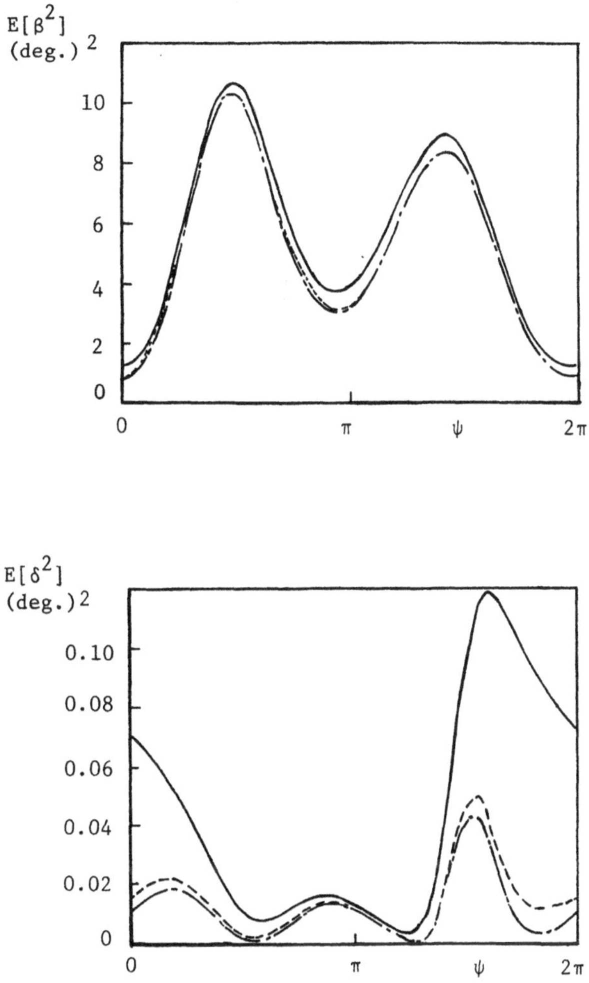

FIG. (8.13) Effect of turbulence spectral levels on flapping and torsion mean squares (C_T = 0.005, γ = 5.0, K_o = 10, ω_δ = 1.2)

——— $D_{11} = D_{22} = 3\pi\times10^{-5}$, $D_{33} = \pi\times10^{-4}$

- - - - - $D_{11} = D_{22} = 3\pi\times10^{-6}$, $D_{33} = \pi\times10^{-5}$

—— - —— $D_{11} = D_{22} = D_{33} = 0$ (Fuh, et al., 1983)

velocity components. Figure (8.13) shows the mean squares of flapping and torsional response under two different sets of turbulent flow levels as compared with the ideal case of no turbulence. Since the mean values of the response are unaffected by the turbulent flow components, the differences between the turbulence and non-turbulence mean square responses represent the variance. It is seen that for the higher turbulence spectral level $D_{\lambda\lambda}=0.0001\pi$ the standard deviation of the flapping angle reaches a maximum value of 0.88° at $\tau=\psi=\pi$. The maximum standard deviation of the torsional angle is 0.3° at $\psi=1.7\pi$ which occurs within the reversed flow region.

8.4.4 FLAPPING RESPONSE TO RANDOM FIELD EXCITATION

The flapping motion of a rotor blade subjected to a random loading, which varies along the blade span with a correlation length less than or equal to the blade length, was investigated by Wan and Lakshmikantham (1974) and Wan (1980). They employed the method of spatial correlation to determine the response statistics. The transverse motion of a flexible blade is governed by the dimensionless partial differential equation

$$\frac{\partial^2 z}{\partial \tau^2} + \frac{\gamma}{6}\left| x + \mu \sin\tau \right| \frac{\partial z}{\partial \tau} + L_{x\tau}[z] = F(x,\tau), \quad 0<x<1, \ \tau>0 \tag{8.68}$$

where

$$L_{x\tau}[\ ] = \frac{EI}{mR^4\Omega^2}\frac{\partial^4}{\partial x^4}[\ ] - \frac{1}{2}(1 - x^2)\frac{\partial^2}{\partial x^2}[\ ] + (x + \frac{\gamma}{6}\mu\cos\tau \left| x + \mu\sin\tau \right|)\frac{\partial}{\partial x}[\ ] \tag{8.69}$$

$$x = r/R, \qquad z = Z/R,$$

and m is the blade mass per unit length.

The external excitation $F(x,\tau)$ is considered to be due to the vertical turbulent inflow and is described by the expression

$$F(x,\tau) = \frac{\gamma}{6}\left| x + \mu\sin\tau \right| \lambda(x,\tau) \tag{8.70}$$

The blade is assumed to experience no transverse motion at time $\tau=0$, i.e.,

$$z(x,0) = \frac{\partial}{\partial \tau} z(x,0) = 0 \tag{8.71}$$

For a hinged blade at the axis of rotation the following boundary conditions are applied

$$z(0,\tau) = \left.\frac{\partial^2 z}{\partial x^2}\right|_{x=0} = \left.\frac{\partial^2 z}{\partial x^2}\right|_{x=1} = \left.\frac{\partial^3 z}{\partial x^3}\right|_{x=1} = 0 \tag{8.72}$$

Equations (8.68,71,72) constitute the initial-boundary value problem of the blade in flapping motion. The random excitation $\lambda(x,\tau)$ is a zero mean process with known statistics. Since equation (8.68) is linear, the response $z(x,\tau)$ will have zero mean, but its second order statistical functions have to be determined. In particular, the following four spatial correlation functions will be considered

$$R_{zz}(x,y,\tau) = E[z(x,\tau)z(y,\tau)], \quad R_{z\dot{z}}(x,y,\tau) = E[z(x,\tau)\dot{z}(y,\tau)]$$
$$R_{\dot{z}z}(x,y,\tau) = E[\dot{z}(x,\tau)z(y,\tau)], \quad R_{\dot{z}\dot{z}}(x,y,\tau) = E[\dot{z}(x,\tau)\dot{z}(y,\tau)] \tag{8.73}$$

where dot denotes partial differentiation with respect to the time parameter τ, and y is the value of x at different azimuth angle. These functions contain the mean square response when $x=y$. On the other hand, they serve as initial conditions for another initial-value problem which leads to determine the autocorrelation function

$$R_z(x_1,\tau_1;x_2,\tau_2) = E[z(x_1,\tau_1)z(x_2,\tau_2)] \tag{8.74}$$

It is possible to derive a set of first order differential equations for the spatial correlation functions of the following type

$$\frac{\partial}{\partial\tau} R_{zz} = E[\dot{z}(x,\tau)z(y,\tau)] + E[z(x,\tau)\dot{z}(y,\tau)] = R_{\dot{z}z} + R_{z\dot{z}}$$
$$\frac{\partial}{\partial\tau} R_{\dot{z}z} = R_{\ddot{z}z} + R_{\dot{z}\dot{z}} \tag{8.75}$$

These relations are obtained by using the property that differentiation and ensemble averaging commute, which is legitimate within the framework of mean square convergence. Eliminating $\ddot{z}$ from (8.75) by using (8.68) gives

$$\frac{\partial}{\partial\tau} R_{\dot{z}z} = R_{\dot{z}\dot{z}} - \frac{\gamma}{6}\left| x + \mu\sin\tau \right| (R_{\dot{z}z} - R_{\lambda z}) - L_{x\tau}[R_{zz}] \tag{8.76}$$

where $R_{\lambda z} = E[\lambda(x,\tau)z(y,\tau)]$

Similarly,

$$\frac{\partial}{\partial\tau} R_{z\dot{z}} = R_{\dot{z}\dot{z}} - \frac{\gamma}{6}\left| y + \mu\sin\tau \right| (R_{z\dot{z}} - \hat{R}_{\lambda z}) - L_{y\tau}[R_{zz}] \tag{8.77}$$

and

$$\frac{\partial}{\partial\tau} R_{\dot{z}\dot{z}} = - L_{x\tau}[R_{z\dot{z}}] - L_{y\tau}[R_{\dot{z}z}] - \frac{\gamma}{6}\left(\left| x + \mu\sin\tau \right| + \left| y + \mu\sin\tau \right|\right)R_{\dot{z}\dot{z}}$$
$$+ \frac{\gamma}{6}\left| x + \mu\sin\tau \right| R_{\lambda\dot{z}}(x,y,\tau) + \frac{\gamma}{6}\left| y + \mu\sin\tau \right| R_{\lambda\dot{z}}(y,x,\tau) \tag{8.78}$$

where

$$\hat{R}_{\lambda z} = E[\lambda(y,\tau)z(x,\tau)], \qquad R_{\lambda\dot{z}} = E[\lambda(x,\tau)\dot{z}(y,\tau)]$$

The differential equations of the spatial correlation functions given by (8.75) through (8.78) must be satisfied in the interior of the semi-infinite unit square column $(0<(x,y)<1,\ \tau > 0)$. On the base of the column, and at $\tau=0$, the initial condition (8.71) can be recast in terms of the spatial correlation functions

$$\underline{R}(x,y,0) = 0, \text{ where } \underline{R} = \begin{bmatrix} R_{zz} & R_{z\dot{z}} \\ R_{\dot{z}z} & R_{\dot{z}\dot{z}} \end{bmatrix}, \quad 0 \le (x,y) \le 1 \tag{8.79}$$

The boundary conditions on the four walls of the column are obtained from (8.72,73)

$$\underline{R}(0,y,\tau) = \frac{\partial^2}{\partial x^2}\underline{R}(0,y,\tau) = \frac{\partial^2}{\partial x^2}\underline{R}(1,y,\tau) = \frac{\partial^3}{\partial x^3}\underline{R}(1,y,\tau)=0,(0\le y<1,\tau>0)$$

$$\underline{R}(x,0,\tau) = \frac{\partial^2}{\partial y^2}\underline{R}(x,0,\tau) = \frac{\partial^2}{\partial y^2}\underline{R}(x,1,\tau) = \frac{\partial^3}{\partial y^3}R(x,1,\tau)=0,(0\le \underline{x}\le 1,\tau>0)$$

(8.80)

It is noticed that the four differential equations (8.75) through (8.78) contain six unknowns. These are the four spatial correlation functions and the response-excitation correlation functions $R_{\lambda z}$ and $R_{\lambda\dot{z}}$. Thus another two equations for $R_{\lambda z}$ and $R_{\lambda\dot{z}}$ must be derived. These two equations will depend on the random properties of the excitation $\lambda(x,\tau)$. At this stage $\lambda(x,\tau)$ is assumed to be generated from the linear filtered white noise equation

$$\dot{\lambda} + \nu\lambda = \sqrt{2\nu}\, W(x,\tau) \tag{8.81}$$

where $W(x,\tau)$ is a temporally uncorrelated random field process with autocorrelation function

$$R_w = E[W(x_2,\tau_2)W(x_1,\tau_1)] = R_s(x_2,x_1)\delta(\tau_2 - \tau_1) \tag{8.82}$$

and the autocorrelation function of the physical random field $\lambda(x,\tau)$ is exponentially correlated in time

$$E[\lambda(x_2,\tau_2)\lambda(x_1,\tau_1)] = R_s(x_2,x_1)\exp\{-\nu\,|\tau_2 - \tau_1|\} \tag{8.83}$$

By using relations (8.28) and (8.31) we write

$$E[W(y,\tau')z(x,\tau)] = \frac{1}{2}\int_0^1 G(x,\tau;x',\tau')\, R_s(y,x')\, dx'$$

$$= 0 \qquad t' \geq \tau > 0, \text{ and } 0 \leq (x,y) \leq 1 \qquad (8.84)$$

which is a consequence of the condition $G(x,t'+;x',t')=0$ as verified by Wan (1972). It is also possible to write

$$E[W(y,\tau')\dot{z}(x,\tau)] = 0 \qquad (8.85)$$

Now the two differential equations for $R_{\lambda z}$ and $R_{\lambda\dot{z}}$ can be written by using (8.78,81,83)

$$\frac{\partial}{\partial\tau} R_{\lambda z}(x,y,\tau) = E[\dot{\lambda}(x,\tau)z(y,\tau)] + E[\lambda(x,\tau)\dot{z}(y,\tau)]$$

$$= -\nu R_{\lambda z} + R_{\lambda\dot{z}} \qquad (8.86)$$

$$\frac{\partial}{\partial\tau} R_{\lambda\dot{z}}(x,y,\tau) = -\left(\nu + \frac{\gamma}{6}\left| y + \mu\sin\tau \right|\right)R_{\lambda\dot{z}} - L_{y\tau}[R_{\lambda z}(x,y,\tau)]$$

$$+ \frac{\gamma}{6}\left| y + \mu\sin\tau \right| R_s(x,y) \qquad (8.87)$$

where $\dot{z}$ and $\dot{\lambda}$ can be eliminated by using equations (8.73,86) and conditions (8.84,85). $R_s(x,y)$ may be expressed by the exponential form

$$R_s(x,y) = \exp\{-\varepsilon(x - y)\} \qquad (8.88)$$

where $\varepsilon \geq 0$ and $1/\varepsilon$ represents the load correlation length.

The flapping random motion of rotor blades was examined by Wan and Lakshmikantham (1974) and Wan (1973b,1974a,b) for three cases of blade rigidity. These were the purely rigid blade case, the zero-stiffness blade case, and the elastic blade case $0 < EI < \infty$.

For the case of a rigid blade flapping motion in hovering flight, the means and mean squares of flapping displacement and velocity were found to decrease as the spanwise correlation length decreases. When the blade was in forward flight, the time and spatial correlation parameters (ν,ε) had a substantial effect on the peaks of flapping mean squares. Any increase in ν or ε was accompanied by a reduction in the mean square response. Wan (1980) reported that the peaks grew in a non-linear fashion with the lock number γ and the growth rate depended significantly on the advance ratio.

The flapping motion of a zero-stiffness blade (string model) is also affected by the spatial correlation parameter ε. Any reduction in ε results in a reduction of the peak value of the displacement mean square. However, the effect is reversed for the velocity mean square response.

For elastic blade $0<EI<\infty$ experiencing random pitching motion, the maximum mean square displacement and velocity were found to decrease as the effective bending stiffness factor ($EI/m\ell^4\Omega^2$) increased. The mean square response parameters reached their limiting value as the effective stiffness parameter increased beyond unity. For blades in practical applications the stiffness factor ranges from 0.01 to 0.6 with a typical value of 0.06. In this case Wan (1973c) indicated that the simplified solution of a rigid blade subjected to spatially uniform random excitation may well be off by 25% or more in the mean square response when the spanwise load correlation length is of the order of the blade length. The difference increases as the correlation length decreases.

8.5 RANDOM FORCED RESPONSE NEAR COMBINATION PARAMETRIC RESONANCE

The influence of combination parametric resonance on the random response of linear systems will be examined in this section. The dynamic response of these systems is usually described by the set of differential equations

$$\ddot{Y}_i + 2\sum_{j=1}^{n}\zeta_{ij}\dot{Y}_j + \omega_i^2[Y_i + \sum_{j=1}^{n}\mu_{ij}Y_j\sin 2\Omega t] = \xi_i(t) \tag{8.89}$$

The external excitations $\xi_i(t)$ are assumed stationary Gaussian random processes. For two degree-of-freedom systems ($i=1,2$) equation (8.89) may exhibit combination resonance of the type $\Omega = |\omega_1 \pm \omega_2|/2$. The following transformation for the response coordinates is introduced

$$\begin{aligned} Y_i(t) &= X_{ic}(t)\cos\nu_i t + X_{is}(t)\sin\nu_i t \\ \dot{Y}_i(t) &= \nu_i\{-X_{ic}(t)\sin\nu_i t + X_{is}(t)\cos\nu_i t\} \end{aligned} \tag{8.90}$$

where X_{ic} and X_{is} are slowly varying amplitudes and the frequencies ν_i satisfy the relations

$$\Omega \cong |\nu_1 \pm \nu_2|/2 \tag{8.91}$$

Carrying out the averaging method for the transformed equations of motion, the following set of state-vector equations is obtained

$$\dot{\underset{\sim}{X}}(t) = \underline{A}\,\underset{\sim}{X} + \underset{\sim}{W}(t) \tag{8.92}$$

where

$$\underline{A} = \begin{bmatrix} -\zeta_{11} & \delta_1 & \gamma_{12} & 0 \\ -\delta_1 & -\zeta_{11} & 0 & \mp\gamma_{12} \\ \pm\gamma_{21} & 0 & \zeta_{22} & \delta_2 \\ 0 & -\gamma_{21} & -\delta_2 & \zeta_{22} \end{bmatrix}, \tag{8.93}$$

$$\underset{\sim}{X}^T = \{X_{1c}, X_{1s}, X_{2c}, X_{2s}\}, \qquad \underset{\sim}{W}^T(t) = \{W_{1c}, W_{1s}, W_{2c}, W_{2s}\}$$

$$\delta_i = (\omega_i - \nu_i)/2\nu_i, \qquad \gamma_{ij} = \omega_i^2\mu_{ij}/4\nu_i$$

W_{ic} and W_{is} are uncorrelated white noise processes whose spectral densities are related to the densities of ξ_i by the relation

$$D_{ic} = D_{is} = D_i = \pi S_{\xi_i \xi_i}(\nu_i)/\nu_i^2 \tag{8.94}$$

The upper sign in matrix $\underline{A}$ refers to a sum combination resonance, while the lower sign is associated with the difference resonance. Introducing the detuning parameter

$$\Delta_i = \omega_i - \nu_i \cong \delta_i \tag{8.95}$$

where Δ_1 and Δ_2 are related by the expression

$$\Delta_1 \pm \Delta_2 = (\omega_1 \pm \omega_2) - 2\Omega \tag{8.96}$$

The stationary solution of system (8.92) may be obtained by assuming that the equilibrium solution $\underset{\sim}{X}=\underset{\sim}{0}$ is stable when $W_{ic}(t)=W_{is}(t)=0$. The stability condition of this solution was given by Dimentberg and Isikov (1983) by the inequality

$$|\gamma_{12}\gamma_{21}| < \zeta_{11}\zeta_{22}\{1 + [(\delta_1 + \delta_2)/(\zeta_{11} + \zeta_{22})]^2\} \tag{8.97}$$

The direct method outlined in section 8.2.1 may be employed to establish the differential equations of covariance and correlation functions of the response

$$\dot{\underline{C}}_x(t) = \underline{A}\underline{C}_x(t) + \underline{C}_x\underline{A}^T + \underline{D}_w \tag{8.98}$$

$$\dot{\underline{R}}_{xx}(t,t_1) = \underline{A}\underline{R}_{xx}(t,t_1), \quad t_1>t, \quad \underline{R}_{xx}(t_0,t_0) = \underline{C}_x(t_0) \tag{8.99}$$

The stationary solution of (8.98) may be obtained by setting the left-hand side to zero. Integrating (8.99) with the stationary solution of (8.98) taken as initial condition results in a complete description for the response statistics.

For zero detuning parameters $\Delta_1=\Delta_2=0$, Dimentberg and Isikov (1983) obtained the analytical solution

$$C_{ic} = C_{is} = \frac{D_i}{2(\zeta_{ii}+\zeta_{ij})}\left\{1+\frac{\zeta_{jj}^2+\gamma_{ij}^2 D_j/D_i}{\zeta_{ii}\zeta_{jj}\mp\gamma_{ij}\gamma_{ji}}\right\}$$

$$C_{1s2s} = \mp C_{1c2c} = \frac{\pm D_1\zeta_{22}\gamma_{21}+D_2\zeta_{11}\gamma_{12}}{2(\zeta_{11}+\zeta_{22})(\zeta_{11}\zeta_{22}\mp\gamma_{12}\gamma_{21})} \tag{8.100}$$

Inspection of (8.100) shows that the difference combination resonance exists only if $\gamma_{12}\gamma_{21}<0$, while the sum resonance takes place when $\gamma_{12}\gamma_{21}>0$. Integrating (8.99) and using (8.100) as initial conditions, the correlation functions can be obtained in a closed form as documented by Dimentberg and Isikov (1983).

8.6 RESPONSE OF NON-LINEAR SYSTEMS

The most difficult task in the area of parametric random vibration is the response analysis of non-linear systems. The existing techniques are limited to special forms of non-linear systems which are not common in most engineering applications. In addition, those problems which meet the conditions of a particular technique may result in difficulties to obtain analytical or numerical solutions. These difficulties include the solution of the FPK equation and the infinite hierarchy of equations governing the response moments.

The literature shows that some progress has been made in developing non-Gaussian closure schemes to resolve the problem of infinite hierarchy. However, the application of these schemes to non-linear random parametric systems is still at its early stage. One of the common and effective non-Gaussian closure schemes is the cumulant-neglect outlined in sections 2.3.7&10. This scheme was originally used in turbulence problems. Random processes such as those encountered in the theory of heterogeneous media and the theory of turbulence may be represented by non-Gaussian probability density functions. Experience has shown that the application of Gaussian closure schemes to such physical problems can lead to unrealistic negative spectral densities of the turbulence (Beran, 1968). In random vibration problems the Gaussian closure may lead to non-stationary response for dynamic systems involving autoparametric interaction (Ibrahim and J. W. Roberts, 1976). However, as will be shown in section 8.6.2, the non-Gaussian closure results in a stationary response for the same problem. For non-linear oscillators under parametric random excitation the Gaussian closure predicts mean square instability at an excitation level below the level predicted by the exact solution by 50% (Ariaratnam, 1980 and Ibrahim and Soundararajan, 1983).

The choice of a proper closure scheme depends primarily on the nature and complexity of the dynamic system under investigation. The scheme is valid if the predicted response is found in agreement with the experimental results. However, experimental verifications may involve difficulties such as the determination of a precise boundary for the onset of motion of the test model. Instead, one may rely on two criteria for the validity of the closure procedure. The first is the non-negativeness of the response mean square and even moments. The second is to examine whether or not the Schwarz inequality is satisfied.

This section deals with the application of the cumulant-neglect closure to solve non-linear dynamic problems under random parametric excitation. Observed discrepancies in the results derived by various methods will be discussed.

8.6.1 SINGLE DEGREE-OF-FREEDOM SYSTEMS

Example 8.3: Determine the stationary response statistics of the non-linear system

$$\ddot{Y} + (2\zeta + \beta Y^2)\dot{Y} + [1 + \xi(\tau)]Y = 0 \tag{i}$$

where, as in example 8.1, dot denotes differentiation with respect to the dimensionless time parameter $\tau = \omega t$. ζ and β are the linear and non-linear damping factors, respectively. $\xi(\tau)$ is a Gaussian wide band random excitation which can be approximated by a physical white noise such that

$$E[\xi(\tau)\xi(\tau+\tau')] = 2D\delta(\tau') \tag{ii}$$

Introducing the coordinate transformation $X_1 = Y$, and $X_2 = \dot{Y}$, equation (i) may be written in the state vector form

$$\begin{aligned} dX_1 &= X_2 d\tau \\ dX_2 &= -\{X_1 + (2\zeta + \beta X_1^2)X_2\}d\tau - \sigma X_1 dB(\tau) \end{aligned} \tag{iii}$$

where $B(\tau)$ is a unit Brownian motion process and $\sigma^2 = 2D$.

Ariaratnam (1980) examined the response of system (i) by using two different approaches: the Stratonovich averaging method and a Gaussian closure scheme. For the sake of comparison, the two solutions are briefly outlined.

i. Stochastic Averaging Solution: Introducing the new variables

$$X_1 = A(\tau)\cos\phi(\tau), \quad X_2 = -A(\tau)\sin\phi(\tau)$$

where $\phi(\tau) = \tau + \theta(\tau)$ (iv)

Following the Stratonovich averaging procedure, it is not difficult to show that the stationary solution of the FPK equation for the amplitude probability density is

$$p(A) = p_o A^{2k-1} \exp\{\beta A^2/2D\} \tag{v}$$

where $p_o = (\beta/2D)^k/\Gamma(k), \quad k = 1 - 4\zeta/D$

provided $k>0$ or

$$D/2\zeta > 2 \tag{vi}$$

The stationary mean square of the amplitude is

$$E[A^2] = \frac{D/2\zeta - 2}{\beta/2\zeta} \tag{vii}$$

and the mean square of the response displacement X_1 is

$$E[X_1^2] = \frac{1}{2}\{D/2\zeta - 2\}/(\beta/2\zeta) \tag{viii}$$

ii. Gaussian Closure Solution (Ibrahim, et al., 1985): The general differential equation of the response joint moments may be derived by using the FPK equation or the Itô stochastic differential formula. Both methods give

$$\dot{m}_{ij} = im_{i-1,j+1} - 2\zeta jm_{ij} - \beta jm_{i+2,j} - jm_{i+1,j-1} + Dj(j-1)m_{i+1,j-2} \tag{ix}$$

where $m_{ij} = E[X_1^i X_2^j]$

It is seen that equation (ix) constitutes a set of infinite coupled equations. If the cumulants of order higher than 2 are neglected, the third and fourth order moments can be expressed in terms of first and second order moments. The stationary solution of first and second order moments is

$$m_{10} = m_{01} = 0$$

$$m_{20} = E[X_1^2] = \{D/2\zeta - 1\}/(\beta/2\zeta) \tag{x}$$

Inspection of solutions (viii) and (x) reveals that the stochastic averaging solution is consistent with the almost sure stability condition, while solution (x) agrees with the mean square stability condition of the linear system.

iii. Non-Gaussian Closure Solution: The first order non-Gaussian closure solution is obtained by generating moment equations up to fourth order:

the first order moment equations are

$$\dot{m}_{10} = m_{01}$$

$$\dot{m}_{01} = -2\zeta m_{01} - \beta m_{21} - m_{10} \qquad \text{(xi)}$$

the second order moment equations are

$$\dot{m}_{11} = m_{02} - 2\zeta m_{11} - \beta m_{31} - m_{20}$$

$$\dot{m}_{20} = 2m_{11} \qquad \text{(xii)}$$

$$\dot{m}_{02} = -4\zeta m_{02} - 2\beta m_{22} - 2m_{11} + 2Dm_{20}$$

the third order moments are

$$\dot{m}_{30} = 3m_{21}$$

$$\dot{m}_{03} = -6\zeta m_{03} - 3\beta m_{23} - 3m_{12} + 6Dm_{21} \qquad \text{(xiii)}$$

$$\dot{m}_{21} = 2m_{12} - 2\zeta m_{21} - \beta m_{41} - m_{30}$$

$$\dot{m}_{12} = m_{03} - 4\zeta m_{12} - 2\beta m_{32} - 2m_{21} + 2Dm_{30}$$

and the fourth order moments are

$$\dot{m}_{40} = 4m_{31}$$

$$\dot{m}_{04} = -8\zeta m_{04} - 4\beta m_{24} - 4m_{13} + 12Dm_{22}$$

$$\dot{m}_{31} = 3m_{22} - 2\zeta m_{31} - \beta m_{51} - m_{40}$$

$$\dot{m}_{13} = m_{04} - 6\zeta m_{13} - 3\beta m_{33} - 3m_{22} + 6Dm_{31}$$

$$\dot{m}_{22} = 2m_{13} - 2\zeta m_{22} - 2\beta m_{42} - 2m_{31} + 2Dm_{40} \qquad \text{(xiv)}$$

It is seen that the last two sets of moment equations contain fifth and sixth order moment terms. In order to close these equations the corresponding fifth and sixth order semi-invariants λ_{23}, λ_{41}, λ_{32}, λ_{24}, λ_{51}, λ_{33}, and λ_{42} are equated to zero. Thus fifth and sixth order moments in (xiii) and (xiv) can be expressed in terms of lower order moments. This procedure results in transforming the linear differential equations (xiii) and (xiv) into a set of non-linear first order moment equations. In view of relations (2.82) the degree of non-linearity reaches sixth order. The resulting fourteen coupled differential equations are solved by numerical integration by using the CSMP (Continuous System Modeling Program) algorithm.

For various sets of small initial conditions, and for excitation levels $D/2\zeta$ well above a certain critical value, the time history response indicates that the response moments of even order grow monotonically with time until they reach a steady state of bounded level governed mainly by the non-linear damping force. A sample of the transient and steady state responses is shown in fig. (8.14). The odd order moments are not displayed and they reach zero level in the steady state region. These results indicate that the system response is stationary and the system oscillates randomly about its equilibrium position.

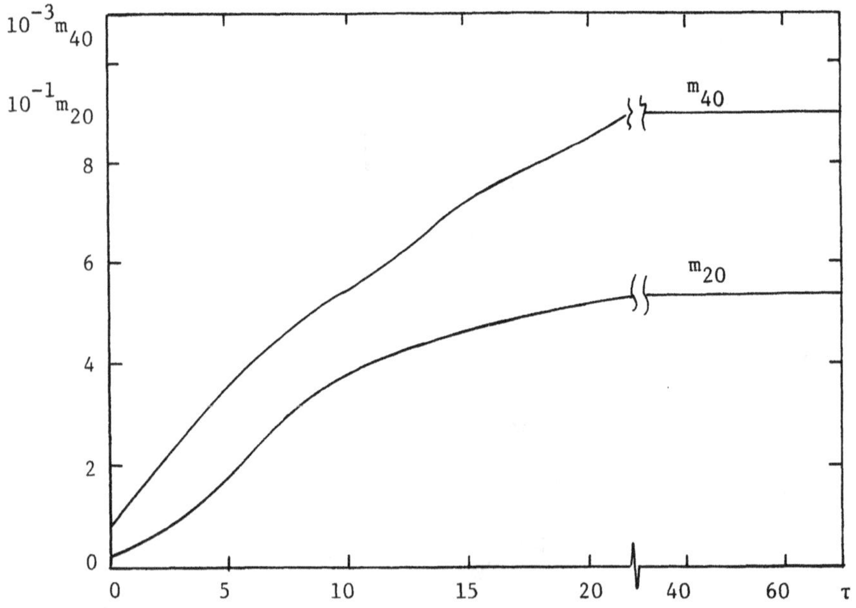

FIG. (8.14) Transient and steady state time history response statistics for $\zeta = 0.02$, $\beta = 0.01$, and $D/2\zeta = 15$

For excitation levels $1<D/2\zeta<(D/2\zeta)_{cr}$, and for any set of small initial conditions, the numerical integration blows up which may imply that real responses are not possible. For excitation levels $D/2\zeta<1$, the system response moments decay to a state of rest. Physically this means that the excitation level is too small to overcome the system damping forces.

The stationary mean square response m_{20} is plotted in figs. (8.15a) and (8.16a) as a function of the excitation level $D/2\zeta$ for two sets of damping constants (ζ,β) = (0.02, 0.01) and (0.05, 0.1), respectively. It is seen that the mean square increases almost linearly with $D/2\zeta > (D/2\zeta)_{cr}$. These figures provide a comparison between the results of three different methods: the non-Gaussian solution (n-G), the Gaussian closure solution (G), and the Stratonovich stochastic averaging solution (S). It is seen that the first order non-Gaussian solution lies between the (G) and (S) solutions. The degree of departure from the G and S solutions is mainly governed by the linear damping ratio ζ. As ζ increases the n-G mean square response becomes closer to the S solution. The limiting effect of the non-linear damping parameter β on the response is common for the three solutions.

Higher order non-Gaussian closure is examined by generating differential equations for the fifth and sixth order moments, and setting seventh and eighth order joint semi-invariants to zero. The results are shown by small circles in fig. (8.15a). It is seen that higher order approximation gives better results.

However, unlike the G and S solutions, which give instability boundaries $D/2\zeta=1$ and 2, respectively, the non-Gaussian solution exhibits a new feature which is manifested by a jump in all even order response statistics (considered in the analysis) at a critical excitation level $(D/2\zeta)_{cr}$. The value of $(D/2\zeta)_{cr}$ is governed by the linear damping ratio ζ.

In order to gain more physical insight of the system response in the neighborhood of the jump point, the analysis is extended to derive the stationary solution by setting the left-hand side of equations (xi-xiv) to zero. The following stationary solution of the resulting algebraic equations is obtained.

$$m_{10} = m_{01} = 0$$

$$m_{11} = m_{21} = m_{31} = 0 \qquad \text{(xva)}$$

From the CSMP results the zero solution of

$$m_{12} = m_{30} = m_{03} = 0 \qquad \text{(xvb)}$$

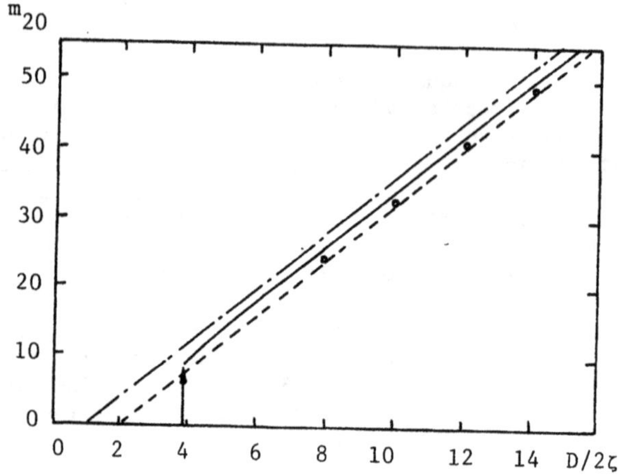

FIG. (8.15a) Mean square response as obtained by

———	1st order non-Gaussian closure
o o	2nd order non-Gaussian closure
—·—	Gaussian closure
- - - -	Stochastic averaging

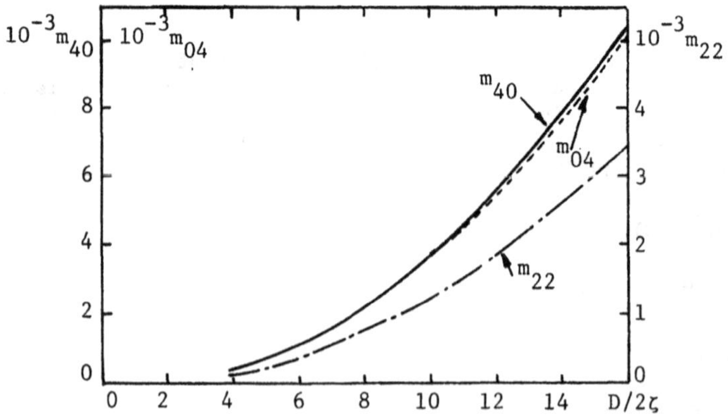

FIG. (8.15b) Fourth order moments of the system response

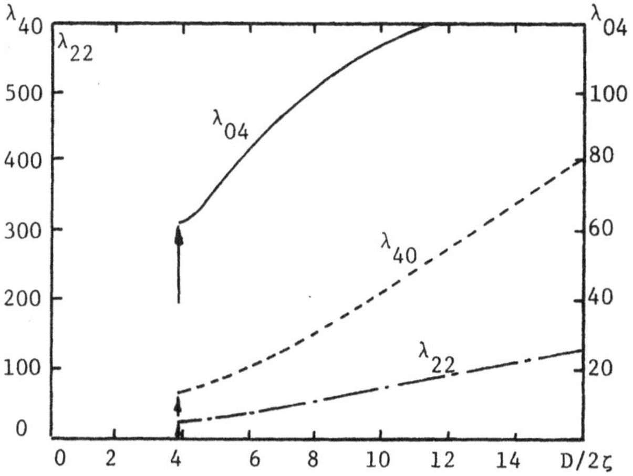

FIG. (8.15c) Fourth order semi-invariants of the system response (ζ = 0.02, β = 0.01)

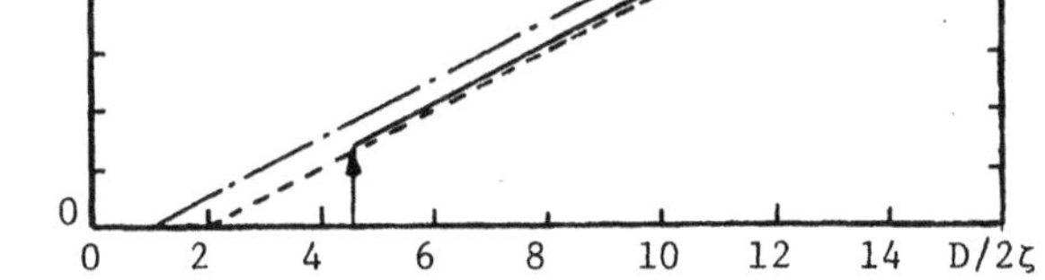

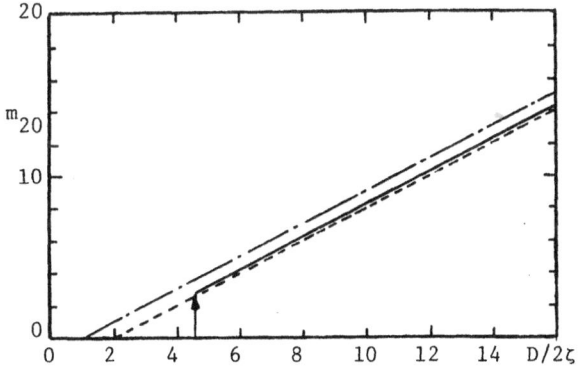

FIG. (8.16a) Mean square response as obtained by
——— non-Gaussian closure
—— - —— Gaussian closure
- - - - - - stochastic averaging

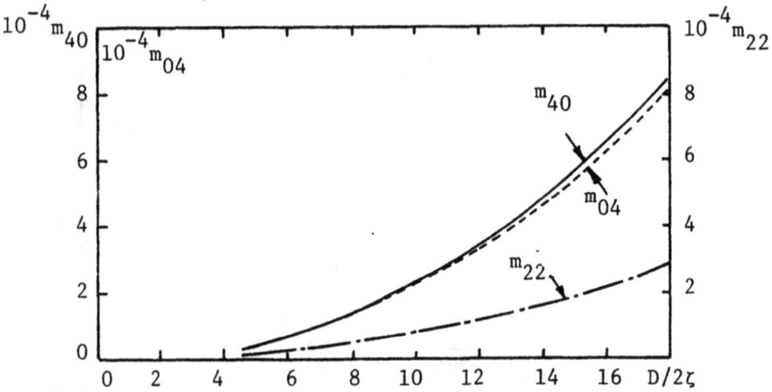

FIG. (8.16b) Fourth order moments of the system response

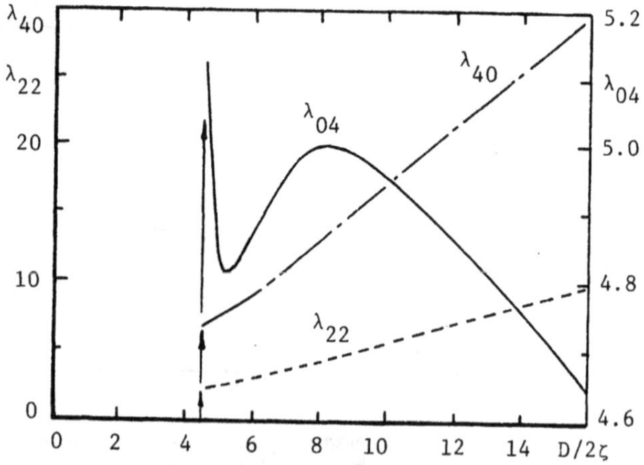

FIG. (8.16c) Fourth order semi-invariants of the system response (ζ = 0.05, β = 0.1)

is enforced, and yields

$$m_{22} = Km_{20}, \qquad K = \frac{D/2\zeta - 1}{\beta/2\zeta}$$

$$m_{40} = 3Km_{20} \qquad m_{02} = m_{20},$$

$$m_{13} = m_{20}\{3\beta m_{20}(3K-2m_{20}) - 2\zeta(3\frac{D}{2\zeta} - 1)K\}, \tag{xvc}$$

where m_{20} is obtained by solving for the roots of the polynomial

$$m_{20}^4 + b_1 m_{20}^3 + b_2 m_{20}^2 + b_3 m_{20} + b_4 = 0 \tag{xvd}$$

where

$$b_1 = -\frac{3}{2}K + \frac{8}{3\beta} \qquad b_2 = \frac{1}{\beta}[-4K\zeta + \frac{K_1}{6} + \frac{1}{9\beta}(K_2 + 1)]$$

$$b_3 = \frac{1}{\beta^2}(-\frac{1}{6}KK_2 + \frac{4}{9}\zeta K_1 - \frac{K}{6}), \qquad b_4 = \frac{1}{54\beta^3}[K_1K_2 + 3K(D - 2\zeta)]$$

$$K_1 = 2\zeta(\frac{3D}{2\zeta} - 1)K, \qquad K_2 = 1 + 12\zeta^2.$$

The roots of equation (xv) are determined numerically by using the IMSL (International Mathematical and Statistical Library) subroutine ZPOLR (Zeros of a Polynomial with Real coefficients). All the roots of (xv) are found complex within an excitation level range defined by the region $1<D/2\zeta<(D/2\zeta)_{cr}$. The imaginary value of the mean square is characterized physically by a non-oscillatory zero amplitude. At $D/2\zeta=1$ the corresponding root is zero and for $D/2\zeta<1$ the roots are negative which implies that the system is completely dormant.

Figures (8.15b) and (8.16b) show the fourth order moments m_{40}, m_{04}, and m_{22} as functions of the excitation level. These moments increase non-linearly with the excitation level $D/2\zeta > (D/2\zeta)_{cr}$. The corresponding fourth order semi-invariants are shown in fig. (8.15c) and (8.16c). The existence of these semi-invariants reflects the fact that the response is non-Gaussian. The response statistic functions displayed in figs. (8.15) and (8.16) are adequate to construct the first order non-Gaussian probability density function.

The instability boundary which defines the location of the jump is plotted in fig. (8.17). Figure (8.17) also shows another two instability boundaries obtained by the Gaussian closure and Stratonovich stochastic averaging solutions.

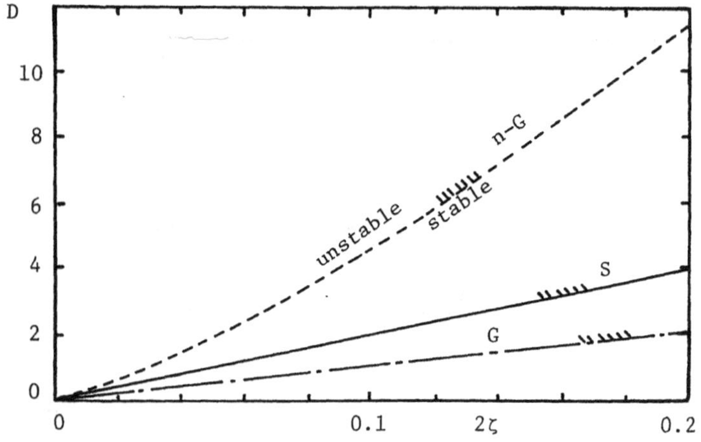

FIG. (8.17) Stability boundaries as obtained by non-Gaussian closure (n-G), stochastic averaging (S), and Gaussian closure (G)

Example 8.4: Examine the random response of the system described by equation (i) in example 8.3 when it is subjected to external and parametric random excitations, that is

$$\ddot{Y} + \left(2\zeta + \beta Y^2\right)\dot{Y} + \left(1 + \xi_1(\tau)\right)Y = \xi_2(\tau) \qquad \text{(i)}$$

This equation is equivalent to the state vector equations

$$dX_1 = X_2 d\tau \qquad \text{(ii)}$$

$$dX_2 = -\left\{X_1 + (2\zeta + \beta X_1^2)X_2\right\} d\tau - X_1\xi_1(\tau) + \xi_2(\tau)$$

i. Stochastic Averaging Solution: Following the same steps as example 8.3 the Itô stochastic differential equation of the amplitude A is

$$dA = \left\{\left(\frac{3}{8} D_1 A + \frac{D_2}{2A}\right) - \frac{\beta}{8} A^3 - \zeta A\right\} d\tau + \left(\frac{1}{4} D_1 A^2 + D_2\right) dB(\tau) \qquad \text{(iii)}$$

where $B(\tau)$ is a unit Brownian motion process, $2D_1$ and $2D_2$ are the spectral densities of $\xi_1(\tau)$ and $\xi_2(\tau)$, respectively.

The stationary solution of the FPK equation for the amplitude probability density function was obtained by Wu and Lin (1984) in the form

$$p(A) = 2(2D_1/\beta)^{-k_2-1} A(A^2 + 4k_1)^{k_2} \exp\{-2\beta(\zeta^2 + k_1)/D_1\} \cdot$$

$$\Gamma^{-1}(k_2 + 1, 2\beta k_1/D_1) \tag{iv}$$

provided $D_1>0$, where $k_1=D_2/D_1$, $K_2=2(\beta k_1 - 2\zeta)/D_1$, and Γ is an incomplete Gamma function subject to the condition that $k_2>0$ when $k_1=0$. The mean square of the amplitude can be obtained by using (iv) as

$$E[A^2] = \int_0^{\infty} A^2 p(A)\, dA$$

$$= \frac{(D_1/2\zeta) - 2}{(\beta/2\zeta)} + \frac{(2D_1/\beta)(2\beta k_1/D_1)^{k_2+1}}{\Gamma(k_2+1, 2\beta k_1/D_1)} \exp(-2\beta k_1/D_1) \tag{v}$$

or

$$E[X_1^2] = \frac{(D_1/2\zeta) - 2}{2(\beta/2\zeta)} + \frac{(D_1/\beta)(2\beta k_1/D_1)^{k_2+1}}{\Gamma(k_2+1, 2\beta k_1/D_1)} \exp(-2\beta k_1/D_1) \tag{vi}$$

ii. Gaussian Closure Solution: The general differential equation of the response joint moment is

$$\dot{m}_{ij} = i m_{i-1,j+1} - 2\zeta j m_{ij} - \beta j m_{i+2,j} - j m_{i+1,j-1} + j(j-1) \cdot$$

$$(D_1 m_{i+2,j-2} + D_2 m_{i,j-2}) \tag{vii}$$

The Gaussian closure solution of (vii) for first and second order moments is obtained by setting cumulants of order greater than 2 to zero in the first and second order moment equations. The resulting solution is

$$m_{20} = m_{02} = \{D_1/2\zeta - 1 + [(D_1/2\zeta - 1)^2 + D_2(\beta/2)]^{1/2}\}/(\beta/\zeta) \tag{viii}$$

iii. Non-Gaussian Closure Solution: Wu and Lin (1984) obtained the first order non-Gaussian closure solution by setting fifth and sixth joint order cumulants to zero. The stationary solution was obtained numerically for all parametric excitation levels which lie in the almost sure stability domain $D_1/2\zeta<2$. Figure (8.18) shows a comparison between the Stratonovich solution S, the non-Gaussian

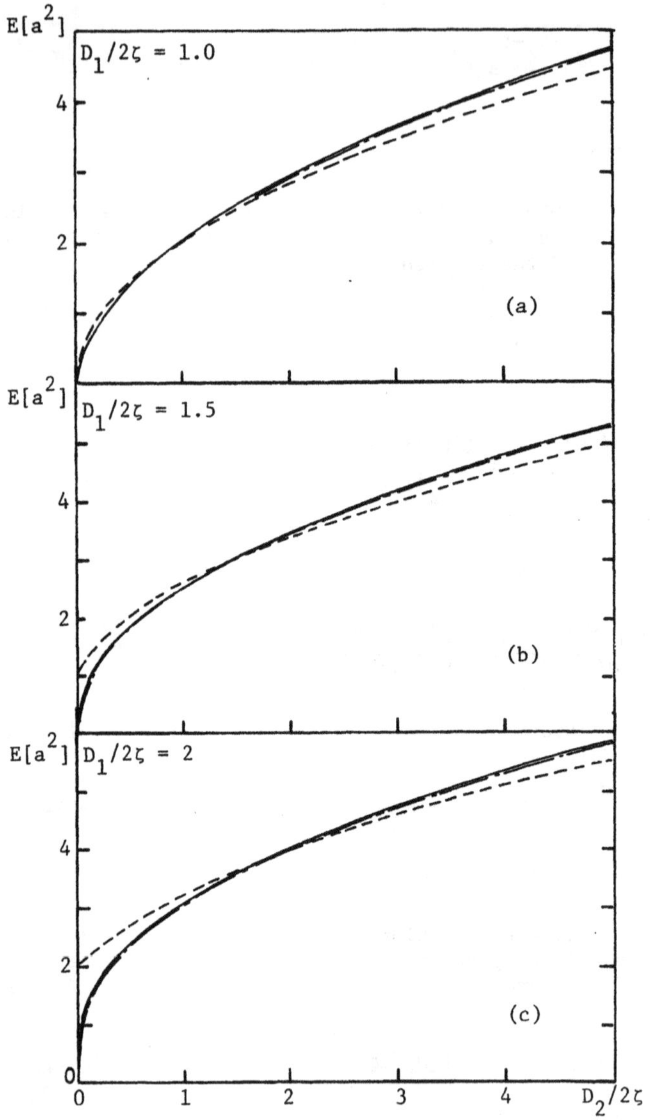

FIG. (8.18) Stationary mean square amplitude of a system with non-linear damping subjected to parametric and external random excitations (non-linear damping coefficient $\beta = 0.1$)

——— stochastic averaging

- - - - - - Gaussian closure

—·—— non-Gaussian closure

solution n-G, and the Gaussian solution G versus the external excitation level $D_2/2\zeta$ for $\alpha = \beta = 0.1$, $2D_1=0.2$. It is clear that the results of the non-Gaussian closure solution are very close to the exact solution.

Example 8.5 (Ibrahim and Soundararajan, 1985): Employ the first order non-Gaussian closure (cumulant-neglect) to determine the response statistics of the liquid free surface motion described in example 6.5.

The general differential equation of the response moment of order $k=i+j$, $m_{ij}=E[X_1^i X_2^j]$, is

$$\dot{m}_{ij} = i m_{i-1,j+1} + j\{-2\zeta m_{ij} + C_1 m_{i+1,j+1} + C_2 m_{i,j+1} + C_3 \zeta m_{i+2,j}$$

$$+ C_4 \zeta m_{i+1,j} + C_5 m_{i+3,j-1} + C_6 m_{i+2,j-1} - m_{i+1,j-1}\}$$

$$+ j(j-1)D\{m_{i+2,j-2} - 2C_6 m_{i+3,j-2} + C_7 m_{i+4,j-2}\} \quad \text{(i)}$$

It is not difficult to show that the Gaussian closure solution gives the mean square response

$$m_{20} = \frac{(D/2\zeta) - 1}{C_{11}(D/2\zeta - C_{12})} \quad \text{(ii)}$$

and

$$m_{02} = m_{20}(1 - 3C_5 m_{20})/(1 + C_1 m_{20}) \quad \text{(iii)}$$

where $C_{11}=6C_5-3C_6^2-C_1$, and $C_{12}=(1/C_{11})(\frac{1}{2} C_3+3C_5)$. This solution is consistent with the condition of asymptotic stability of mean square $D/2\zeta<1$. The mean squares of the response are always positive for values of $D/2$ greater than 1. Furthermore, this solution depends on the excitation parameter $D/2\zeta$ and the non-linear coefficients.

Following the same steps of examples 8.3&4, the first order non-Gaussian closure solution can be established by setting the joint cumulants of fifth and sixth orders to zero in the third and fourth order moment equations generated from (i). In view of the non-linearity introduced by the cumulant-neglect closure, the stationary solution will have more than one solution. Setting to zero all derivatives of m_{ij} on the left-hand side of the moment equations results in fourteen algebraic equations.

These equations were solved numerically by using the IMSL (International Mathematical and Statistical Library) Subroutine ZSPOW. The numerical algorithm results in more than one solution depending on the initial guessing values. Different sets of guessing solutions were introduced and five distinct stationary responses were obtained. All solutions gave positive mean squares for the response displacement and velocity. However, two of these solutions were discarded because the associated higher even moments (such as m_{40} or m_{22}) were negative over the whole range of the excitation level $D/2\zeta$. A third solution exhibited positive mean squares and higher even moments over a limited range of excitation level $1.2<D/2\zeta<27$. This solution was associated with numerical instability in the higher order moments and the algorithm indicated no convergence in the solution. The fourth solution demonstrates successful numerical iteration with very high degree of accuracy over excitation level $D/2\zeta > 1.86$. However, this solution did not satisfy Schwarz's inequality. The fifth solution is believed to be the most realistic one since it satisfies all moment properties and Schwarz's inequality over a range of excitation level greater than a threshold value which is governed by the linear damping factor.

The influence of the damping ratio on the response statistical functions, as obtained by the fifth solution, is examined and the numerical results are plotted in figs. (8.19a-c) for damping ratios $\zeta = 0.1$, 0.05, and 0.25. It is found that the damping factor has no significant effect on the bounded response moments, as they are mainly governed by the non-linearity of the system in addition to the non-linearity that arises from the non-Gaussian closure relations. These figures reveal an important feature, not predicted by other methods, that the response moments exhibit the jump phenomenon at a threshold excitation level. This feature is similar, to a great extent, to the behavior of deterministic response of non-linear systems. The damping factor affects the critical excitation level $(D/2\zeta)_{cr}$ at which a jump occurs for all even response statistics. A stability boundary can then be constructed on the $(D,2\zeta)$ plane to identify the occurrence of the jump phenomenon. This boundary is shown by the curve n-G in fig. (8.20). For small damping factor $\zeta < 0.2$ the critical excitation level $(D/2\zeta)_{cr}$ is almost constant and is found to be in the neighborhood of $D/2\zeta \cong 1.5$. For higher damping ratio $\zeta > 0.2$, the stability boundary becomes non-linear on the $(D,2\zeta)$ plane. Figure (8.20) also provides a comparison between the mean square stability boundaries as obtained by the Gaussian closure and Stratonovich stochastic averaging. It is seen that the non-Gaussian closure stability boundary is very close to the Stratonovich boundary up to $2\zeta \approx 0.8$ above which the two boundaries diverge.

Figure (8.21) shows the displacement mean square as obtained by three different methods: Gaussian closure, Stratonovich stochastic averaging, and non-Gaussian closure for damping ratio $\zeta = 0.1$, respectively. The results indicate that the mean square as obtained by the non-Gaussian closure is remarkably less than the other two solutions by almost 20%. This noticeable reduction is attributed to the fact that the non-Gaussian closure increases the degree of non-linearity of the response moment equations.

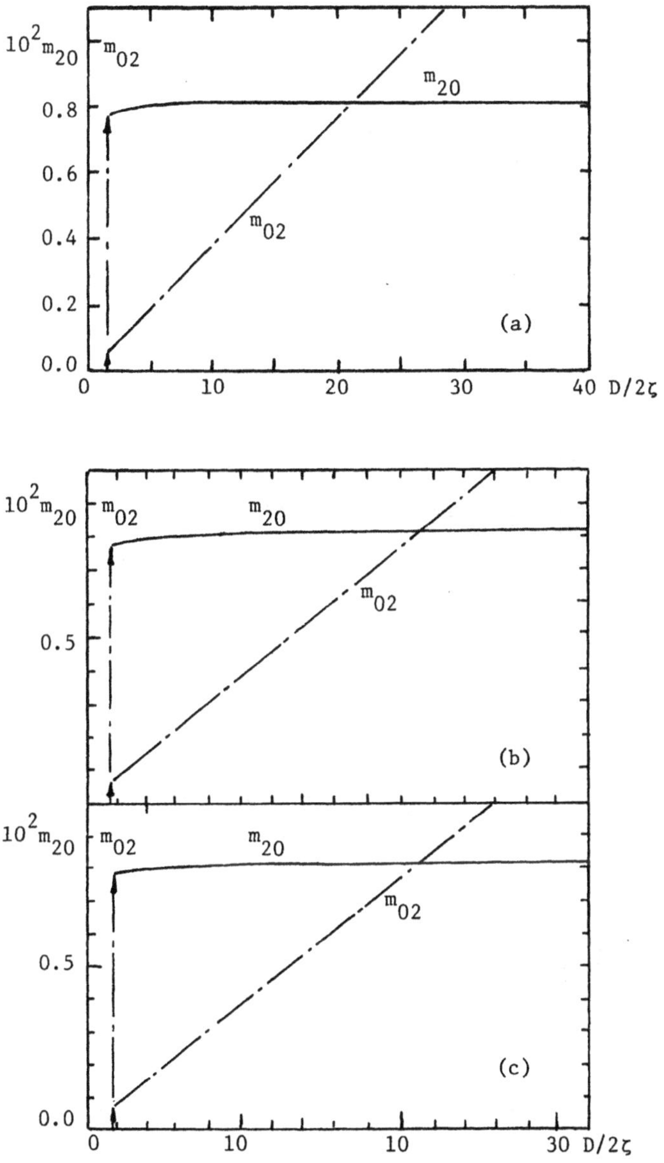

FIG. (8.19) Mean square responses of liquid free surface under vertical random excitation for linear damping ratio (a) $\zeta = 0.1$, (b) $\zeta = 0.05$, and (c) $\zeta = 0.25$

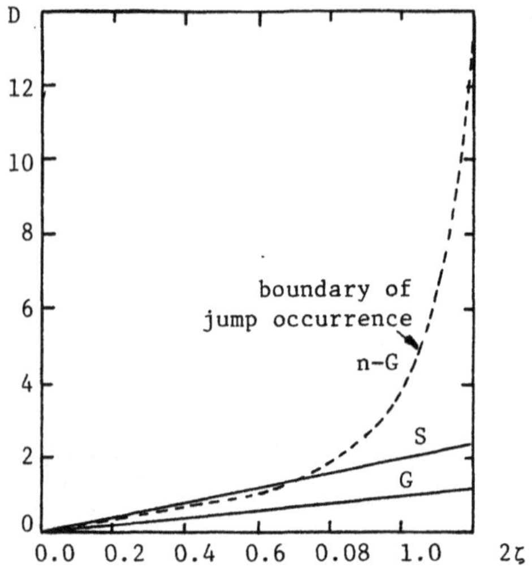

FIG. (8.20) **Stability boundaries as obtained by (n-G) non-Gaussian closure, (G) Gaussian closure, and (S) stochastic averaging**

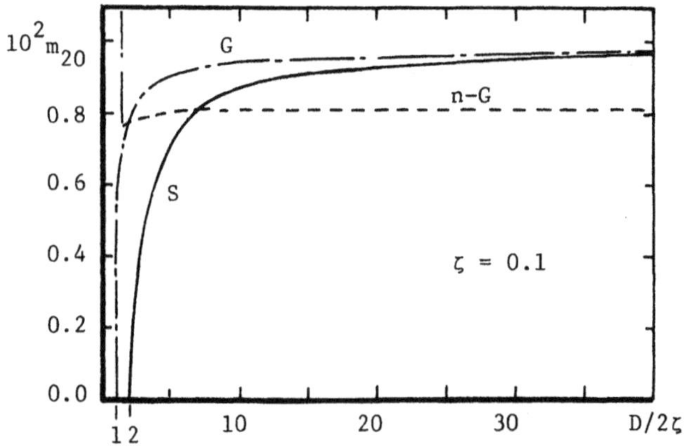

FIG. (8.21) **Comparison of mean square responses as derived by (n-G) non-Gaussian closure, (G) Gaussian closure, and (S) stochastic averaging**

8.6.2 SYSTEMS WITH AUTOPARAMETRIC COUPLING

Example 8.6 (Ibrahim and Heo, 1985): With reference to fig. (8.22) the autoparametric random interaction of a system of coupled beams (k_1 and k_2) with end masses (m_1 and m_2) to a random support motion will be investigated. The system resembles an analytical model of aeroelastic structures such as aircraft wing with fuel store. Under random support acceleration $\ddot{\xi}(t)$ the mass m_1 moves vertically (q_1) and, under the conditions of internal resonance, the mass m_2 moves laterally (q_2). The mathematical modeling can be derived via the Lagrangian formulation. Both the axial and lateral components of the velocity of the wing and the fuel storage are included in determining the kinetic energy and, by using the static deformation curve of the cantilever, these components are found to be in the ratio $6q_i/5\ell_i$, where ℓ_i is the length of beam i. The equations of motion in terms of the generalized coordinates q_i are (Barr and Ashworth, 1977)

$$\begin{bmatrix} m_1 + m_2\left(1+2.25(\ell_2/\ell_1)^2\right) & 1.5m_2\ell_2/\ell_1 \\ 1.5m_2\ell_2^2/\ell_1 & m_2 \end{bmatrix}\begin{Bmatrix} \ddot{q}_1 \\ \ddot{q}_2 \end{Bmatrix} + \begin{bmatrix} k_1 & 0 \\ 0 & k_2 \end{bmatrix}\begin{Bmatrix} q_1 \\ q_2 \end{Bmatrix}$$

$$= -\ddot{\xi}(t)\begin{Bmatrix} m_1 + m_2 \\ 0 \end{Bmatrix} - \ddot{\xi}(t)\begin{bmatrix} 2.25m_2\ell_2/\ell_1^2 & 1.5m_2/\ell_1 \\ 1.5m_2/\ell_1 & 1.2m_2/\ell_2 \end{bmatrix}\begin{Bmatrix} q_1 \\ q_2 \end{Bmatrix} - m_2\begin{Bmatrix} \psi_1 \\ \psi_2 \end{Bmatrix} \tag{i}$$

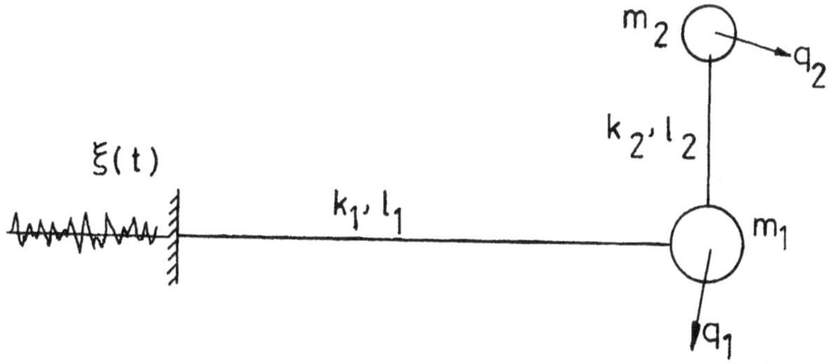

FIG. (8.22) Schematic diagram of coupled beams with autoparametric resonance

where

$$\psi_1 = 0.9\,\frac{\ell_2^2}{\ell_1^2}\,q_1\ddot{q}_1 + 0.45\,\frac{\ell_2^2}{\ell_1^2}\,\dot{q}_1^2 + \frac{1.2}{\ell_2}\,(q_2\ddot{q}_2 + \dot{q}_2^2) + \frac{0.3}{\ell_1}\,q_1\ddot{q}_2 + \frac{3}{\ell_1}$$

$$(\ddot{q}_1 q_2 + \dot{q}_1\dot{q}_2)$$

$$\psi_2 = \frac{0.3}{\ell_1}\,q_1\ddot{q}_1 - \frac{1.2}{\ell_1}\,\dot{q}_1^2 + \frac{1.2}{\ell_2}\,q_2\ddot{q}_1 \tag{ii}$$

$$k_i = 3E_iI_i/\ell_i^3$$

It is seen that the left-hand side of equations (i) represents the linear conservative part of the equations of motion. This part involves dynamic coupling. The first term on the right-hand side is the non-homogeneous random excitation $\ddot{\xi}(t)$, the second term constitutes the parametric effect of the excitation, and ψ_1 and ψ_2 in the third expression include all quadratic non-linearities. The linear eigenvalues (ω_1, ω_2) and eigenvectors of system (i) are determined by setting the right-hand side to zero. The eigenvectors are used in establishing the linear transformation into the principal coordinates y_i, i.e.,

$$\{q\} = [R]\{y\} \tag{iii}$$

where $[R]$ is the modal matrix. Premultiplying equations (i) by $[R]^{-1}\,[m]^{-1}$, where $[m]$ is the mass matrix, and introducing transformation (iii) gives

$$\begin{bmatrix} 1 & 0 \\ 0 & 1 \end{bmatrix}\begin{Bmatrix} Y_1'' \\ Y_2'' \end{Bmatrix} + \begin{bmatrix} 2\zeta_1 & 0 \\ 0 & 2\zeta_2 r \end{bmatrix}\begin{Bmatrix} Y_1' \\ Y_2' \end{Bmatrix} + \begin{bmatrix} 1 & 0 \\ 0 & r^2 \end{bmatrix}\begin{Bmatrix} Y_1 \\ Y_2 \end{Bmatrix} =$$

$$\xi''(\tau)\begin{Bmatrix} a_1 \\ b_1 \end{Bmatrix} + \varepsilon\begin{Bmatrix} \bar{\psi}_1 \\ \bar{\psi}_2 \end{Bmatrix} + \varepsilon\xi''(\tau)\begin{bmatrix} a_2 & a_3 \\ b_2 & b_3 \end{bmatrix}\begin{Bmatrix} Y_1 \\ Y_2 \end{Bmatrix} \tag{iv}$$

where a linear viscous damping is incorporated, and $r = \omega_2/\omega_1$ is the ratio of the normal mode frequencies. The non-dimensional principal coordinates Y_1 and Y_2 are related to the dimensional principal coordinates y_1 and y_2 through the relationship

$$\{Y_1, Y_2\} = \{y_1, y_2\}/q_1^o \tag{v}$$

where q_1^o is the response root mean square of the system when the length of the vertical beam shrinks to zero, i.e., the response of the wing beam with end mass $(m_1 + m_2)$. A prime denotes differentiation with respect to the time parameter $\tau = \omega_1 t$. The non-linear functions ψ_1 and ψ_2 are:

$$\bar{\psi}_1 = a_4 Y_1 Y_1'' + a_5 Y_1 Y_2'' + a_6 Y_2 Y_1'' + a_7 Y_2 Y_2'' + a_8 Y_1'^2 + a_9 Y_1' Y_2' + a_{10} Y_2'^2$$

$$\bar{\psi}_2 = b_4 Y_1 Y_1'' + b_5 Y_1 Y_2'' + b_6 Y_2 Y_1'' + b_7 Y_2 Y_2'' + b_8 Y_1'^2 + b_9 Y_1' Y_2' + b_{10} Y_2'^2$$

$$\varepsilon = q_1^o / \ell_1 \qquad \text{(vi)}$$

The coefficients a_i and b_i are related to the original system parameters. Expressions (vi) include all quadratic non-linear terms which can be divided into two classes: non-linear terms of the same mode and autoparametric terms such as $Y_1 Y_2''$. The autoparametric terms give rise to the internal resonance condition $\omega_2 = \omega_{1/2}$.

The random acceleration $\xi''(\tau)$ is assumed to be Gaussian wide band random process with zero mean and a smooth spectral density 2D up to some frequency which is higher than any characteristic frequency of the system. If the acceleration terms are removed from the non-linear part of equations (iv) by successive elimination, equations (iv) may be approximated by a set of Itô's equations and the response coordinates constitute a Markov process. Introducing the coordinate transformation

$$\{Y_1, Y_2, Y_1', Y_2'\} = \{X_1, X_2, X_3, X_4\} \qquad \text{(vii)}$$

equations (iv) can be written by the following set of Itô's equations:

$$X_1' = X_3$$

$$X_2' = X_4$$

$$X_3' = -X_1 - 2\zeta_1 X_3 - a_4 X_1^2 - (a_6 + r^2 a_5) X_1 X_2 - r^2 a_7 X_2^2 - 2\zeta_1 a_4 X_1 X_3$$

$$-2\zeta_2 r a X_1 X_4 - 2\zeta_1 a_6 X_2 X_3 - 2\zeta_2 r a_7 X_2 X_4 + a_8 X_3^2 + a_9 X_3 X_4 + a_{10} X_4^2$$

$$-(A_1 + A_2 X_1 + A_3 X_2 + A_4 X_1^2 + A_5 X_1 X_2 + A_6 X_2^2) W(\tau)$$

$$X_4' = -r^2X_2 - 2\zeta_2 rX_4 - b_4X_1^2 - (b_6 + r^2b_5)X_1X_2 - r^2b_7X_2^2 - 2\zeta_1 b_4X_1X_3$$

$$-2\zeta_2 rb_5X_1X_4 - 2\zeta_1 b_6X_2X_3 - 2\zeta_2 rb_7X_2X_4 + b_8X_3^2 + b_9X_3X_4 + b_{10}X_4^2$$

$$- (B_1 + B_2X_1 + B_3X_2 + B_4X_1^2 + B_5X_1X_2 + B_6X_2^2)W(\tau) \qquad \text{(viii)}$$

where the coefficients A_i and B_i are related to a_i and b_i.

In equations (viii) the random acceleration has been replaced by the white noise process $W(\tau)$. The autocorrelation function of $W(\tau)$ is defined by the relation

$$R_W(\tau') = E[W(\tau)W(\tau + \tau')] = 2D\delta(\tau') \qquad \text{(ix)}$$

where 2D is the spectral density. In view of the complexity of the state equations (viii) it is not expected to obtain a stationary solution for the corresponding Fokker-Planck equation. Instead, it is possible to generate a general differential equation for all possible moments by using the Itô stochastic calculus or the Fokker-Planck equation. It is not difficult to show that the differential equation of the response joint moments is given in the form

$$m_{ijk\ell}' = im_{i-1,j,k+1,\ell} + jm_{i,j-1,k,\ell+1} + k[-m_{i+1,j,k-1,\ell} - 2\zeta_1 m_{ijk\ell}$$

$$- a_4m_{i+2,j,k-1,\ell} - (r^2a_5 + a_6)m_{i+1,j+1,k-1,\ell} - r^2a_7m_{i,j+2,k-1,\ell}$$

$$- 2\zeta_1 a_4m_{i+1,jk\ell} - 2\zeta_2 ra_5m_{i+1,j,k-1,\ell+1} - 2\zeta_1 a_6m_{i,j+1,k\ell}$$

$$- 2\zeta_2 ra_7m_{i,j+1,k-1,\ell+1} + a_8m_{ij,k+1,\ell} + a_9m_{ijk,\ell+1}$$

$$+ a_{10}m_{ij,k-1,\ell+2}] + \ell[-r^2m_{i,j+1,k,\ell-1} - 2\zeta_2 rm_{ijk\ell} - b_4m_{i+2,jk,\ell-1}$$

$$-(r^2b_5 + b_6)m_{i+1,j+1,k,\ell-1} - r^2b_7m_{i,j+2,k,\ell-1} - 2\zeta_1 b_4m_{i+1,j,k+1,\ell-1}$$

$$- 2\zeta_2 rb_5m_{i+1,jk\ell} - 2\zeta_1 b_6m_{i,j+1,k+1,\ell-1} - 2\zeta_2 rb_7m_{i,j+1,k\ell}$$

$$+ b_8m_{ij,k+2,\ell-1} + b_9m_{ij,k+1,\ell} + b_{10}m_{ijk,\ell+1}]$$

$$+k(k - 1)[DA_1^2m_{ij,k-2,\ell} + 2DA_1A_2m_{i+1,j,k-2,\ell} + 2DA_1A_3m_{i,j+1,k-2,\ell}$$

$$+ D(2A_1A_4 + A_2^2)m_{i+2,j,k-2,\ell} + 2D(A_1A_5 + A_2A_3)m_{i+1,j+1,k-2,\ell}$$

$$+ D(A_3^2 + 2A_1A_6)m_{i,j+2,k-2,\ell}] + k\ell[2DA_1B_1m_{ij,k-1,\ell-1}$$

$$+ 2D(A_1B_2 + A_2B_1)m_{i+1,j,k-1,\ell-1} + 2D(A_1B_3 + A_3B_1)m_{i,j+1,k-1,\ell-1}$$

$$+ 2D(A_1B_4 + A_4B_1 + A_2B_2)m_{i+2,j,k-1,\ell-1}$$

$$+ 2D(A_1B_5 + A_5B_1 + A_2B_3 + A_3B_2)m_{i+1,j+1,k-1,\ell-1}$$

$$+ 2D(A_3B_3 + A_1B_6 + A_6B_1)m_{i,j+2,k-1,\ell-1}]$$

$$+ \ell(\ell - 1)[DB_1^2m_{ijk,\ell-2} + 2DB_1B_2m_{i+1,jk,\ell-2} + 2DB_1B_3m_{i,j+1,k,\ell-2}$$

$$+ D(2B_1B_4 + B_2^2)m_{i+2,jk,\ell-2} + 2D(B_1B_5 + B_2B_3)m_{i+1,j+1,k,\ell-2}$$

$$+ D(B_3^2 + 2B_1B_6)m_{i,j+2,k,\ell-2}] \tag{x}$$

where the definition $m_{ijk\ell} = \int\ldots\int X_1^i\, X_2^j\, X_3^k\, X_4^\ell\, p(\underset{\sim}{X},\tau)dX_1\ldots dX_4$

$$= E[X_1^i\, X_2^i\, X_3^k\, X_4^\ell] \text{ is adopted.}$$

It is seen that a moment equation of order $n=i+j+k+\ell$ contains moments of order n and n+1. In order to solve for the steady state response the moment equations must be closed. The response moments will be determined by using Gaussian and non-Gaussian closure schemes.

i. Gaussian Closure Solution

From the general differential equation (x) one can generate four equations for the first order moments and ten equations for the second order moments. These equations are, however, coupled with third order moment terms. In this section the fourteen equations will be closed by making the assumption that the system nonlinearities are too small to the extent that the response can be regarded as "nearly" Gaussian. In this case the cubic semi-invariants vanish and third order moment terms can be written in terms of lower order moments.

The closed fourteen coupled equations are solved numerically by using the IMSL DVERK Routine (Runge-Kutta-Verner fifth and sixth order numerical integration method). The transient and steady state responses of the system mean square displacements $E[Y_1^2]$ and $E[Y_2^2]$ are plotted in fig. (8.23) for internal resonance ratio r = 0.5, mass

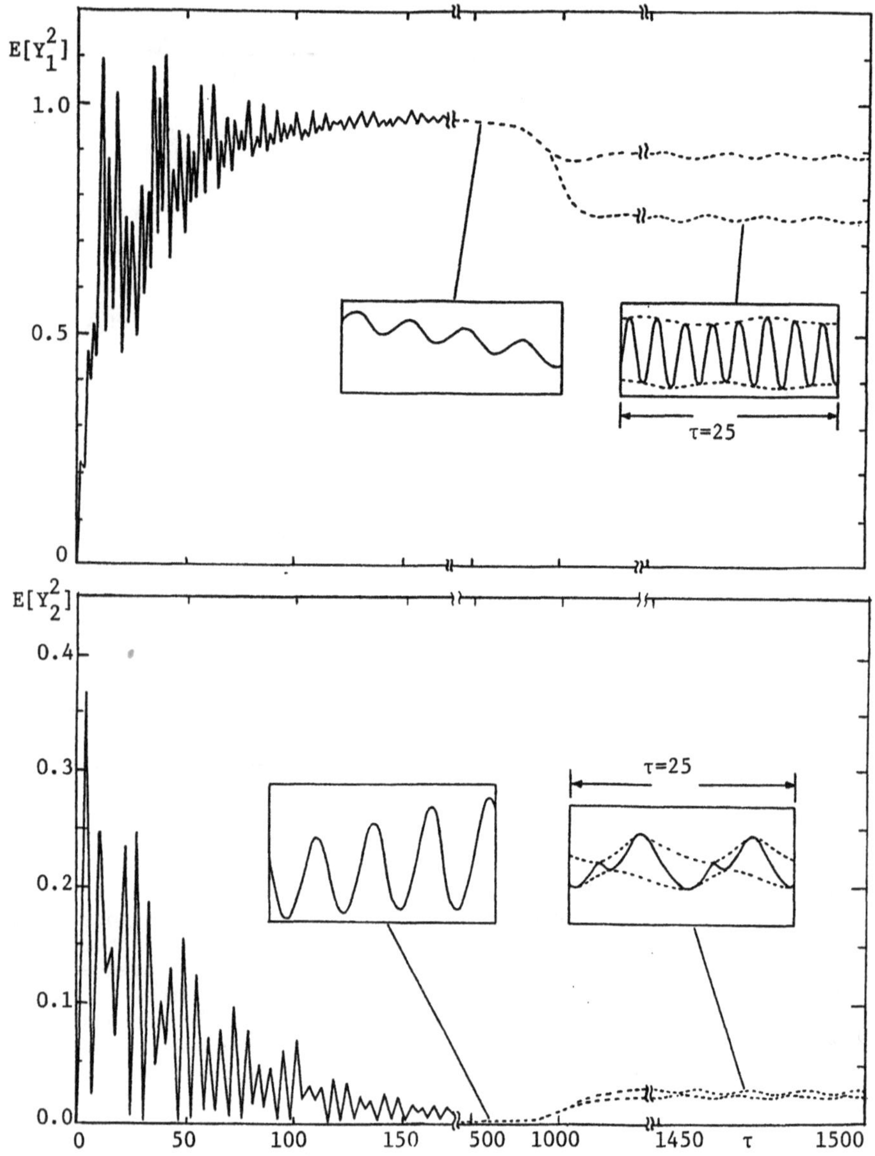

FIG. (8.23) Transient and steady-state responses based on Gaussian closure solution
($r = 0.5$, $\zeta_1 = \zeta_2 = 0.02$, $m_2/m_1 = 0.2$, $\varepsilon = 0.02$)

ratio $m_2/m_1 = 0.2$, beams length ratio $\ell_2/\ell_1 = 0.6$, and $\varepsilon = 0.02$. It is seen that the steady state response fluctuates between two boundaries which will be referred to as lower and upper limits. Repeating the numerical integration for various values of the internal resonance parameter $r = 0.5 \pm \varepsilon$, one can examine the effect of the system parameters upon the response mean squares.

The effect of damping ratios ζ_1 and ζ_2 is shown in fig. (8.24a). It is seen that the region of autoparametric interaction becomes wider as the damping ratios decrease. Figure (8.24b) shows the influence of the non-linear coupling ε. For very small ε the system does not reflect any autoparametric coupling for the whole range of internal resonance ratio. As ε increases the system enters the region of autoparametric interaction. This region becomes wider as ε increases. The effect of the mass ratio is shown in fig. (8.24c).

ii. Non-Gaussian Closure Solution

Since the system is non-linear, the response process is not Gaussian distributed and the corresponding third and higher order semi-invariants will not vanish. These higher semi-invariants give a measure to the deviation of the response from normality. However, their contribution diminishes as their order increases if the process is slightly deviated from Gaussian. Thus one can establish a better approximation if fifth and higher order semi-invariants will be equated to zero.

From equation (x) one can generate moment equations of order up to four. This will result in 69 equations which are coupled and contain fifth order moment terms. Replacing fifth moment terms in terms of lower order moments, the 69 equations will be closed. The resulting 69 coupled differential equations are solved numerically by using the IMSL DVERK subroutine. Figure (8.25) exemplifies the time history response of the displacement mean squares for internal resonance ratio $r = 0.5$ and damping ratios $\zeta_1 = \zeta_2 = 0.02$. During the transient period the mean square of the first normal mode displacement grows until it reaches a peak value at $\tau = 60$ then drops to a lower level at $\tau = 150$. The mean square of the second normal mode displacement grows more slowly until it reaches its peak at $\tau = 150$, which is the time at which the first normal mode mean square reaches its minimum value. This feature reflects the fact that the two normal modes exchange energy during a transient period after which each mode shows a complete stationary response. Unlike the Gaussian closure solution, the non-Gaussian closure solution brings the system into a stationary state.

The stationarity of the solution is confirmed by setting the left-hand sides of the closed 69 equations to zero. The resulting non-linear algebraic equations were solved numerically by using the ZSCNT subroutine which is basically the Secant method for simultaneous non-linear equations. The algebraic solution is found to be identical to the stationary solution obtained by numerical integration.

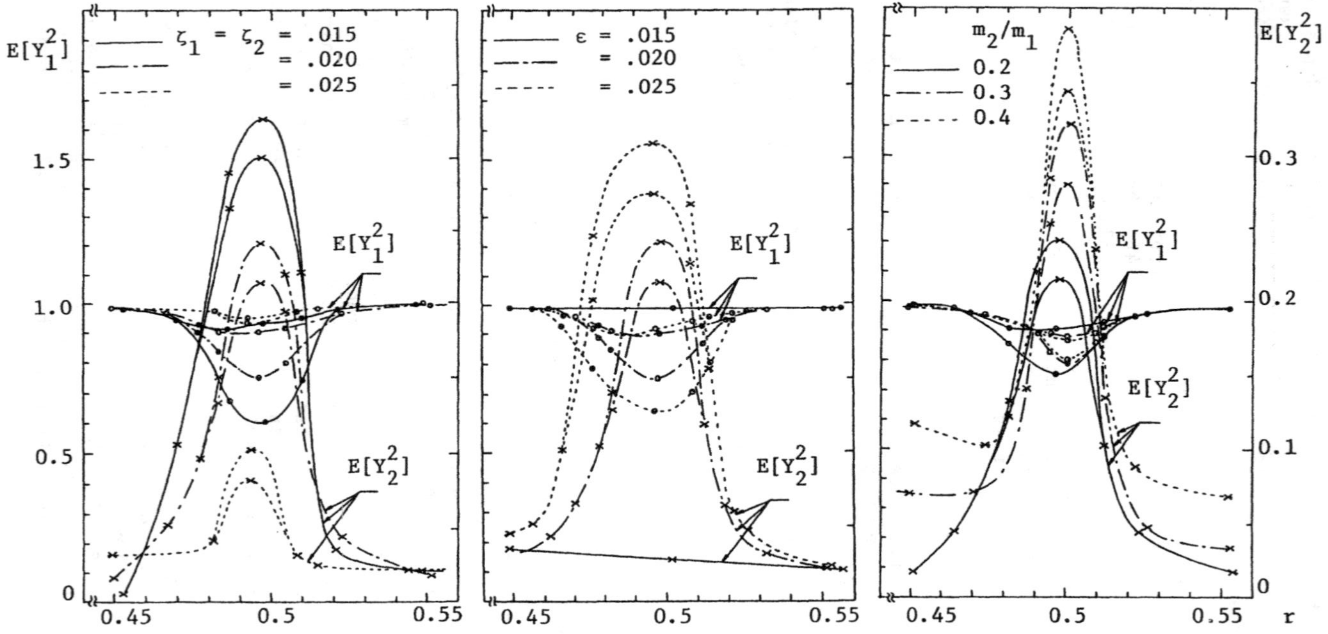

(a) $m_2/m_1 = 0.2$, $\varepsilon = 0.02$

(b) $\zeta_1 = \zeta_2 = 0.02$
$m_2/m_1 = 0.2$

(c) $\zeta_1 = \zeta_2 = 0.002$
$\varepsilon = 0.02$

FIG. (8.24) Gaussian closure solution for various system parameters

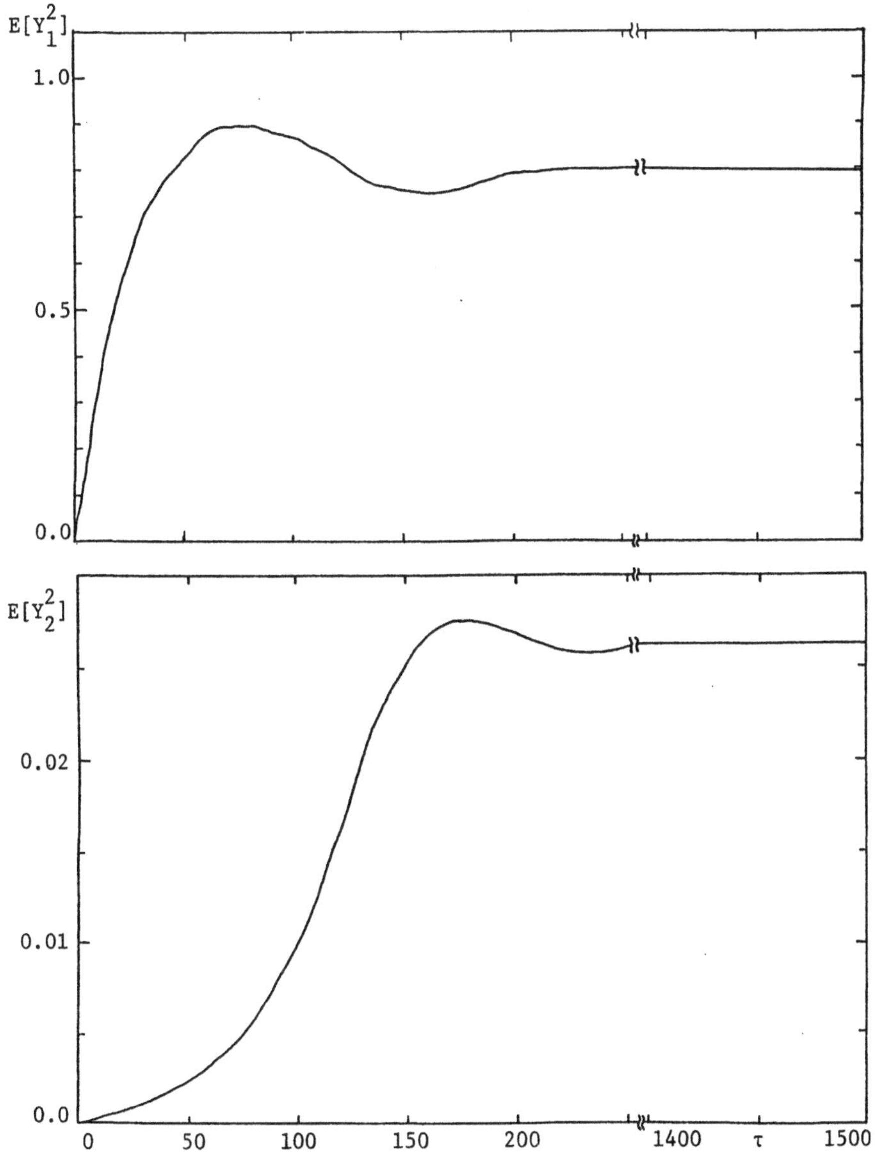

FIG. (8.25) Transient and steady-state responses based on non-Gaussian closure solution
($r = 0.5$, $\zeta_1 = \zeta_2 = 0.02$, $m_2/m_1 = 0.2$, $\varepsilon = 0.02$)

Originally, an algebraic solution for the 14 equations closed by the Gaussian closure scheme was attempted. However, the solution did not converge for all possible guessing values. This shows that the Gaussian closure scheme is not adequate to model the system non-linearity and thus results in a non-stationary solution. The validity of the stationarity was previously verified by Schmidt (1977a) who obtained a stationary solution of the Fokker-Planck equation of a non-linear two degree-of-freedom system via the stochastic averaging method.

The numerical integration of the 69 closed differential equations is repeated for various values of internal resonance ratio $r = 0.5 \pm \varepsilon$. The influence of the damping ratios, non-linear coupling parameter ε, and mass ratio are shown in figs. (8.26a) through (8.26c), respectively. The effect of these parameters on the response mean squares is similar to their effect in the Gaussian solution curves, but the response curves have one branch which is located within the limiting curves of the Gaussian solution.

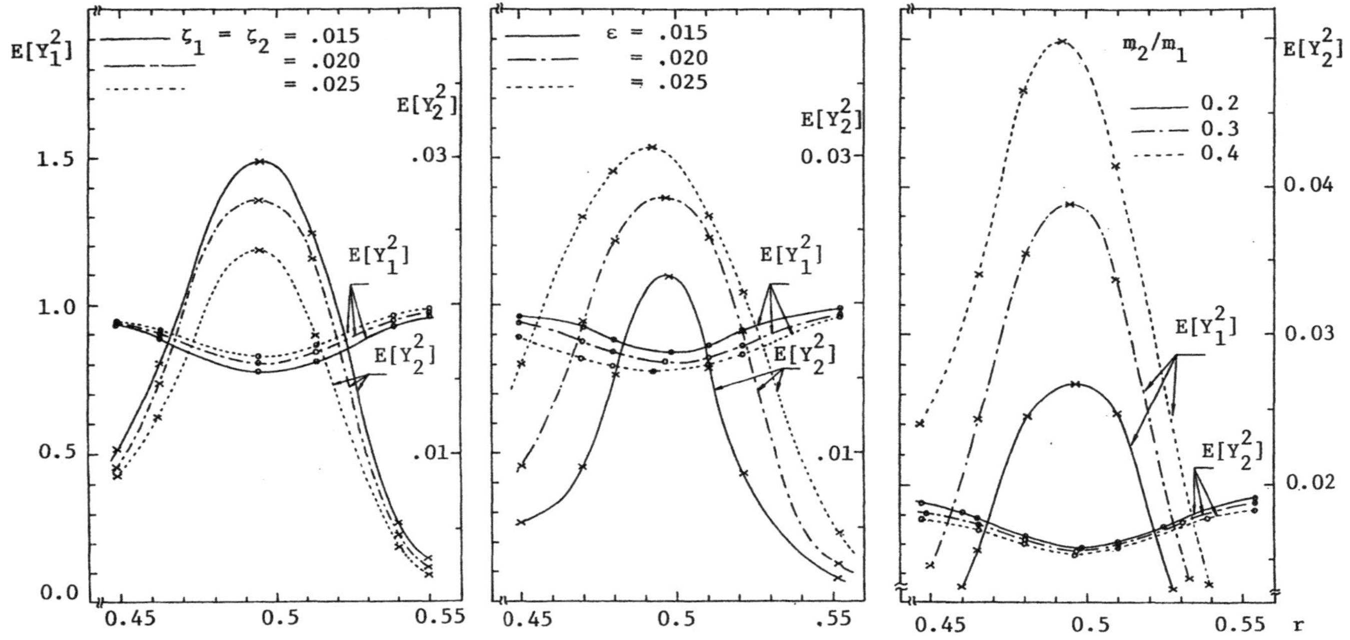

(a) $m_2/m_1 = 0.2$, $\varepsilon = 0.02$

(b) $\zeta_1 = \zeta_2 = 0.02$
$m_2/m_1 = 0.2$

(c) $\zeta_1 = \zeta_2 = 0.02$
$\varepsilon = 0.02$

FIG. (8.26) Non-Gaussian closure solution for various system parameters

CHAPTER 9

Experimental Results

9.1 INTRODUCTION

The experimental investigations pertaining to the behavior of dynamic systems under parametric random excitations are very important not only to verify analytical results but also to explore response characteristics not predicted by the available analytical methods. However, the literature shows very few and sporadic attempts. The lack of experimental results is mainly attributed to the difficulties encountered in measurements of small random parameters. Bolotin (1984, p. 273) and J. W. Roberts (1980) discussed various technical difficulties in measuring stochastic stability boundaries which involve unrealizable infinitesimal numbers.

The earliest experimental work is believed to be done by Bogdanoff and Citron (1965a,b) in an attempt to verify the theoretical conclusion that an inverted pendulum may be stabilized under second order, stationary random parametric excitation having a discrete-power spectral density. They found that all differences among excitation frequencies must be large to ensure the stability of the pendulum in the inverted position. Other experimental investigations were then followed. Dalzell (1967) conducted a series of tests to explore the behavior of fluid in a cylindrical tank under relatively low frequency random longitudinal excitation. Baxter and Evan-Iwanowski (1975) conducted laboratory experiments to observe and measure the response amplitude and frequency of a simply supported elastic column under axial random forces. Although these investigations are valuable, none of them provided any correlations with analytical results. The first attempt towards that goal was done by J. W. Roberts (1980) who conducted comprehensive analytical and experimental investigations to determine the mean square stability of a two degree-of-freedom system under wide band random excitations.

In general, experimental investigations involve estimations of the response spectral distribution, the total mean square and the probability distribution function. In most cases, experimental data reduction of recorded samples is performed by using Analog-Digital methods. The experimental techniques, results, and observations reported in the literature will be discussed in the following sections.

9.2 RANDOM BEHAVIOR OF LIQUID FREE SURFACE

The large amplitude of liquid free surface motion under longitudinal random acceleration was investigated by Dalzell (1967) on a circular cylindrical container. The experiments were conducted for three classes of random accelerations: ultra narrow band, narrow band, and wide band excitations.

i. Ultra Narrow Band Excitation: This excitation is essentially random modulated sine wave of frequency equal to twice that of the first axisymmetric sloshing mode. The free surface was observed to fluctuate from almost negligible movement to very large amplitude. Occasionally, a spike on the tank axis grew until it broke up into globules. An important feature in this set of experiments was that the variation of the excitation spectral level did not affect the estimated response peak spectral density. In other words, the excitation root mean square (rms) acceleration level defining the onset of the half-subharmonic response was not bracketed. Figure (9.1) shows the excitation and response spectra. The spectral density of the excitation acceleration $S_a(\omega)$ is given in $g^2/\Delta\omega$. The spectral density of the fluid elevation on the tank axis of symmetry $S_\eta(\omega)$ is measured in $(\text{diameter})^2/\Delta\omega$. Figure (9.1) also shows a consistent fluid elevation spectral density peak centered on the excitation frequency. This situation is similar to the harmonic response observed under sinusoidal excitation. It was observed that as the excitation level decreases, the contributions at frequencies other than the first axisymmetric mode decrease.

ii. Narrow Band Excitation: This case involved excitation bandwidths 4 and 16 times that of ultra narrow band excitation. The excitation is centered on twice the frequency of the first axisymmetric sloshing mode. For small bandwidth both harmonic and half-subharmonic responses were observed as shown in fig. (9.2). Figure (9.3) shows the excitation and response spectra for larger bandwidth $\Delta\omega = 0.68$. It was reported that subharmonic behavior was experienced, but transition from harmonic to subharmonic behavior was not identified.

iii. Wide Band Excitation: A passband excitation was adjusted so that the energy content at the first axisymmetric mode was about 1/50 that at twice this frequency. For low excitation level the fluid response was harmonic, and an abrupt transition to subharmonic was observed when the excitation level increases. Figure (9.4) shows the excitation and fluid response spectra for a relatively high excitation level.

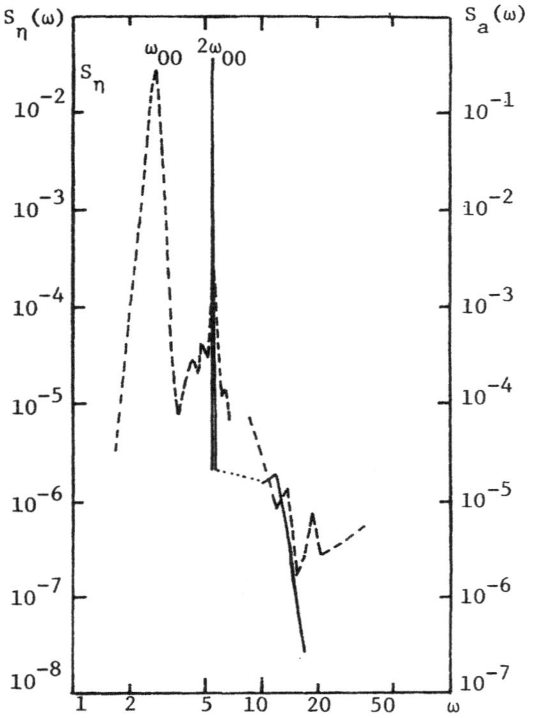

FIG. (9.1) **Spectra: ultra narrow band excitation $\Delta\omega = 0.04$**

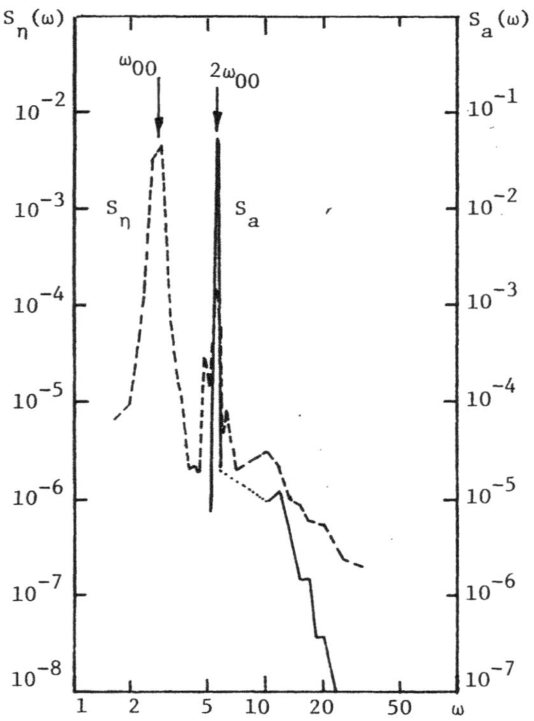

FIG. (9.2) **Spectra: narrow band excitation $\Delta\omega = 0.17$**

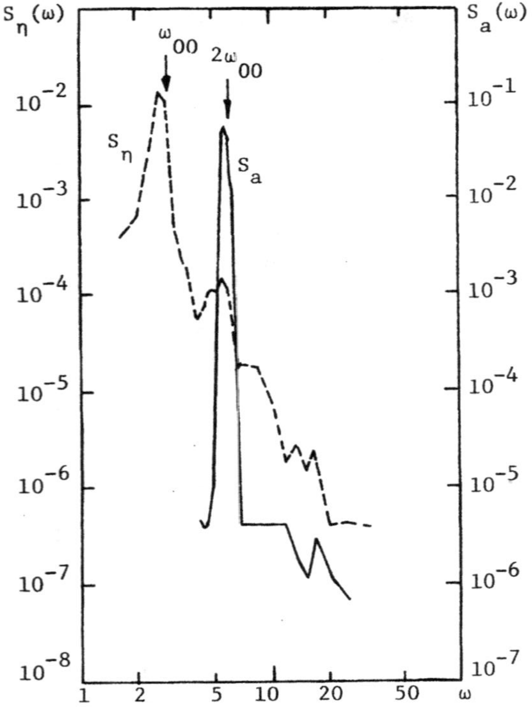

FIG. (9.3) Spectra: narrow band excitation $\Delta\omega = 0.68$

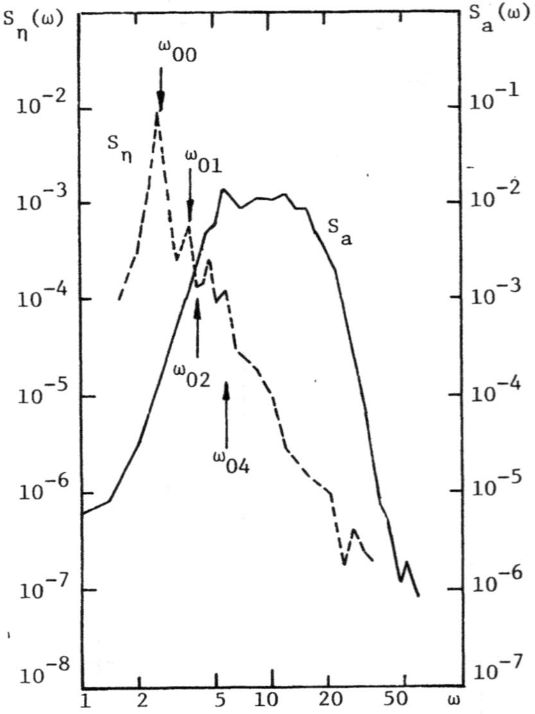

FIG. (9.4) Spectra: wide band excitation of variable energy

Probability Distribution of the Response: The excitations in the above three cases were essentially Gaussian distributed. The probability distributions of the responses were found to be non-Gaussian especially for the large amplitude half-subharmonic response. However, samples of harmonic responses were found almost normal distributed. Figure (9.5) shows the response probability distributions plotted on a normal probability paper. Dalzell conducted a least square fitting procedure and found that the shape of an extreme value distribution (Cramér, 1946) was very similar to that of the estimated cumulative distribution of liquid free surface elevations. The extreme value distribution, referred to as double exponential distribution, is defined by the relationship

$$F(Y) = \exp\{-\exp(-Y)\} \qquad (9.1)$$

where the variate $Y = \frac{\pi}{\sqrt{6}} X + \gamma$

γ = Euler constant = 0.577215665

X = standardized variate = $(\eta - \bar{\eta})/\hat{\sigma}_\eta$

η = fluid elevation, $\bar{\eta}$ = estimated mean value of fluid elevation,

$\hat{\sigma}_\eta$ = r.m.s. fluid elevation.

The probability density can be determined from the definition

$$p(Y) = dF(Y)/dY$$

$$= \exp\{-Y - \exp(-Y)\} \qquad (9.2)$$

Figure (9.6) shows a possible probability density for half-subharmonic fluid elevation response to ultra narrow band excitation.

9.3 RANDOM BEHAVIOR OF ELASTIC COLUMNS

Baxter and Evan-Iwanowski (1975) conducted a series of laboratory experiments to observe and measure the non-linear parametric response of an oscillatory elastic column. The parametric excitations had five possible power spectral density bandwidths. These included a sinusoidal excitation and four random excitations with bandwidths 3Hz, 10Hz, 30Hz, and 100Hz. Typical time history response curves for the five excitations are shown in fig. (9.7). It is seen that the response of the column to random excitation has the form of a narrow band random process. The average response amplitude differs between the various bandwidths, but the response frequency remains consistently near the natural frequency of the column for all excitation bandwidths.

The response of the column was characterized by a temporary but complete decay of the response amplitude. At those times, the column

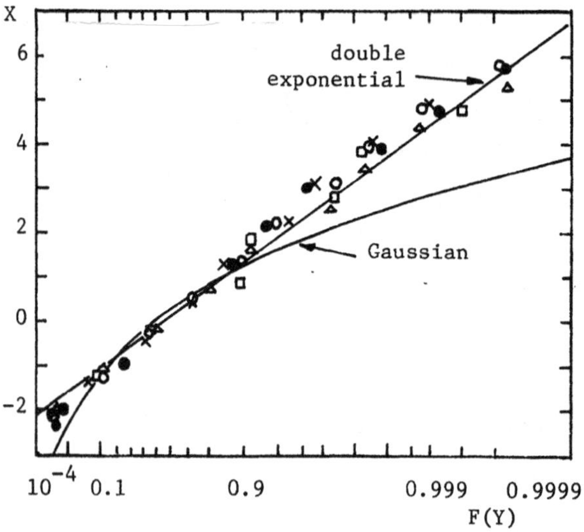

FIG. (9.5) Probability distribution plotted on double exponential paper (● ultra narrow band, x,o,Δ narrow bands with different bandwidths, □ broad band) (Dalzell, 1967)

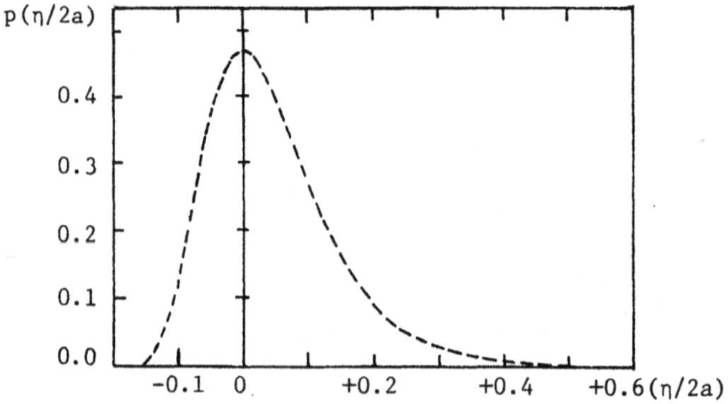

FIG. (9.6) Possible probability density for half-sub-harmonic, under ultra narrow band random excitation (η - fluid elevation on the tank axis, 2a - tank diameter) (Dalzell, 1967)

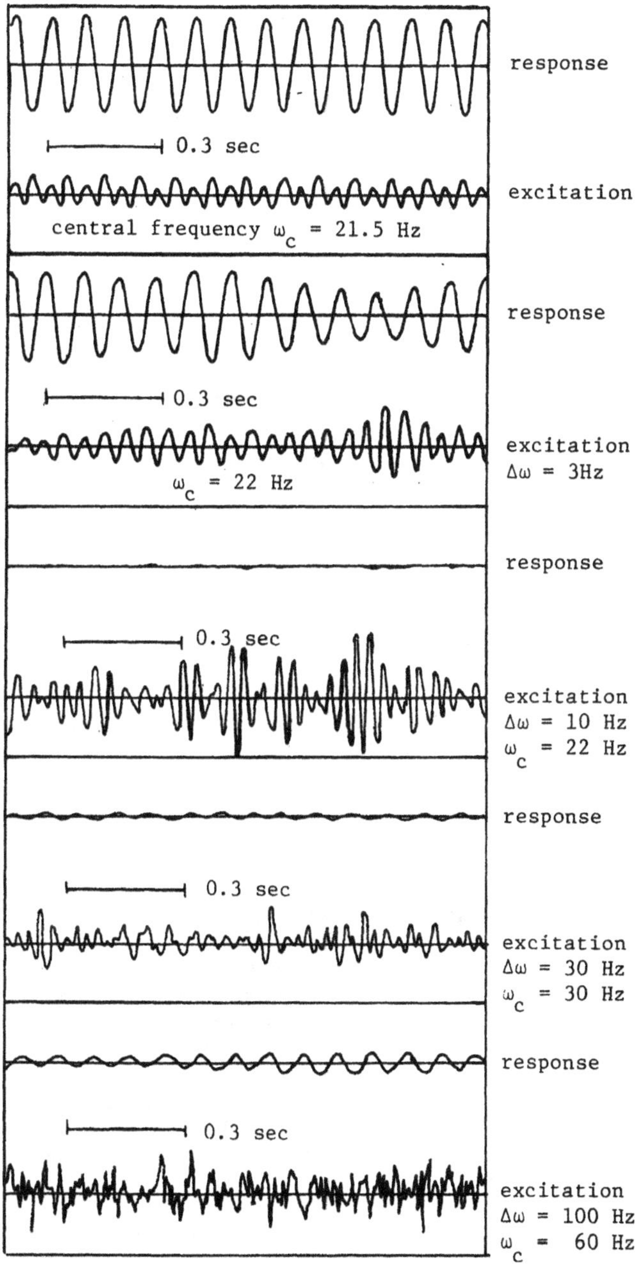
response
0.3 sec
excitation
central frequency ω_c = 21.5 Hz
response
0.3 sec
excitation
$\Delta\omega$ = 3Hz
ω_c = 22 Hz
response
0.3 sec
excitation
$\Delta\omega$ = 10 Hz
ω_c = 22 Hz
response
0.3 sec
excitation
$\Delta\omega$ = 30 Hz
ω_c = 30 Hz
response
0.3 sec
excitation
$\Delta\omega$ = 100 Hz
ω_c = 60 Hz

FIG. (9.7)

would stop all lateral oscillations even though the excitation was statistically well within the fundamental parametric resonance region. The absence of the lateral motion appeared more frequently as the excitation bandwidth increased but in any case, it could occur at any frequency within the resonance region.

The probability distribution of the column response for 3Hz narrow band random excitation was determined from a recorded trace of the response amplitude. The results of these measurements at seven amplitude levels (indicated by small circles) are shown in fig. (9.8a). The corresponding probability density is shown in fig. (9.8b).

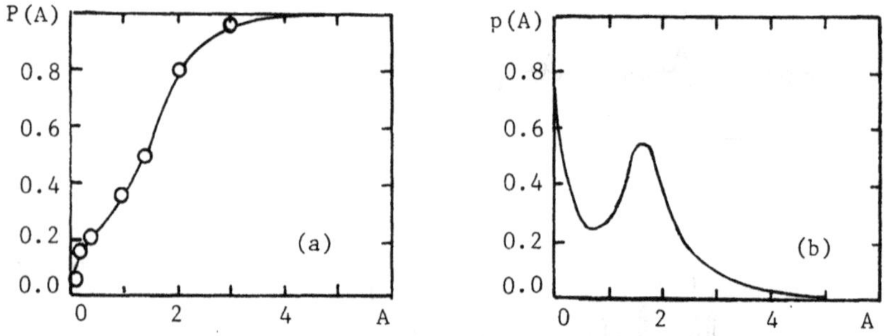

FIG. (9.8) Approximate probability distribution and probability density functions as measured by Baxter and Evan-Iwanowski (1975) (for central excitation frequency ω_c = 20.2 Hz, natural frequency of the column ω_n = 9.6 Hz)

9.4 SYSTEMS WITH AUTOPARAMETRIC INTERACTION

The random behavior of two degree-of-freedom dynamic systems with autoparametric coupling was examined on the model shown in fig. (9.9) by J. W. Roberts (1980). The system consists of a horizontal beam, of equivalent mass M, excited at one point. The coupled vertical beam with tuning mass m can oscillate in a direction perpendicular to the plane of the figure if the system frequency tuning ratio r is in the neighborhood of 0.5, ($r = \omega_2/\omega_1$, ω_1 is the fundamental frequency of the horizontal beam including the tuning mass, and ω_2 is the fundamental frequency of the vertical beam). The system was excited by a wide band random forcing excitation with flat spectral density up to a frequency which is four times the fundamental horizontal beam frequency.

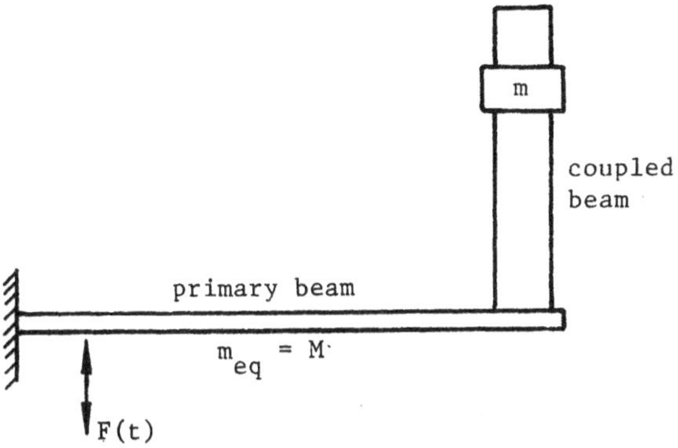

FIG. (9.9) Schematic diagram of the experimental model with autoparametric coupling (J. W. Roberts, 1980)

Roberts carried out a series of tests to determine a transition boundary of the onset of random motions of the coupled vertical beam under forcing excitation of the horizontal beam. Under various excitation levels he observed three distinct characteristic forms of the coupled beam behavior. These were:

i. Zero motion over the whole observation period.

ii. Partially developed random motion with periods of zero motion. This regime tended to dominate with reducing excitation.

iii. Continuous random motion.

Considerable difficulty was reported in establishing a precise determination of changeover from partially developed regime to zero motion region. However, a procedure was adopted to measure the stability boundary. This was done by increasing the excitation level to a point where definite random motions of the vertical beam were observed. The excitation level was reduced in stages and the corresponding mean square response of the coupled beam was determined. Finally, a graph of these measurements was extrapolated to locate an excitation level representing the transition to unimodal random response of the system with zero motion of the coupled cantilever. It is obvious that this procedure has one reservation in that the extrapolation will not predict any possibility for mean square jump or collapse at a critical excitation level as predicted in examples 8.3 and 8.5. According to this procedure, Roberts obtained the system stability boundaries for unimodal response which are shown by small circles in figs. (9.10a,b). These figures provide comparisons between the experimental and predicted stability

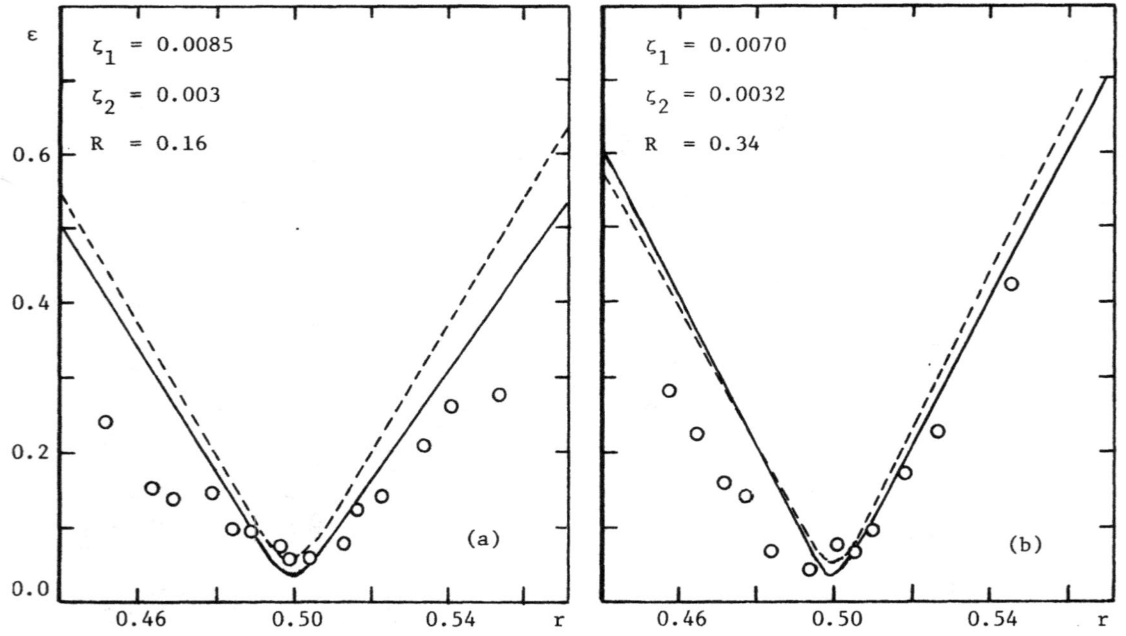

FIG. (9.10) Comparisons of measured and predicted stability boundaries o experimental, ——— perturbation solution, - - - - - - Gaussian closure, ε = excitation level parameter (J. W. Roberts, 1980)

boundaries. Two theoretical stability boundaries are shown. The first was determined by the Gaussian closure scheme and the second was evaluated by employing a perturbation method. It is seen that the three boundaries have typical V-shape curves centered in the vicinity of the internal resonance condition $r = 0.5$. For $r<0.5$ there is remarkable deviation from the experimental boundary which gives a wider instability region than predicted by the analytical methods. It has been indicated in example (7.6) that the mass ratio $R = [m/(m+M)]$ has a negligible effect on the stability boundary over the whole range of the internal detuning ratio r. However, the experimental boundaries indicate that the mass ratio has no effect at $r = 0.5$ but not elsewhere.

References

Abramowitz, M. and Stegen, I. A. (1972). Handbook of Mathematical Functions with Formulas, Graphs and Mathematical Tables. Dover Publications, New York.

Adomian, G. (1971). The closure approximation in the hierarchy equations. J. Stat. Phys. 3(2), 127-133.

Adomian, G. and Malakian, K. (1979). Closure approximation error in the mean solution of stochastic differential equations by the hierarchy method. J. Stat. Phys. 21(2), 181-189.

Ahmadi, G. (1977a). On the stability of a class of continuous systems with non-stationary random coefficients. Int. J. System Science 8, 1201-1207.

Ahmadi, G. (1977b). On the stability of systems of coupled partial differential equations with random excitation. J. Sound Vib. 52, 27-35.

Ahmadi, G. (1979). On the mean square stability of a class of non-stationary coupled partial differential equations. Ing-Arch. 48, 213-219.

Ahmadi, G. and Glockner, P. G. (1982). Approximate stability criteria for some second order linear differential equations with stationary Gaussian random coefficients. ASME J. Appl. Mech. 49, 648-649.

Akhmetkaliev, T. (1965). Connection between the stability of stochastic difference and differential systems. Differents. Urvan. 1(8), 1016-1026.

Akhmetkaliev, T. (1966). On the stability of stochastic difference systems under constantly acting disturbances. Differents. Urvan. 2(9), 1161-1169.

Alekseyev, V. M. and Valeev, K. G. (1971). Analysis of the oscillations of a linear system with random coefficients. Izv. Vyssh. Uchebn. Zaved., Radiofiz. 14(12).

Andronov, A. A., Pontryagin, L. S., and Witt, A. A. (1933). On the statistical investigation of dynamical systems. J. Exper. & Theor. Phys. 3, 165-180.

Ariaratnam, S. T. (1967); in Proc. Int. Conf. Dynamic Stability of Structures. (Edited by G. Hermann). 255-265, Pergamon Press. Dynamic stability of a column under loading.

Ariaratnam, S. T. (1972); in Proc. IUTAM Symp. Stability of Stochastic Dynamical Systems. Lecture Notes in Mathematics 294. (Edited by R. F. Curtain). 291-302, Springer-Verlag, Berlin. Stability of mechanical systems under parametric excitation.

Ariaratnam, S. T. (1980); in New Approaches in Non-linear Problems in Dynamics. (Edited by P. J. Holmes). 470-474, SIAM, Philadelphia. Bifurcation in non-linear stochastic systems.

Ariaratnam, S. T. and Graefe, P. W. U. (1965a). Linear systems with stochastic coefficients, (I). Int. J. Contr. 1(3), 239-250.

Ariaratnam, S. T. and Graefe, P. W. U. (1965b). Linear systems with stochastic coefficients, (II). Int. J. Contr. 2(2), 161-170.

Ariaratnam, S. T. and Srikantaiah, T. K. (1978). Parametric instabilities in elastic structures under stochastic loading. J. Struc. Mech. 6(4), 349-365.

Ariaratnam, S. T. and Tam, D. S. F. (1976). Parametric random excitation of a damped Mathieu oscillator. Z. Angew. Math. Mech. (ZAMM) 56, 449-452.

Ariaratnam, S. T. and Tam, D. S. F. (1979). Random vibration and stability of a linear parametrically excited oscillator. Z. Angew. Math. Mech. (ZAMM) 59, 79-84.

Arnold, L. (1974). Stochastic Differential Equations: Theory and Applications. John Wiley & Sons, New York.

Assaf, Sh. A. and Zirkle, L. D. (1976). Approximate analysis of non-linear stochastic systems. Int. J. Contr. 23(4), 477-492.

Atkinson, J. D. and Caughey, T. K. (1968). First order piecewise linear systems with random parametric excitation. Int. J. Non-linear Mech. 3(4), 399-411.

Auslender, E. I. and Mil'shtein, G. N. (1982). Asymptotic expansions of the Liapunov index for linear stochastic systems with small noise. Appl. Math. Mech. (Prikl. Mat. Mekh., PMM) 46(3), 277-286.

Barr, A. D. S. (1980); in Proc. Int. Conf. Recent Advances in Structural Dynamics. 545-568, Institute of Sound and Vibration Research, University of Southampton, England. Some developments in parametric stability and non-linear vibration.

Barr, A. D. S. and Ashworth, R. P. (1977). Final Report, AFOSR 74-2723 (EOARD), University of Dundee, Scotland. Parametric and non-linear mode interaction behavior in the dynamics of structures.

Baxter, G. K. (1971). Ph.D. Thesis, University of Syracuse. The non-linear response of mechanical systems to parametric random excitation.

Baxter, G. K. and Evan-Iwanowski, R. M. (1975). Response of a column in random vibration tests. ASCE J. Struct. Div. 101, 1749-1761.

Beaman, J. J. and Hedrick, J. K., Jr. (1981a). Improved statistical linearization for analysis and control of non-linear stochastic systems, part I: an extended statistical linearization technique. Trans. ASME J. Dynamic Systems, Measurements, and Control 103, 14-21.

Beaman, J. J. and Hedrick, J. K., Jr. (1981b). Improved statistical linearization for analysis and control of non-linear stochastic systems, part II: application to control system design. Trans. ASME J. Dynamic Systems, Measurements, and Control 103, 22-27.

Belinfante, J. G., Kolman, B., and Smith, H. A. (1966). Introduction to Lie groups and Lie algebra with applications. SIAM Rev. 8(1), 11-46.

Beliveau, J. G., Vaicaitis, R., and Shinozuka, M. (1977). Motion of suspension bridge subject to wind loads. ASCE J. Struct. Div. 103(ST6), 1189-1205.

Bellman, R. E. (1953). Stability Theory of Differential Equations. McGraw-Hill, New York.

Bellman, R. E. (1954). Limit theorems for non-commutative operations, I. Duke Math. J. 21(3), 491-500.

Bellman, R. E. (1960). Introduction to Matrix Analysis. McGraw-Hill, New York.

Bellman, R. E. and Richardson, J. M. (1968). Closure and preservation of moment properties. J. Math. Anal. Appl. 23, 639-644.

Beran, M. J. (1968). Statistical Continuum Theories. Interscience, John Wiley & Sons, New York.

Bergen, A. R. (1960). Stability of systems with randomly time-varying parameters. IRE Trans. Automatic Control 5, 265-269.

Bertram, J. E. and Sarachik, P. E. (1959); in Proc. Int. Symp. Circuits and Information Theory. 260-270, IRE Trans. CT-6, Los Angeles. Stability of circuits with randomly time-varying parameters.

Bharucha, B. H. (1961). Ph.D. Thesis, Univ. of California, Berkeley. On the stability of randomly varying systems.

Bharucha-Reid, A. T. (1972). Random Integral Equations. Academic Press, New York.

Blankenship, G. L. (1975). Lie theory and the moment stability problem in stochastic differential equations. IFAC 1975, Cambridge, Mass.

Blankenship, G. L. (1977). Stability of linear differential equations with random coefficients. IEEE Trans. Auto. Control AC-22, 834-838.

Blankenship, G. L. and Papanicolaou, G. C. (1978). Stability and control of stochastic systems with wide band noise disturbances, I. SIAM J. Appl. Math. 34, 437-476.

Bogdanoff, J. L. (1962). Influence on the behavior of a linear dynamical system of some imposed rapid motions of small amplitude. J. Acoust. Soc. Am. 34(8), 1055-1062.

Bogdanoff, J. L. and Citron, S. J. (1965a); in Proc. 9th Midwestern Mech. Conf. Development in Mechanics. 3(1), 3-15, John Wiley, New York. On the stabilization of the inverted pendulum.

Bogdanoff, J. L. and Citron, S. J. (1965b). Experiments with an inverted pendulum subjected to random parametric excitation. J. Acoust. Soc. Am. 38(9), 447-452.

Bogdanoff, J. L. and Kozin, F. (1962). Moments of the output of linear random systems. J. Acoust. Soc. Am. 34(8), 1063-1068.

Bogoliubov, N. N. and Mitropol'skii, Y. A. (1961). Asymptotic Methods in the Theory of Non-linear Oscillation. Gordon & Breach, New York.

Bolotin, V. V. (1964). The Dynamic Stability of Elastic Systems. Holden-Day, Inc., San Francisco.

Bolotin, V. V. (1972); in Study No. 6 On Stability. Chapter 11, 385-422, University of Waterloo. Reliability theory and stochastic stability.

Bolotin, V. V. (1984). Random Vibration of Elastic Systems. Martinus and Nijhoff Publishers, The Hague, The Netherlands.

Bolotin, V. V. and Moskvin, V. G. (1972). On parametric resonances in stochastic systems. Mech. Solids (Mekh. Tver. Tela) 7(4), 77-82.

Bolotin, V. V. and Moskvin, V. G. (1973). Excitation of parametric vibrations in stochastic systems with two degrees-of-freedom. Mech. Solids (Mekh. Tver. Tela) 8(3), 30-37.

Bover, D. C. C. (1978a). Ph.D. Thesis, Australian National University. A computational study of several problems in stochastic modeling.

Bover, D. C. C. (1978b). Moment equations for non-linear stochastic systems. J. Math Anal. Appl. 65(2), 306-320.

Brissaud, A. and Frisch, U. (1974). Solving linear stochastic differential equations. J. Math. Phys. 15(5), 524-534.

Brockett, R. W. (1976). Parametrically stochastic linear differential equations. Math. Program. Study 5, 8-21.

Bunke, H. (1970). On the stability condition in the mean of a system of differential equations with stochastic parameters. (in German) Monatsb. Deutsch. Akad. Wiss., Berlin, 12, 734-740.

Bunke, H. (1971). On the stability of ordinary differential equations under persistent random disturbances. (in German) Z. Angew. Math. Mech. (ZAMM) 51, 543-546.

Bunke, H. (1972). Ordinary Differential Equations with Random Parameters. (in German) Akademie-Verlag, Berlin.

Caughey, T. K. (1960). Comment on 'On the stability of random systems and the stabilization of deterministic systems with random noise,' by J. C. Samuels. J. Acoust. Soc. Am. 32, p. 1356.

Caughey, T. K. (1963a). Derivation and application of the Fokker-Planck equation to discrete non-linear systems subjected to white random excitation. J. Acoust. Soc. Am. 35(11), 1683-1692.

Caughey, T. K. (1963b). Equivalent linearization technique. J. Acoust. Soc. Am. 35(11), 1706-1711.

Caughey, T. K. (1971); in Advances in Applied Mechanics. (Edited by C. S. Yih). Vol. 11, 209-243, Academic Press, New York. Non-linear theory of random vibration.

Caughey, T. K. and Dienes, J. K. (1961). Analysis of a non-linear first order system with white noise input. J. Appl. Phys. 32, 2476-2479.

Caughey, T. K. and Dienes, J. K. (1962). The behavior of linear systems with random parametric excitation. J. Math. Phys. 41, 300-318.

Caughey, T. K. and Dickerson, J. R. (1967). Stability of linear dynamic systems with narrow band parametric excitation. ASME J. Appl. Mech. 34, 709-713.

Caughey, T. K. and Gray, A. H., Jr. (1965). On the almost sure stability of linear dynamic systems with stochastic coefficients. ASME J. Appl. Mech. 32, 365-372.

Caughey, T. K. and Ma, F. (1982). The steady-state response of a class of dynamical systems to stochastic excitation. ASME J. Appl. Mech. 49, 629-632.

Caughey, T. K. and Payne, H. J. (1967). On the response of a class of self-excited oscillators to stochastic excitation. Int. J. Non-Lin. Mech. 2, 125-151.

Cesari, L. (1963). Asymptotic Behavior and Stability Problems in Ordinary Differential Equations. Academic Press, New York.

Chelpanov, I. B. (1962). Vibration of a second order system with a randomly varying parameter. Appl. Math. Mech. (Prikl. Mat. Mekh., PMM) 26(4), 1145-1152.

Chen, W. L. and Huang, T. C. (1984); in Random Vibrations. (Edited by T. C. Huang and P. D. Spanos). ASME WAM, AMD-Vol. 65, 25-34. The stability and response of a randomly excited hanging string in a fluid.

Chow, P. L. and Chiou, K. L. (1981). Asymptotic stability of randomly perturbed linear periodic systems. SIAM J. Appl. Math. 40(2), 315-326.

Clarkson, B. L. (1977). Stochastic Problems in Dynamics. Pitman, London.

Coppel, W. (1975). Stability and Asymptotic Behavior of Differential Equations. Heath, Boston, Mass.

Cramér, H. (1946). Mathematical Methods of Statistics. Princeton University Press, Princeton.

Crandall, S. H. (1963). Perturbation technique for random vibration of non-linear systems. J. Acoust. Soc. Am. 35(11), 1700-1705.

Crandall, S. H. (1966); in Applied Mechanics Surveys. (Edited by H. N. Abramson, H. Liebowitz, J. M. Crowly, and S. Juhasz). 681-689, Spartan Books, Washington, D. C. Random vibration.

Crandall, S. H. (1980). Non-Gaussian closure for random vibration of non-linear oscillators. Int. J. Non-linear Mech. 15, 303-313.

Crandall, S. H. (1981); in Proc. IX International Conf. on Non-linear Oscillations. Kiev, Aug. 30-Sept. 5, 1981. Non-Gaussian closure techniques for stationary random vibration.

Crandall, S. H. and Mark, W. D. (1963). Random Vibration in Mechanical Systems. Academic Press, New York.

Crandall, S. H. and Zhu, W. Q. (1983). Random vibration: a survey of recent developments. ASME J. Appl. Mech., 50th Anniversary Issue 50, 953-962.

Cumming, I. G. (1967). Derivation of the moments of a continuous stochastic system. Int. J. Contr. 5(1), 85-90.

Curtain, R. F. (1972). Stability of Stochastic Systems. Lecture Notes in Mathematics 294, Springer-Verlag, Berlin.

Dalzell, J. F. (1967). Southwest Res. Inst., San Antonio, Tech. Rept. 1. Exploratory studies of liquid behavior in randomly excited tanks: longitudinal excitation.

Danilin, A. B. and Yadykin, I. B. (1981). Exponential stability of stochastic finite-difference systems. Automation & Remote Control (Avtom. i Telemekh.) 42(4) pt. 1, 440-445.

Darkhovskii, B. S. and Leibovich, V. C. (1971). Statistical stability and output moments of a certain class of systems with a random varying structure. Automation & Remote Control (Avtom. i Telemekh.) 32(10), 1560-1567.

Dash, P. and Iyenger, R. N. (1982). Analysis of randomly time-varying systems by Gaussian closure technique. J. Sound Vib. 83(2), 241-251.

Dashevskii, M. L. (1967). Approximate analysis of the accuracy on non-stationary, non-linear systems, using the method of semi-invariants. Automation and Remote Control (Avtom. i Telemekh.) 28(11), 1673-1690.

Dashevskii, M. L. and Liptser, R. Sh. (1967). Application of conditional semi-invariants in problems of non-linear filtering of Markov processes. Automation and Remote Control (Avtom. i Telemekh.) 28(6), 63-74.

Davenport, W. B., Jr. and Root, W. L. (1958). An Introduction to the Theory of Random Signals and Noise. McGraw-Hill, New York.

Dickerson, J. R. and Caughey, T. K. (1969). Stability of continuous dynamic systems with parametric excitation. ASME J. Appl. Mech. 36(2), 212-216.

Dimentberg, M. F. (1967). Subharmonic resonance in a system with a randomly varying natural frequency. Appl. Math. Mech. (Prikl. Mat. Mekh., PMM) 31(4), 761-762.

Dimentberg, M. F. (1980a). Oscillations of a system with non-linear stiffness under simultaneous external and parametric random excitations. Mech. Solids (Mekh. Tver. Tela) 15(5), 42-45.

Dimentberg, M. F. (1980b). Non-linear Stochastic Problems of Mechanical Vibrations. (in Russian) Izdatel'stvo Nauka, Moscow.

Dimentberg, M. F. (1980c). Exact solution of a particular problem of oscillations of a system with random parametric excitation. Appl. Math. Mech. (Prikl. Mat. Mekh., PMM) 44(6), 816-818.

Dimentberg, M. F. (1982). An exact solution to a certain non-linear random vibration problem. Int. J. Non-linear Mech. 17(4), 231-236.

Dimentberg, M. F. and Gorbunov, A. A. (1975). Certain problems of identification of an oscillatory system with random parametric excitation. Soviet Appl. Mech. (Prikl. Mekh.) 11(4), 401-404.

Dimentberg, M. F. and Isikov, N. E. (1977). Oscillations of systems having periodically varying parameters when experiencing random action. Mech. Solids (Mekh. Tver. Tela) 12(4), 66-72.

Dimentberg, M. F. and Isikov, N. E. (1983). Combination resonances in systems with periodic parametric and random external excitation. Mech. Solids (Mekh. Tver. Tela) 18(1), 21-25.

Dimentberg, M. F., Isikov, N. E., and Model, R. (1981). Vibration of a system with cubic non-linear damping and simultaneous periodic and random parametric excitation. Mech. Solids (Mekh. Tver. Tela) 16(8), 19-21.

Dimentberg, M. F. and Menyailov, A. I. (1979). Response of a single-mass vibroimpact system to white-noise random excitation. Z. Angew. Math. Mech. (ZAMM) 59, 709-716.

Dimentberg, M. F. and Sidorenko, A. S. (1978). Interaction between oscillations that arise in a linear system when external and parametric random perturbations are present. Mech. Solids (Mekh. Tver. Tela) 13(3), 1-4.

Doob, J. L. (1953). Stochastic Processes. John Wiley & Sons, New York.

Dowell, E. H. (1984). Observation and evolution of chaos for an autonomous system. ASME J. Appl. Mech. 51, 664-673.

Drexler, J. and Kropác, O. (1968); in Proc. Conf. Dynamics, 101-109. Contribution to random vibration of one class of non-linear parametrically excited two-mass systems.

Drexler, J. and Kropác, O. (1969); in 12th IUTAM Cong. Applied Mechanics, 179-191. One class of non-linear stochastic differential equations characterized by random excitation.

Einstein. A. (1905). Über die von der molekularkinetischen theorie der Wärne geforderte bewegung von in ruhenden flüssigkeiten susspendierich teilchen. Ann. Physik 17, 549-560. Also (1956); collected in Investigations of the Theory of Brownian Movement. Dover Publication, New York.

Evan-Iwanowski, R. M. (1976). Resonance Oscillations in Mechanical Systems. Elsevier Science, New York.

Fedosov, Y. A., Podvintsev, Yu. V., Sebryakov, G. G., and Chernyshev, A. V. (1968); in Automatic Control and Computer Technology. (in Russian) No. 9, Mashgiz, Moscow. Certain problems of the correlation theory of dynamic accuracy of systems with random parameters.

Fontenot, L. L., McDonough, G. F., and Lomen, D. O. (1965); in 6th Int. Symp. Space Technology and Science. 199-209, Tokyo. Liquid-free surface instability resulting from random vertical acceleration.

Friedman, A. (1975). Stochastic Differential Equations and Applications. Vol. 1, Academic Press, New York.

Froude, W. (1863). Remarks on Mr. Scott Russell's paper on rolling. Trans. Instit. Naval Res. 4, 232-275.

Fuh, J. S., Hong, C. Y. R., Lin, Y. K., and Prussing, J. E. (1983). Coupled flap-torsional response of a rotor blade in forward flight due to atmospheric turbulence excitations. J. Amer. Helicopter Soc. 28(7), 3-12.

Fujimori, Y. (1978). Ph.D. Thesis, University of Illinois at Urbana, Champaign. Effect of atmospheric turbulence on the stability of a lifting rotor blade.

Fujimori, Y., Lin, Y. K., and Ariaratnam, S. T. (1979). Rotor blade stability in turbulent flow, Part II. AIAA J. 17(7), 673-678.

Gantmacher, F. R. (1959a). Application of the Theory of Matrices. Interscience Publishers, Inc., New York.

Gantmacher, F. R. (1959b). Theory of Matrices. Chelsea Publishing Co., New York.

Gaonkar, G. H. (1971a). Dynamic systems with random initial state. J. Eng. Math. 5(3), 171-178.

Gaonkar, G. H. (1971b). Interpolation of aerodynamic damping of lifting rotors in forward flight from measured response variance. J. Sound Vib. 18(3), 381-389.

Gaonkar, G. H. (1971c). Linear systems with non-stationary random inputs. Int. J. Conr. 14(1), 161-174.

Gaonkar, G. H. (1971d). Computational aspects of state correlation matrix and threshold crossings of variable systems with canonical expansion of input random vectors. Int. J. Contr. 14(3), 401-415.

Gaonkar, G. H. (1972). A general method with shaping filters to study random vibration statistics of lifting rotors with feedback controls. J. Sound Vib. 21(2), 213-225.

Gaonkar, G. H. (1974a). Peak statistics and narrow band feature of coupled torsion flapping rotor blade vibrations to turbulence. J. Sound Vib. 34(1), 35-52.

Gaonkar, G. H. (1974b). A study of lifting rotor flapping response peak distribution in atmospheric turbulence. J. Aircraft 11(2), 104-111.

Gaonkar, G. H. (1977). Random vibration peaks in rotorcraft and the effect of non-uniform gusts. J. Aircraft 14(1), 68-76.

Gaonkar, G. H. (1980); in Proc. AIAA/ASME/ASCE/AHS 21st Structures, Structural Dynamics and Materials Conference. 938-956, Seattle, Washington. Review of the non-stationary gust responses of flight vehicles.

Gaonkar, G. H. (1981). Gust response of rotor propeller systems. J. Aircraft 18(5), 389-396.

Gaonkar, G. H. and Hohenemser, K. H. (1969). Flapping response of lifting rotor blades to atmospheric turbulence. J. Aircraft 6(6), 496-503.

Gaonkar, G. H. and Hohenemser, K. H. (1971). Stochastic properties of turbulent excited rotor blade vibration. AIAA J. 9(3), 419-424.

Gaonkar, G. H. and Hohenemser, K. H. (1972). An advanced stochastic model for threshold cressing studies of rotor blade vibrations. AIAA J. 10(8), 1100-1101.

Gaonkar, G. H., Hohenemser, K. H. and Yin, S. K. (1972). Random gust response statistics for coupled torsion-flapping rotor blade vibrations. J. Aircraft 9, 726-729.

Gaonkar, G. H. and Subramanian, A. K. (1977). A study of feedback, blade and hub parameters on flap bending due to non-uniform rotor disk turbulence. J. Sound Vib. 51(4), 501-515.

Geman, S. (1982). Almost sure stable oscillations in a large system of randomly coupled equations. SIAM J. Appl. Math. 42(4), 695-703.

Gikhman, I. I. and Skorokhod, A. V. (1965). Introduction to the Theory of Random Processes. W. B. Saunders Comp., Philadelphia.

Gikhman, I. I. and Skorokhod, A. V. (1972). Stochastic Differential Equations. (Band 72), Springer-Verlag, Berlin.

Gopalsamy, K. (1976). On a class of linear systems with random coefficients. Z. Angew. Math. Mech. (ZAMM) 56(11), 453-459.

Gorbunov, A. A. and Dimentberg, M. F., Jr. (1974). Some diagnostic problems for an oscillating system with periodic parametric excitation. Mech. Solids (Mekh. Tver. Tela) 9(2), 43-46.

Gradshteyn, I. S. and Ryzhik, I. M. (1980). Tables of Integrals, Series and Products. (Edited by A. Jefferey). Academic Press, New York.

Graefe, P. W. U. (1966). Stability of a linear second order system under random parametric excitation. Ing. Arch. 35, 202-205.

Grant, B. E. (1966). A method for measuring aerodynamic damping of helicopter rotors in forward flight. J. Sound Vib. 3(3), 407-421.

Gray, A. H., Jr. (1967). Frequency-dependent almost sure stability conditions for a parametrically excited random vibrational system. ASME J. Appl. Mech. 34, 1017-1019.

Gray, A. H., Jr. and Caughey, T. K. (1965). A controversy in problems involving random parametric excitation. J. Math. Phys. 44(3), 288-296.

Haddara, M. R. (1975); in Proc. Int. Conf. on Stability of Ships and Ocean Vehicles. University of Strathclyde, Glasgow. A study of the stability of the mean and variance of rolling motion in random waves.

Hahn, W. (1963). Theory and Application of Liapunov's Direct Method. Prentice-Hall, Inc., Englewood Cliffs, N. J.

Haines, C. W. (1967). Hierarchy methods for random vibrations of elastic strings and beams. J. Eng. Math. 1, 293-305.

Hatwal, H., Mallik, A. K., and Ghosh, A. (1983a). Forced non-linear oscillations of an autoparametric system, part I: periodic responses. ASME J. Appl. Mech. 50, 657-662.

Hatwal, H., Mallik, A. K., and Ghosh, A. (1983b). Forced non-linear oscillations of an autoparametric system, part II: chaotic response. ASME J. Appl. Mech. 50, 663-668.

Hemp, G. W. and Sethna, P. R. (1968). On dynamical systems with high frequency parametric excitation. Int. J. Non-linear Mech. 3, 351-365.

Hennig, K. (1983). Random Vibration and Reliability. Akademie-Verlag, Berlin.

Hohenemser, K. H. and Crews, S. T. (1973). Model tests on unsteady rotor wake effects. J. Aircraft 10(1), 58-60.

Holmes, P. J. and Moon, F. C. (1983). Strange attractors and chaos in non-linear mechanics. ASME J. Appl. Mech. 50th Anniversary Issue 50, 1021-1032.

Howe, M. S. (1974). The mean square stability of an inverted pendulum subjected to random parametric excitation. J. Sound Vib. 32(3), 407-421.

Hsu, C. S. and Lee, T. H. (1969); in Proc. IUTAM Symp. Instability of Continuous Systems. (Edited by H. Leipholz), 112-118. A stability study of continuous systems under parametric excitation via Liapunov's direct method.

Ibrahim, R. A. (1978a). Parametric vibration, part III: current problems (1). The Shock Vib. Digest 10(3), 41-57.

Ibrahim, R. A. (1978b). Parametric vibration, part IV: current problems (2). The Shock Vib. Digest 10(4), 19-47.

Ibrahim, R. A. (1978c). Stationary response of a randomly parametric excited non-linear system. ASME J. Appl. Mech. 45, 910-916.

Ibrahim, R. A. (1981). Parametric vibration, part VI: stochastic problems (2). Shock Vib. Digest 13(9), 23-35.

Ibrahim, R. A. (1982). Self-excited vibration of a non-linear system with random parameters. Shock Vib. Bulletin 53(1), 135-144.

Ibrahim, R. A. and Barr, A. D. S. (1978a). Parametric vibration, part I: mechanics of linear problems. The Shock Vib. Digest 10(1), 15-29.

Ibrahim, R. A. and Barr, A. D. S. (1978b). Parametric vibration, part II: mechanics of non-linear problems. The Shock Vib. Digest 10(2), 9-24.

Ibrahim, R. A. and Heo, H. (1985). Autoparametric vibration of coupled beams under random support motion. Paper submitted for presentation at the ASME Vibration Conference, Cincinnati, Ohio.

Ibrahim, R. A. and Roberts, J. W. (1976). Broad band random excitation of a two degree-of-freedom system with autoparametric coupling. J. Sound Vib. 44(3), 335-348.

Ibrahim, R. A. and Roberts, J. W. (1977). Stochastic stability of the stationary response of a system with autoparametric coupling. Z. Angew. Math. Mech. (ZAMM) 57, 643-649.

Ibrahim, R. A. and Roberts, J. W. (1978). Parametric vibration, part V: stochastic problems. Shock Vib. Digest 10(5), 17-38.

Ibrahim, R. A. and Soundararajan, A. (1983). Non-linear parametric liquid sloshing under wide band random excitation. J. Sound Vib. 91(1), 119-134.

Ibrahim R. A. and Soundararajan, A. (1985). An improved approach for random parametric response of dynamic systems with non-linear inertia. Int. J. Non-linear Mech. (in print).

Ibrahim, R. A., Soundararajan, A., and Heo, H. (1985). Stochastic response of non-linear dynamic systems based on a non-Gaussian closure. ASME J. Appl. Mech. (in print).

Infante, E. F. (1968). On the stability of some linear non-autonomous random systems. ASME J. Appl. Mech. 35, 7-12.

Infante, E. F. and Plaut, R. H. (1969). Stability of a column subjected to a time-dependent axial load. AIAA J. 7(4), 766-768.

Irwin, H. P. A. H. and Schuyler, G. D. (1978); in ASCE Spring Convention and Exhibit, Reprint 3268, Pittsburgh, Penn. Wind effect on a full aeroelastic bridge model.

Itô, K. (1944). Stochastic integral. Proc. Imperial Academy, Tokyo 20, 519-524.

Itô, K. (1951a). On a formula concerning stochastic differentials. Nagoya Math. J. 3, 55-65.

Itô, K. (1951b). Stochastic differential equations. Mem. Amer. Soc. 4.

Itô, K. and Nisio, M. (1964). On stationary solutions of a stochastic differential equation. J. Math. Kyoto Univ. 4, 1-75.

Ivovich, V. A. (1969). Vibration of a plane grid subjected to random parametric excitation. Soviet Appl. Mech. (Prikl. Mekh.) 5(3), 92-99.

Iyenger, R. N. and Dash, P. K. (1976). Random vibration analysis of stochastic time-varying systems. J. Sound Vib. 45(1), 69-89.

Iyenger, R. N. and Dash, P. K. (1978). Study of the random vibration of non-linear systems by the Gaussian closure technique. ASME J. Appl. Mech. 45, 393-399.

Jaeger, L. G. and Barr, A. D. S. (1966); in Proc. Symp. Design for Earthquake Loadings, VII-1, McGill University, Canada. Parametric instability in structures subjected to prescribed support motion.

Jazwinski, A. H. (1970). Stochastic Processes and Filtering Theory. Academic Press, New York.

Johnson, W. (1980). Helicopter Theory. Princeton University Press, Princeton, New Jersey.

Kats, I. I. (1964). On the stability of stochastic systems in the large. Appl. Math. Mech. (Prikl. Mat. Mekh., PMM) 28, 449-456.

Kats, I. I. and Krasovskii, N. N. (1960). On the stability of systems with random parameters. Appl. Math. Mech. (Prikl. Mat. Mekh., PMM) 24(5), 1225-1246.

Katsnel'son, A. N., Kolovskii, M. Z., and Troitskaya, Z. V. (1971). Stochastic stability of linear systems. Mech. of Solids (Mekh. Tver. Tela) 6(3), 56-62.

Keller, J. B. (1964). Stochastic equations and wave propagation in random media. Amer. Math. Soc. XVI (annual Symposia in Applied Mathematics).

Kendall, M. and Stuart, A. (1969). The Advanced Theory of Statistics, Vol. 1, Distribution Theory. 3rd Edition, Charles Griffin, London.

Khas'miniskii, R. Z. (1962). On the stability of the trajectories of Markov processes. Appl. Math. Mech. (Prikl. Mat. Mekh., PMM) 26, 1552-1565.

Khas'miniskii, R. Z. (1964). Operation of a conservative system with small friction and small random noise. Appl. Math. Mech. (Prikl. Mat. Mekh., PMM) 28(5), 1126-1130.

Khas'miniskii, R. Z. (1966). A limit theorem for solution of differential equations with random right-hand side. Theory Probab. Appl. 11(3), 390-406.

Khas'miniskii, R. Z. (1967). Necessary and sufficient conditions for the asymptotic stability of linear systems. Theory Probab. Appl. 11, 144-147.

Khas'miniskii (Has'miniskii) R. Z. (1980). Stochastic Stability of Differential Equations. Sijthoff & Noordhoff, Alphen aan den Rijn, The Netherlands.

Kistner, A. (1977); in Proc. IUTAM Symp. Stochastic Problems in Dynamics. (Edited by B. L. Clarkson). 37-53, Pitman, London. On moments of linear systems excited by a colored noise process.

Kistner, A. (1978). Ph.D. Thesis, University of Stuttgart. Strict statement about the solution, moment and stability of linear system with parametric color noise. (in German).

Kliatskin, V. I. (1975). Statistical Description of Dynamic Systems with Fluctuating Parameters. (in Russian) Nauka, Moscow.

Kliatskin, V. I. (1980). Stochastic Equations and Waves in Randomly Inhomogeneous Media. (in Russian) Nauka, Moscow.

Kolmogorov, A. N. (1933). Grundbegriffe der Wahrscheinkeitsrechung. Springer-Verlag, Berlin.

Kolomiets, V. G. (1967); in Proc. 4th Int. Conf. Non-linear Oscillations. 181-185, Prague. Random oscillations of non-autonomous quasilinear systems.

Kolomiets, V. G. (1972); in Proc. IUTAM Symp. Stochastic Stability of Dynamical Systems. Lecture Notes in Mathematics 294. (Edited by R. F. Curtain). 317-323, Springer-Verlag, Berlin. Application of averaging principle in non-linear oscillatory stochastic systems.

Kolovskii, M. Z. and Troitskaya, Z. V. (1972). On the stability of linear systems with random parameters. Appl. Math. Mech. (Prikl. Mat. Mekh., PMM) 36, 202-207.

Konstantinov, V. M. (1970). On the stability of stochastic difference systems. Problems of Information Transmission (Probl. Peredachi Inf.) 6(1), 81-86.

Kozin, F. (1963). On almost sure stability of linear systems with random coefficients. J. Math. Phys. 42, 59-67.

Kozin, F. (1965a). On almost sure asymptotic sample properties of diffusion processes defined by stochastic differential equations. J. Math. Kyoto Univ. 4, 515-528.

Kozin, F. (1965b). On relations between moment properties and almost sure Liapunov stability for linear stochastic systems. J. Math. Anal. Appl. 10(2), 342-352.

Kozin, F. (1966). Discussion on: "On the almost sure stability of linear dynamic systems with stochastic coefficients," by T. K. Caughey and A. H. Gray. ASME J. Appl. Mech. 33, 234-235.

Kozin, F. (1969). A survey of stability of stochastic systems. Automatica 5, 95-112.

Kozin, F. (1972); in Proc. IUTAM Symp. Stability of Stochastic Dynamical Systems. Lecture Notes in Mathematics 294. (Edited by R. F. Curtain). 186-229, Springer-Verlag, Berlin. Stability of the linear stochastic system.

Kozin, F. and Milstead, R. M. (1979). The stability of a moving elastic strip subjected to random parametric excitation. ASME J. Appl. Mech. 46(2), 404-410.

Kozin, F. and Prodromou, S. (1971). Necessary and sufficient conditions for almost sure sample stability of linear Itô equations. SIAM J. Appl. Math. 21(3), 413-424.

Kozin, F. and Sugimoto, S. (1977); in Proc. IUTAM Symp. Stochastic Problems in Dynamics. (Edited by B. L. Clarkson). 8-35, Pitman, London. Decision criteria for stability of stochastic system from observed data.

Kozin, F. and Wu, C. M. (1973). On the stability of linear stochastic differential equations. ASME J. Appl. Mech. 40, 87-92.

Kraichnan, R. H. (1962). The closure problem of turbulence theory. Amer. Math. Soc. 13, 199-225.

Krasovskii, N. N. (1961). On mean square optimum stabilization of damped random perturbations. Appl. Math. Mech. (Prikl. Mat. Mekh., PMM) 25(5), 1212-1227.

Kropáč, O. (1971). On some qualitative characteristics of one class of non-linear parametrically excited stochastic differential equations. J. Sound Vib. 14(2), 241-249.

Kropáč, O. and Drexler, J. (1967); in Proc. IV Int. Conf. on Non-linear Oscillations. 349-360, Prague. An analog study of random parametric vibration of a non-linear dynamic second order system.

Kul'terbaev, Kh. P. (1979). Random parametric oscillations of cylindrical hollow shells. Soviet Appl. Mech. (Prikl. Mekh.) 14(9), 938-942.

Kurnik, W. and Tylikowski, A. (1983). Stochastic stability and non-stability of a cylindrical shell. Ing.-Atch. 53, 363-369.

Kushner, H. J. (1965). On the construction of stochastic Liapunov functions. Trans. IEEE AC-10, p. 477.

Kushner, H. J. (1967). Stochastic Stability and Control. Academic Press, New York.

Kuznetsov, P. I., Stratonovich, R. L., and Tikhonov, V. I. (1960). Quasi-moment functions in the theory of random processes. Theory Probab. Appl. 5, 80-97.

Landa, P. S. and Stratonovich, R. L. (1962). Theory of fluctuations of different transition systems with one stationary state with respect to the other. Vestnik MGU Ser. Fiz. Astron., Series III, No. 1, Moscow University.

Laning, J. H. and Battin, R. H. (1956). Random Processes in Automatic Control. McGraw-Hill, New York.

La Salle, J. and Lefschetz, S. (1961). Stability by Liapunov's Direct Method with Applications. Academic Press, New York.

Ledwich, M. A. (1974). Ph.D. Thesis, University of Cambridge. The application of Hermite polynomials to system identification.

Leibowitz, M. A. (1963). Statistical behavior of linear systems with randomly varying parameters. J. Math. Phys. 4, 852-858.

Lepore, J. A. and Shah, H. C. (1968). Dynamic stability of axially loaded columns subjected to stochastic excitations. AIAA J. 6(8), 1515-1521.

Lepore, J. A. and Shah, H. C. (1970). Dynamic stability of circular plates under stochastic excitation. J. Spacecraft Rockets 7, 582-587.

Lepore, J. A. and Stoltz, R. A. (1971a); in Proc. AIAA/ASME 12th Structural Dynamics and Materials Conf., Anaheim, Calif. Dynamic stability of cylindrical shells under axial and radial stochastic excitations.

Lepore, J. A. and Stoltz, R. A. (1971b); in Developments of Mechanics. (Edited by L. H. N. Lee and A. A. Szewczyk). 571-585, University of Notre Dame Press, Notre Dame, Indiana. Stability of linear dynamic systems under stochastic parametric excitations.

Lepore, J. A. and Stoltz, R. A. (1972); in Proc. IUTAM Symp. Stability of Stochastic Dynamical Systems. Lecture Notes in Mathematics 294. (Edited by R. F. Curtain). 239-251, Springer-Verlag, Berlin. Stability of linear cylindrical shells subjected to stochastic excitations.

Lepore, J. A. and Stoltz, R. A. (1973). Stability of a stochastically excited non-linear cylindrical shell. AIAA J. 11(6), 801-806.

Lepore, J. A. and Stoltz, R. A. (1974). Stability of cylindrical shells under random excitations. ASCE J. Eng. Mech. Div. EM3 100, 531-546.

Levit, M. V. and Yakubovich, V. A. (1970). Algebraic criterion for stochastic stability of linear systems with parametric action of the white noise type. Appl. Math. Mech. (Prikl. Mat. Mekh., PMM) 36(1), 130-136.

Lin, Y. K. (1967). Probabilistic Theory of Structural Dynamics. McGraw-Hill, New York.

Lin, Y. K. (1969). Random processes. ASME Appl. Mech. Rev. 22(8), 825-831.

Lin, Y. K. and Ariaratnam, S. T. (1980). Stability of bridge motion in turbulent winds. J. Struct. Mech. 8(1), 1-15.

Lin, Y. K., Fujimori, Y., and Ariaratnam, S. T. (1979). Rotor blade stability in turbulent flows, part I. AIAA J. 17(6), 545-552.

Lin, Y. K. and Holmes, P. J. (1978). Stochastic analysis of wind-loaded structures. ASCE J. Eng. Mech. Div. 104(EM2), 421-440.

Lin, Y. K. and Prussing, J. E. (1982). Concepts of stochastic stability in rotor dynamics. J. Amer. Helicopter Soc. 27(2), 73-74.

Loeve, M. (1963). Probability Theory. Van Nostrand, Princeton, New Jersey.

Ly, B. L. (1974). Ph.D. Thesis, University of Waterloo, Ontario. Topics in the stability of stochastic systems.

Lyubarskii, G. Y. and Robotnikov, Y. L. (1963). On differential equations with random coefficients. Theory Probab. Appl. 8, 290-298.

McKean, H. P., Jr. (1969). Stochastic Integrals. Academic Press, New York.

McLachlan, N. W. (1947). Theory and Application of Mathieu Functions. Oxford University Press, New York.

McShane, E. I. (1970). Stochastic Calculus and Stochastic Models. Academic Press, New York.

Mehr, C. B. and Wang, P. K. C. (1966). Discussion on "On the almost sure stability of linear dynamic systems with stochastic coefficients" by T. K. Caughey and A. H. Gray. ASME J. Appl. Mech. 33, 234-236.

Meirovitch, L. (1970). Methods of Analytical Dynamics. McGraw-Hill, New York.

Melsa, J. L. and Sage, A. P. (1973). An Introduction to Probability and Stochastic Processes. Prentice-Hall, Inc., New Jersey.

Milstead, R. M. (1975). Ph.D. Thesis, Polytechnic Institute of New York, New York. The dynamic stability of an axially moving thin elastic strip subjected to random parametric excitation.

Mirkina, A. S. (1975). Response of linear systems to transient parametric perturbation. Soviet Appl. Mech. (Prikl. Mekh.) 11(10), 1097-1103.

Mirkina, A. S. (1977). Determining the second instability region for equations with random coefficients. Mech. Solids (Mekh. Tver. Tela) 12(6), 80-85.

Mitchell, R. R. (1968a). NASA-CR-98009, Washington, D. C. Stochastic stability of the liquid-free surface in vertically excited cylinders.

Mitchell, R. R. (1968b). NASA-CR-980016, Washington, D. C. Stability of a simply supported rod subjected to a random longitudinal force.

Mitchell, R. R. (1970). Ph.D. Thesis, Purdue University, Lafayette, Indiana. Necessary and sufficient conditions for sample stability of second order stochastic differential equations.

Mitchell, R. R. (1972). Stability of the inverted pendulum subjected to almost periodic and stochastic base motion: an application to the method of averaging. Int. J. Non-linear Mech. 7, 101-123.

Mitchell, R. R. and Kozin, F. (1974). Sample stability of second order linear differential equations with wide band noise coefficients. SIAM J. Appl. Math. 27(4), 571-605.

Model, R. (1978). Combination resonance of stochastic oscillatory systems. (in German) Z. Angew. Math. Mech. (ZAMM) 58, 377-382.

Moon, F. C. (1980). Experiments on chaotic motions of a forced nonlinear oscillator: strange attractors. ASME J. Appl. Mech. 47, 638-644.

Moon, F. C. and Holmes, P. J. (1979). A magnetoelastic strange attractor. J. Sound Vib. 65(2), 276-296.

Morozan, T. (1967a). Stability of controlled systems with random parameters. (in Russian) Rev. Roumaine Math. Pures Appl. 12, 545-552.

Morozan, T. (1967b). Stability of some linear stochastic systems. J. Diff. Equa. 3, 153-169.

Morozan, T. (1967c). Stability of linear systems with random parameters. J. Diff. Equa. 3, 170-178.

Morozan, T. (1968). Stability of differential systems with random parameters. J. Math. Anal. Appl. 24, 669-676.

Morozan, T. (1969). Stability of Systems with Random Parameters. (in Romanian) Editura Academiei Republicii Socialiste Roumaine, Bucharest.

Mortensen, R. E. (1969). Mathematical problems of modeling stochastic non-linear dynamic systems. J. Statis. Phys. 1(2), 271-296.

Moskvin, V. G. and Smirnov, A. I. (1975). On the stability of linear stochastic systems. Mech. Solids. (Mekh. Tver. Tela) 10(4), 58-61.

Muhuri, P. K. (1980). A study of the stability of the rolling motion of a ship in an irregular seaway. Int. Shipbuilding Progr. 27, 139-142.

Nakamizo, T. (1970). On the state estimation for non-linear dynamic systems. Int. J. Contr. 11(4), 683-695.

Nakamizo, T. and Sawaragi, Y. (1972); in Proc. IUTAM Symp. Stability of Stochastic Dynamical Systems. Lecture Notes in Mathematics 294. (Edited by R. F. Curtain). 173-185, Springer-Verlag, Berlin. Analytical study on n-th order linear system with stochastic coefficients.

Nayfeh, A. H. and Mook, D. T. (1979). Non-Linear Oscillations. Wiley-Interscience Publications, New York.

Nemat-Nasser, S. (1972); in Study No. 6 On Stability. Chapter 10, 351-384, University of Waterloo, Ontario. On stability under non-conservative loads.

Ness, D. J. (1967). Small oscillations of a stabilized inverted pendulum. Amer. J. Phys. 35, 964-967.

Nevel'son, M. B. (1966). Some remarks concerning the stability of a linear stochastic system. Appl. Math. Mech. (Prikl. Mat. Mekh., PMM) 30(6), 1332-1335.

Nevel'son, M. B. (1967). Behavior of a linear system under small random excitation of its parameters. Appl. Math. Mech. (Prikl. Mat. Mekh., PMM) 31(3), 552-555.

Nevel'son, M. B. and Khas'miniskii, R. Z. (1966a). Stability of stochastic systems. Prob. Information Transm. 2(3), 61-74.

Nevel'son, M. B. and Khas'miniskii, R. Z. (1966b). Stability of a linear system with random perturbations of its parameters. Appl. Math. Mech. (Prikl. Mat. Mekh., PMM) 30(2), 487-493.

Nevel'son, M. B. and Khas'miniskii, R. Z. (1973). Stochastic Approximation and Recursive Estimation. Vol. 47, American Math. Soc., Providence, Rhode Island.

Newland, D. E. (1965). Energy sharing in random vibration of non-linearly coupled modes. J. Inst. Math. Appl. 1(3), 199-207.

Newland, D. E. (1975). An Introduction to Random Vibration and Spectral Analysis. Longman, London.

Nigam, N. C. (1983). Introduction to Random Vibrations. The MIT Press, Cambridge, Massachusetts.

Ormiston, R. A. and Hodges, D. H. (1972). Linear flap-lag dynamics of hingeless helicopter rotor blades in hover. J. Amer. Helicopter Soc. 17(2), 2-14.

Palmer, J. T. (1966). Ph.D. Thesis, Purdue University, Lafayette, Indiana. Sufficient conditions for almost sure Liapunov stability for a class of linear systems.

Papanicolaou, G. G. and Kohler, W. (1974). Asymptotic theory of mixing stochastic ordinary differential equations. Commun. Pure Appl. Math. 27, 641-668.

Papoulis, A. (1965). Probability, Random Variables, and Stochastic Processes. McGraw-Hill, New York.

Parthasarathy, A. (1972). Ph.D. Dissertation, Syracuse University, Syracuse, New York. Deterministic and stochastic stability of a non-autonomous oscillatory system.

Parthasarathy, A. and Evan-Iwanowski, R. M. (1978). On the almost sure stability of linear stochastic systems. SIAM J. Appl. Math 34(4), 643-656.

Parzen, E. (1960). Modern Probability Theory and Its Applications. John Wiley, New York.

Parzen, E. (1962). Stochastic Processes. Holden-Day, San Francisco.

Paulling, J. R. and Rosenberg, R. M. (1959). On unstable ship motions resulting from non-linear coupling. J. Ship Res. 3(1), 36-46.

Peters, D. A. (1975a); NASA TM X-62, 425, April 1975. An approximate closed-form solution for lead-lag damping of rotor blades in hover.

Peters, D. A. (1975b). Flap-lag stability of helicopter rotor blades in forward flight. J. Amer. Helicopter Soc. 20(4), 2-13.

Phillis, Y. A. (1982). Entropy stability of continuous dynamic systems. Int. J. Control 35, 323-340.

Pinsky, M. A. (1974). Stochastic stability and the Dirichlet problem. Commun. Pure Appl. Math. 27, 311-350.

Plaut, R. H. and Infante, E. F. (1970). On the stability of some continuous systems subjected to random excitation. ASME J. Appl. Mech. 37(3), 623-628.

Podvintsev, Yu. V. (1971). The stability of vibrating systems in the case of random parametric excitation. Soviet Appl. Mech. (Prikl. Mekh.) 7(8), 904-907.

Price, W. G. (1975). A stability analysis of the roll motion of a ship in an irregular seaway. Int. Shipbuilding Progress 22, 103-112.

Price, W. G. and Bishop, R. E. D. (1974). Probability Theory of Ship Dynamics. Chapman and Hall, London.

Prodromou, S. E. (1970). Ph.D. Thesis, Polytechnic Institute of Brooklyn, N. Y. Necessary and sufficient conditions for stability of stochastic systems.

Prussing, J. E. (1981). Stabilization of an unstable linear system by parametric white noise. ASME J. Appl. Mech. 48(1), 198-199.

Prussing, J. E. and Lin, Y. K. (1982). Rotor blade flap-lag stability in turbulent flows. J. Amer. Helicopter Soc. 27(2), 51-57.

Prussing, J. E. and Lin, Y. K. (1983). A closed-form analysis of rotor blade flap-lag stability in hover and low-speed forward flight in turbulent flow. J. Amer. Helicopter Soc. 28(7), 42-46.

Rabotnikov, Iu. L. (1964). On the impossibility of stabilizing a system in the mean square by random perturbation of its parameters. Appl. Math. Mech. (Prikl. Mat. Mekh., PMM) 28(5), 1131-1136.

Richardson, J. M. (1964). The application of truncated hierarchy techniques in the solution of a stochastic differential equation. Amer. Math. Soc. 16, 290-302.

Richardson, J. M. and Levitt, L. C. (1967). Linear closure aproximation method for classical statistical mechanics. J. Math. Phys. 8, 1707-1715.

Roberts, J. B. (1978). The energy envelope of a randomly excited non-linear oscillator. J. Sound Vib. 60(2), 177-185.

Roberts, J. B. (1981). Response of non-linear mechanical systems to random excitation, part 2: equivalent linearization and other methods. The Shock Vib. Digest 13(5), 15-29.

Roberts, J. B. (1982). The effect of parametric excitation on ship rolling in random waves. J. Ship Res. 26(4), 246-253.

Roberts, J. W. (1980). Random excitation of a vibratory system with autoparametric interaction. J. Sound Vib. 69(1), 101-116.

Robson, J. D. (1963). An Introduction to Random Vibration. Edinburgh University Press, Scotland.

Rosenbloom, A., Heilfron, J., and Trautman, D. L. (1955); in IRE Convention Record, Computers, Information Theory, Automatic Control 4, 106-113, New York. Analysis of linear systems with randomly varying inputs and parameters.

Sagirow, P. S. (1976). Zur abscliebung der moment engleichungen linearer systemme mit stochastischer parameterregung. Z. Angew. Math. Mech. (ZAMM) 56, T75-T76.

Samuels, J. C. (1959); in Proc. Int. Symp. Circuit and Information Theory. IRE CT-6, 248-259. On the mean square stability of random linear systems.

Samuels, J. C. (1960). On the stability of random systems and the stabilization of deterministic systems with random noise. J. Acoust. Soc. Amer. 32, 594-601.

Samuels, J. C. (1961). Theory of stochastic linear systems with Gaussian parameter variations. J. Acoust. Soc. Amer. 33(12), 1782-1786.

Samuels, J. C. and Eringen, A. C. (1959). On stochastic linear systems. J. Math. Phys. 38, 83-103.

Sancho, N. G. F. (1968). Moment equation of a stochastic system with generalized Poisson parameters. Int. J. Contr. 8, 417-421.

Sancho, N. G. F. (1969). Moment equations of a stochastic system with two different random parameters. Int. J. Contr. 9(1), 83-88.

Sancho, N. G. F. (1970a). Technique for finding the moment equations of a non-linear stochastic system. J. Math. Phys. 11(3), 771-774.

Sancho, N. G. F. (1970b). On the approximate moment equations of a non-linear stochastic differential equation. J. Math. Anal. Appl. 29, 384-391.

Sancho, N. G. F. (1970c). Non-linear stochastic differential equations containing random parameters with small and large correlation time. J. Math Phys. 11(4), 1283-1287.

Sawaragi, Y., Nakamizo, T., and Ohe, Y. (1967); in Proc. 16th Japan Natl. Cong. Applied Mechanics. 342-347, Tokyo. Mean square stability of linear systems with random parametric excitation.

Scanlan, R. H., Beliveau, J. G., and Budlong, K. S. (1974). Indicial aerodynamic functions for bridge decks. ASCE J. Eng. Mech. Div. 100(EM4), 657-672.

Schmidt, G. (1975). Parametric Oscillations. (in German) VEB Deutscher Verlag der Wissenschaften, Berlin.

Schmidt, G. (1976); in Proc. VII Int. Conf. Non-linear Oscillations. 341-359, Berlin. Non-linear systems under random and periodic parametric excitation.

Schmidt, G. (1977a); in Proc. IUTAM Symp. Stochastic Problems in Dynamics. (Edited by B. L. Clarkson). 197-213, Pitman, London. Probability densities of parametrically excited random vibrations.

Schmidt, G. (1977b); in Proc. 14th Int. Cong. Theoretical and Applied Mechanics. (Edited by W. T. Koiter). 439-450, North-Holland Publishing Co., Amsterdam. Vibrating mechanical systems with random parametric excitation.

Schmidt, G. (1979); in Proc. VIII Int. Conf. Non-linear Oscillations. (Edited by L. Püst). 633-638, Academia, Publishing House of the Czechoslovak Academy of Sciences, Prague. Forced and parametrically excited non-linear random vibration.

Schmidt, G. (1981). Vibrations caused by simultaneous random forced and parametric excitations. (in German) Z. Angew. Math. Mekh. 60(9), 409-419.

Schmidt, G. and Schulz, R. (1983); in Proc. IUTAM Symp. Random Vibrations and Reliability. (Edited by K. Hennig). 307-315, Akademię-Verlag, Berlin. Non-linear random vibration of systems with several degrees of freedom.

Sethna, P. R. (1972); in Proc. IUTAM Symp. Stability of Stochastic Dynamical Systems. Lecture Notes in Mathematics 294. (Edited by R. F. Curtain). 273-282, Springer-Verlag, Berlin. Ultimate behavior of a class of stochastic differential systems dependent on parameter.

Sethna, P. R. (1973). Method of averaging for systems bounded for positive time. J. Math. Anal. Appl. 41, 69-96.

Sethna, P. R. and Orey S. (1980). Some asymptotic results for a class of stochastic systems with parametric excitations. Int. J. Non-linear Mech. 15(6), 431-441.

Shannon, C. E. and Weaver, W. (1949). The Mathematical Theory of Communications. University of Illinois Press, Illinois.

Simiu, E. and Scanlan, R. H. (1978). Wind Effects on Structures, An Introduction to Wind Engineering. John Wiley & Sons, New York.

Sissingh, G. J. (1968). Dynamics of rotors operating at high advance ratios. J. Amer. Helicopter Soc. 13(3), 56-63.

Sissingh, G. J. and Kuczynski, W. A. (1970). Investigations on the effect of blade torsion on the dynamics of flapping motion. J. Amer. Helicopter Soc. 15, 2-9.

Skorokhod, A. V. (1965). Studies in the Theory of Random Processes. Addison-Wesley, Mass.

Soeda, T. and Umeda, K. (1966). Stability of randomly time-varying control systems by the second method of Liapunov. Bull. Fac. Eng., Tokushima Univ. 3, p. 43.

Soeda, T. and Umeda, K. (1970). On the stability of control systems with randomly time-varying characteristics. Int. J. Contr. 11, 361-373.

Soong, T. T. (1973). Random Differential Equations in Science and Engineering. Academic Press, New York.

Spanos, P. D. (1978). Energy analysis of structural vibrations under modulated random excitation. ASCE J. Struct. Mech. 6(3), 289-302.

Spanos, P. D. (1981). Stochastic linearization in structural dynamics. ASME Appl. Mech. Rev. 34, 1-8.

Sperling, L. (1979). Analysis of stochastically excited non-linear systems using linear differential equations for generalized quasi-moment functions. (in German) Z. Angew. Math. Mech. (ZAMM) 59(4), 169-176.

Stratonovich, R. L. (1963). Topics in the Theory of Random Noise. Vol. I, Gordon & Breach, New York.

Stratonovich, R. L. (1966). A new representation for stochastic integrals and equations. SIAM J. Contr. 4, 362-371.

Stratonovich, R. L. (1967). Topics in the Theory of Random Noise. Vol. II, Gordon & Breach, New York.

Stratonovich, R. L. and Romanovskii, Yu. A. (1965); in Non-linear Transformations of Stochastic Processes. (Edited by P. I. Kuznetsov, R. L. Stratonovich, and V. I. Tikhonov). Paper 26, 322-326. Parametric effect of a random force on linear and non-linear oscillating systems.

Sunahara, Y., Asakura, T., and Morita, Y. (1977); in Proc. IUTAM Symp. Stochastic Problems in Dynamics. (Edited by B. L. Clarkson). 138-175, Pitman, London. On the asymptotic behavior of non-linear stochastic dynamical systems considering the initial states.

Sveshnikov, A. A. (1966). Applied Methods of the Theory of Random Functions. Pergamon Press, New York.

Sveshnikov, A. A. (1978). Problems in Probability Theory, Mathematical Statistics and Theory of Random Functions. Dover Publications, Inc., New York.

Syski, R. (1967); in Stochastic Differential Equations. (Edited by J. L. Saaty). Chapter 8, 346-456, McGraw-Hill, New York. Modern non-linear equations.

Szopa, J. (1976). Application of Volterra stochastic integral equations of the II-nd kind to the analysis of dynamical systems of variable inertia. J. Tech. Phys. 17(4), 423-433.

Szopa, J. (1977). Method of determining a function of moments of the stochastic output of a linear time varying system. (in Polish) Arch. Automatics and Telemech. 22(4), 359-374.

Szopa, J. (1979); in Proc. VIIIth Int. Conf. Non-linear Oscillations. 677-682, Institute of Thermomechanics, Czechoslovakia Academy of Sciences, Prague, 1978. The analysis of dynamic systems of variable inertia described by non-linear stochastic differential equations.

Szopa, J. (1981). Response of stochastic linear systems. ASCE J. Eng. Mech. Div. 107 (EM1), 1-11.

Szopa, J. (1982). Response of a multi-degree-of-freedom system of variable coefficients to random excitations. Z. Angew. Math. Mech. (ZAMM) 62, 321-328.

Szopa, J. and Wojtylak, M. (1979); in Proc. IV Conf. Computer Methods in Mechanics of Structures. (in German), 323-330. Numerical problems of calculating the probabilistic characteristics of response in stochastic dynamical systems with variable inertia.

Thomas, J. B. (1971). An Introduction to Applied Probability and Random Processes. John Wiley, New York.

Tikhonov, V. I. (1958). Fluctuation action in the simplest parametric systems. Autom. Rem. Control (Avtom. i Telem.) 19(8), 705-711.

Tondl, A. (1978). On the Interaction Between Self-Excitation and Parametric Vibrations. Monograph No. 25, National Research Institute for Machine Design, Bechovice, Prague.

Tylikowski, A. (1978). Stability of a non-linear rectangular plate. ASME J. Appl. Mech. 45, 583-585.

Tylikowski, A. (1979); in Proc. VIIIth Int. Conf. Non-linear Oscillations. 715-720, Institute of Thermomechanics, Czechoslovak Academy of Sciences, Prague. Dynamic stability of a non-linear rectangular plate.

Tylikowski, A. (1984). Dynamic stability of a non-linear cylindrical shell. ASME J. Appl. Mech. 51, 852-856.

Valeev, K. G. (1971). Dynamic stabilization of unstable systems. Solids Mech. (Mekh. Tver. Tela) 16(4), 9-16.

Valeev, K. G. and Dolya, V. V. (1974). On the dynamic stabilization of oscillations of a pendulum. Soviet Appl. Mech. (Prikl. Mekh.) 10(2), 88-93.

Vanmarcke, E. (1983). Random Fields: Analysis and Synthesis. The MIT Press, Cambridge, Massachusetts.

Van Trees, H. L. (1968). Detection, Estimation and Modulation Theory. John Wiley & Sons, New York.

Vrkoc, I. (1966). Extension of the averaging method to stochastic equations. Czech. Math. J. 16(91), 518-544.

Wan, F. Y. M. (1972). Linear partial differential equations with random forcing. Studies Appl. Math. (J. Math. Phys.) 11(2), 163-178.

Wan, F. Y. M. (1973a). Non-stationary response of linear-varying dynamical systems to random excitation. ASME J. Appl. Mech. 40, 442-448.

Wan, F. Y. M. (1973b). A direct method for linear dynamical problems in continuous mechanics with random loading. Studies Appl. Math. (J. Math. Phys.) LII 5(3), 259-276.

Wan, F. Y. M. (1973c). An in-core finite difference method for separable boundary value problems on a rectangle. Studies in Appl. Math. (J. Math. Phys.) LII 5(2), 103-113.

Wan, F. Y. M. (1974a). Effect of spanwise load-correlation on rotor blade flapping. AIAA paper 74-418, Las Vegas.

Wan, F. Y. M. (1974b). Dynamical problems of continuous media with random boundary data. Int. J. Solids Struct. 10, 35-44.

Wan, F. Y. M. (1980). Flapping response of lifting rotor blade to spanwise non-uniform random excitation. J. Eng. Math. 14(4), 241-261.

Wan, F. Y. M. and Lakshmikantham, C. (1973). Rotor blade response to random loads: a direct time-domain approach. AIAA J. 11(1), 24-28.

Wan, F. Y. M. and Lakshmikantham, C. (1974). Spatial correlation method and a time-varying flexible structure. AIAA J. 12(5), 700-707.

Wang, P. K. C. (1965). On the almost sure stability of linear time-lag systems with stochastic parameters. Int. J. Contr. 2(5), 433-440.

Wang, P. K. C. (1966). On the almost sure stability of linear stochastic distributed parameter dynamical systems. ASME J. Appl. Mech. 33, 182-186.

Watt, D. and Barr, A. D. S. (1983). Stability boundary for psuedo-random parametric excitation of a linear oscillator. ASME J. Vib. Acoust. Stress & Reliab. in Design 105, 326-331.

Wax, N. (1955). Noise and Stochastic Process. Dover Publications, New York.

Wedig, W. (1969). Doctoral Thesis, Universität Karlsruhe, Karlsruhe. (in German). Stability conditions of vibrating systems with random parametric excitation.

Wedig, W. (1972a); in Proc. IUTAM Symp. Stability of Stochastic Dynamical Systems. Lecture Notes in Mathematics 294. (Edited by R. F. Curtain). 160-172, Springer-Verlag, Berlin. Regions of instability for a linear system with random parametric excitation.

Wedig, W. (1972b). Stability conditions of parametric vibrating systems under wide band excitation. (in German) Z. Angew. Math. Mech. (ZAMM) 52, T77-T79.

Wedig, W. (1972c). Stability conditions of oscillations under parametric filtered noise excitation. (in German) Z. Angew. Math. Mech. (ZAMM) 52(3), 161-166.

Wedig, W. (1973). Instability regions of first and second type of vibrating systems under random excitation. (in German) Z. Angew. Math. Mech. (ZAMM) 53(4), T248-T250.

Wei, J. F. (1978); Ph.D. Thesis, Washington University, St. Louis, Mo. Flap-lag stability of helicopter and windmill rotor blades in powered flight and autorotation by a perturbation method.

Weidenhammer, F. (1964). Stability conditions for vibrating systems with random parametric excitations. (in German) Ing. Arch. 33, 404-415.

Wiener, N. (1923). Differential space. J. Math. Phys. 2, 131-174.

Wiener, N. (1958). Non-linear Problems in Random Theory. John Wiley, New York.

Wilcox, R. M. and Bellman, R. (1970). Truncation and preservation of moment properties for Fokker-Planck moment equations. J. Math. Anal. Appl. 32, 532-542.

Willems, J. L. (1975a). Stability of higher order moments for linear stochastic systems. Ing. Arch. 44(2), 123-129.

Willems, J. L. (1975b). Criteria for moment stability of linear stochastic systems. Z. Angew. Math. Mech. (ZAMM) 55, 532-533.

Willems, J. L. (1977); in Proc. IUTAM Symp. Stochastic Problems in Dynamics. (Edited by B. L. Clarkson). 67-89, Pitman, London. Moment stability of linear white noise and colored noise systems.

Willems, J. L. and Aeyels, D. (1976). An equivalent result for moment stability criteria for parametric stochastic systems and Itô equations. Int. J. Syst. Sci. 7, 577-590.

Willsky, A. S., Marcus, S. I., and Martin, D. N. (1975). On the stochastic stability of linear systems containing colored multiplicative noise. IEEE Trans. Automat. Contr. AC-20, 711-713.

Wong, E. and Zakai, M. (1965). On the relation between ordinary and stochastic differential equations. Int. J. Eng. Sci. 3(2), 213-229.

Wonham, W. M. (1966). Liapunov criteria for weak stochastic stability. J. Diff. Equa. 2(2), 195-209.

Wu, C. M. (1971). Ph.D. Thesis, Polytechnic Institute of Brooklyn. On the almost sure stability of stochastic systems.

Wu, W. F. and Lin, Y. K. (1984). Cumulant-neglect closure for nonlinear oscillators under parametric and external excitations. Int. J. Non-linear Mech. 19(4), 349-362.

Yaglom, A. M. (1962). An Introduction to the Theory of Stationary Random Functions. Prentice-Hall, New Jersey.

Yakubovich, V. A. and Starzhinskii, V. M. (1975). Linear Differential Equations with Periodic Coefficients (Two Volumes). John Wiley & Sons, New York, Israel Program for Scientific Translations.

Yudaev, G. S. (1979). On the stability of stochastic difference systems. Izv. Vyssh. Uchebn. Zaved., Mat. 8, 74-78.

Zhu, W. Q. (1983a); in Proc. IUTAM Symp. on Random Vibration and Reliability. 347-357, Akademie-Verlag, Berlin. Stochastic averaging of the energy envelope of nearly Liapunov systems.

Zhu, W. Q. (1983b); in Proc. Int. Workshop on Stochastic Structural Mechanics, Innsbruck. On the method of stochastic averaging of energy envelope.

Unreferenced Literature

Ahmadi, G. and Sattaripour, A. (1976). Dynamic stability of a column subjected to an axial random load. Indus. Math. 26, 67-77.

Ariaratnam, S. T. (1967); in Proc. Canadian Congress of Appl. Mech. 3.163-3.189, Univ. Laval. Dynamic stability under random excitation.

Ariaratnam, S. T. (1971); in Proc. IUTAM Symp. Instability of Continuous Systems. (Edited by H. Leipholz). 78-84, Springer-Verlag, Berlin. Stability of structures under stochastic disturbances.

Ariaratnam, S. T. and Graefe, P. W. U. (1965). Linear systems with stochastic coefficients, (III). Int. J. Contr. 2(3), 205-210.

Ariaratnam, S. T. and Tam, D. S. F. (1974); in Stochastic Problems in Mechanics. Study No. 10, 183-192, University of Waterloo Press. Stability of weakly stochastic linear systems.

Ariaratnam, S. T. and Tam, D. S. F. (1977); in Proc. IUTAM Symp. Stochastic Problems in Dynamics. (Edited by B. L. Clarkson). 90-105, Pitman, London. Moment stability of coupled linear systems under combined harmonic and stochastic excitation.

Benderskii, M. M. (1972). On the asymptotic behavior of the moments of the solutions of a linear system with random coefficients. (in Russian) Trudy FTINT. Matem. Fiz. Funk. Anal. (Har'kov) 3, 15-21.

Benderskii, M. M. and Pastur, L. A. (1973). Asymptotical behavior of the solutions of a second order equation with random coefficients. (in Russian) Teori. Funk. Anal. (Har'kov) 22, 3-14.

Bourret, R. C., Frisch, U., and Pouquet, A. (1973). Brownian motion of harmonic oscillator with stochastic frequency. Physica 65(2), 303-320.

Bunke, H. (1963a). Stability of a system of stochastic differential equations. (in German) Z. Angew. Math. Mech. (ZAMM) 43, 63-70.

Bunke, H. (1963b). On almost sure stability of linear stochastic stability. (in German) Z. Angew. Math. Mech. (ZAMM) 43, 533-535.

Bunke, H. (1963c). On the determination of stability in probability of a stochastic system. (in German) Monatsb. Deutsch. Akad. Wiss., Berlin, 5, 277-280.

Bunke, H. (1965). On the asymptotic behavior of the solution of linear stochastic differential equations. (in German) Z. Angew. Math. Mech. (ZAMM) 45, 1-9.

Bunke, H. (1967). Remarks on the asymptotic behavior of the solution of linear stochastic differential equations. (in German) Z. Angew. Math. Mech. (ZAMM) 47, p. 64.

Bunke, H. (1968a). On the asymptotic behavior of the solution of linear stochastic differential equations in the mean square. (in German) Z. Angew. Math. Mech. (ZAMM) 48, 70-71.

Bunke, H. (1968b); in Proc. IV Int. Conf. Non-linear Oscillations. 131-135, Prague, Sept. 1967. On the stability of stochastic systems which depend on Markov processes.

Bunke, H. (1968c). Stability of linear differential equations with Markov coefficients. (in German) Monatsb. Deutsch. Akad. Wiss., Berlin, 10, 406-411.

Bunke, H. (1972); in Proc. IUTAM Symp. Stability of Stochastic Dynamical Systems. Lecture Notes in Mathematics 294. (Edited by R. F. Curtain). 283-290, Springer-Verlag, Berlin. Stable periodic solutions of weakly non-linear stochastic differential equations.

Car'kov, E. F. (1976). Asymptotic exponential stability in the mean square of the trivial solution of stochastic functional differential equations. (in Russian) Teor. Veroya. 21, 871-875.

Chan, S. Y. and Chuang, K. (1966). A study of linear time-varying systems subjected to stochastic disturbances. Automatica 4, 31-48.

Dickerson, J. R. (1967). Ph.D. Thesis, California Institute of Technology, Pasadena, California. Stability of parametrically excited differential equations.

Dickerson, J. R. (1976). Response of linear dynamic system with random coefficients. Advances in Eng. Sci. 2, 741-746.

D'iakov, Iu. E. (1960). Forced vibration in circuits with a randomly varying capacitance. Radio and Elect. (Radiotekh. i Elektron.) 5(5) 228-231.

Dienes, J. K. (1961). Ph.D. Thesis, Calif. Inst. of Tech., Pasadena, California. Some applications of the theory of continuous Markov processes to random oscillation problems.

Dimentberg, M. F. (1966a). Resonance properties of a one degree-of-freedom system with randomly varying resonance frequency. Mech. Solids (Mekh. Tver. Tela) 1(1), 24-27.

Dimentberg, M. F. (1966b). Amplitude-frequency characteristics for a system with randomly varying parameters. Mech. Solids (Mekh. Tver. Tela) 1(2), 127-129.

Dimentberg, M. F. (1975). Vibrations of a linear system with periodic parametric and external random excitation. Sov. Phys. Dokl. 20 (11), p. 743.

Dimentberg, M. F. (1976). Response of a non-linearly damped oscillator to combined periodic parametric and random external excitations. Int. J. Non-linear Mech. 11, 83-87.

Dimentberg, M. F. (1982). Methods of moments in problems of dynamics of systems with randomly varying parameters. Appl. Math. Mech. (Prikl. Mat. Mekh., PMM) 46(2), 161-166.

Dimentberg, M. F. (1983); in Proc. IUTAM Symp. Random Vibration and Reliability. (Edited by K. Hennig). 245-252, Akademie-Verlag, Berlin. Response of systems with randomly varying parameters to external excitation.

Dimentberg, M. F. and Frolov, K. V. (1966). Vibrating systems with one degree-of-freedom acted upon by periodic and variation of eigenfrequency according to some random law. (in Russian) Mashin. 3, 3-10.

Dimentberg, M. F. and Frolov, K. V. (1971); in collection Non-linear and Optimal Systems. (in Russian) Nauka, Moscow. Some parametric problems of statistical dynamics.

Evlanov, L. G. and Konstantinov, V. M. (1976). Systems with Random Parameters. (in Russian) Nauka, Moscow.

Fedosov, Y. A. and Sebryakov, G. G. (1972); in Spacecraft Control Systems. 2-14, Joint Publications Research Service, Arlington, Virginia. Method of analyzing random parametric effects in spacecraft control systems.

Friedman, A. (1972); in Proc. IUTAM Symp. Stability of Stochastic Systems. Lecture Notes in Mathematics 294. (Edited by R. F. Curtain). 14-20, Springer-Verlag, Berlin. Stability and angular behavior of solutions of stochastic differential equations.

Friedman, A. and Pinsky, M. P. (1973). Asymptotic behavior of solutions of linear stochastic differential systems. Trans. Amer. Math. Soc. 181, 1-22.

Frisch, U. (1966). On the solution of stochastic differential equations with Markovian coefficients. (in French) C. R. Akad. Sci., Paris, series A-B 262, A762-A765.

Frolov, K. V. (1972). Parametric machine dynamics problems. (in Russian) Revue Roumaine des Sci Tech., Serie de Mecan. Appl. 17(2), 265-290.

Gabasov, R. (1965). On the stability of stochastic systems with a small parameter multiplying the derivatives. (in Russian) Uspekhi Mat. Nauk 20(1-121), 189-196.

Gikhman, I. I. (1964); in Winter School on Theory of Probability and Math. Statis. 41-86, Izdat. Akad. Nauk Ukrain. SSR. Differential equations with random functions.

Gikhman, I. I. (1973); in Selected Transl. Statis. and Probab. 12, 125-154, Amer. Math. Soc., Providence, R. I. Stability of stochastic differential equations.

Gray, A. H., Jr. (1964). Ph.D. Thesis, California Institute of Technology, Pasadena. Stability and related problems in randomly excited systems.

Gray, A. H., Jr. (1965). Behavior of linear systems with random parametric excitation. J. Acoust. Soc. Amer. 37(2), 235-239.

Harris, C. J. (1972); in Proc. IUTAM Symp. Stability of Stochastic Dynamical Systems. Lecture Notes in Mathematics 294. (Edited by R. F. Curtain). 230-238, Springer-Verlag, Berlin. The Fokker-Planck-Kolmogorov equation in the analysis of non-linear feedback stochastic systems.

Huang, W-h. (1982). Vibration of some structures with periodic random parameters. AIAA J. 20(7), 1001-1008.

Ibrahim, R. A. and Roberts, J. W. (1979); in Proc. VIII Int. Conf. on Non-linear Oscillations. (Edited by L. Pust). 355-360, Academia, Publishing House of Czech. Academy of Sciences, Prague. Broad band random parametric excitation of a non-linear system.

Infante, E. F. (1967); in Proc. Int. Symp. on Differential Equations and Dynamical Systems. Academic Press, New York. Stability criteria for n-th order, homogeneous linear differential equations.

Insarov, E. F., Kislitsyn, V. A., and Kofman, V. D. (1968). Analysis of the precision of non-stationary dynamic systems containing random parameters. (in Russian) Izv. Vys. Uch. Zav. Avia. Tekh. 3, 3-11.

Iudaev, G. S. (1977). On stability of stochastic differential equations. Appl. Math. Mech. (Prikl. Mat. Mekh., PMM) 41(3), 427-433.

Ivovich, V. A. (1971). Dynamic instability of an elastic system under a random parametric excitation. Stroit. Mekhanika i Raschet Sooruzhenii 4, 28-31. (English translation by Air Force Systems Command, Wright-Patterson AFB, Ohio, Foreign Tech. Div., NTIS 1974).

James, D. J. G. (1972); in Proc. IUTAM Symp. Stochastic Stability of Dynamical Systems. Lecture Notes in Mathematics 294. (Edited by R. F. Curtain). 147-159, Springer-Verlag, Berlin. Stability of mode-reference systems with random inputs.

Jirina, M. (1962). Ordinary differential or difference equations with random coefficients and random right-hand side. Czech. Math. J. 12(87), 457-474.

Jirina, M. (1963). Harmonisable solutions of ordinary differential equations with random coefficients and random right-hand side. Czech. Math. J. 13(88), 360-371.

Kats, I. I. (1965). On the stability of the first approximation of a system with random parameters. (in Russian) Ural. Gos. Univ. Mat. Zap. 3(1), 43-52.

Khas'miniskii, R. Z. (1963). The behavior of a self-oscillating system acted upon by slight noise. Appl. Math. Mekh. (Prikl. Mat. Mekh., PMM) 27(4), 1035-1044.

Khas'miniskii, R. Z. (1966). On the stability of non-linear stochastic systems. Appl. Math. Mech. (Prikl. Mat. Mekh., PMM) 30, 1082-1089.

Khas'miniskii, R. Z. (1967). Stability in the first approximation for stochastic systems. Appl. Math. Mech. (Prikl. Mat. Mekh., PMM) 31, 1025-1030.

Kheifets, M. I. (1978). On single-rotor correctable gyrocompass motion with random parametric disturbances. Mech. of Solids (Mekh. Tver. Tela) 13(2), 50-56.

Khrisanov, S. M. (1976). Stability of a linear oscillatory system parametrically excited by a random process of one class. Ukr. Mat. Zh. 28(5), 699-703.

Kistner, A. (1975); in Mech. and System Theory. 147-160, Stuttgart University, West Germany, (in German). Approximation methods for investigation of moment stability of linear systems with random parametric excitation.

Kistner, A. (1979a); in Stochastic Control Theory and Stochastic Differential Systems. Lecture Notes in Control and Information Sciences, No. 16. (Edited by M. Kohlmann and W. Vogel). 447-455, Springer-Verlag, Berlin. On the solution and the moments of linear systems with randomly distributed parameters.

Kistner, A. (1979b). Strong relation between the moments of linear systems and random parameters. (in German) Z. Angew. Math. Mech. (ZAMM) 59, T132-T134.

Kolomiets, V. G. (1963). On the parametric random effect on linear and non-linear oscillating systems. Ukr. Mat. Zh. 15(2), 199-205.

Kolomiets, V. G. (1964). Parametric random oscillations in linear and non-linear systems. Sb. Dokl. Taskent. Politekn. Insf. 6, 49-59.

Kolomiets, V. G. and Korenevskii, D. G. (1967). The stability of linear systems with random perturbations. Soviet Appl. Mech. (Prikl. Mekh.) 3(8), 82-84.

Kolovskii, M. Z., Sablin, A. D., and Troitskaya, Z. V. (1971). Oscillations in non-linear systems with variable or random parameters. Mech. Solids (Mekh. Tver. Tela) 6(4), 18-24.

Kozin, F. and Bogdanoff, J. L. (1963). Comment on: "The behavior of linear systems with random parametric excitation," by T. K. Caughey and J. K. Dienes. J. Math. Phys. 42, 336-337.

Kropác, O. (1981a); in Proc. 13th Int. Conf. Machine Dynamics Interdynamics 81. 251-258, Warszawa. Solution of stochastic dynamic problems using conditional characteristics.

Kropác, O. (1981b); in Proc. IX Int. Conf. on Non-linear Oscillations, Kiev. Response of systems with random external and parametric excitation: a conditional probability approach.

Kropác, O. and Drexler, J. (1975); in Proc. VII Int. Conf. Non-linear Oscillations, Berlin. On a useful approach to the solutions of some non-linear stochastic vibration problems.

Kropác, O. (1979); in Proc. VIII Int. Conf. Non-linear Oscillations. (Edited by L. Pust). 413-422, Academia, Publishing House of the Czechoslavak Academy of Sciences, Prague. Non-stationary random vibration in non-linear systems.

Kushner, H. J. (1967); in Proc. National Academic Society 53(8). On the stability of stochastic dynamical systems.

Kushner, H. J. (1972); in Proc. IUTAM Symp. Stability of Stochastic Dynamical Systems. Lecture Notes in Mathematics 294. (Edited by R. F. Curtain). 92-124, Springer-Verlag, Berlin. Stochastic stability.

Kuz'ma, V. K. (1965). Simulation of an oscillating system wih randomly varying parameters. Soviet Appl. Mech. (Prikl. Mekh.) 1(1), 125-127.

Ladde, G. (1977). Logarithmic norm and stability of linear systems with random parameters. Int. J. Syst. Sci. 8(9), 1057-1066.

Lambert, L. (1979). Determination of the moments of stochastic parametric systems. (in German) Z. Angew. Math. Mech. (ZAMM) 59, 397-398.

Levit, M. V. (1972). Algebraic criteria for the stochastic stability of linear systems with the parametric action of correlated white noise. Appl. Math. Mech. (Prikl. Mat. Mekh., PMM) 36(3), 516-521.

Lin, Y. K. and Hong, C. Y. R. (1981); in Proc. Structures and Materials Conference. ASME Winter Annual Meeting, Washington, D. C. Turbulence-excited flapping motion of a rotor blade in hovering flight.

Lin, Y. K. and Shih, T. Y. (1980). Column response to vertical-horizontal earthquakes. ASCE J. Eng. Mech. (EM6) 106, 1099-1109.

Lin, Y. K. and Shih, T. Y. (1982). Vertical seismic load effect on building response. ASCE J. Eng. Mech. (EM2) 108, 331-343.

Loginov, V. M. (1980). Behavior of an oscillator under a periodic parametric action with chaotic modulation. Mech. Solids (Mekh. Tver. Tela) 15(5), 35-41.

Ma, F. and Caughey, T. K. (1981). On the stability of stochastic systems. Int. J. Non-linear Mech. 16, 139-153.

Ma, F. and Caughey, T. K. (1982). Mean stability of stochastic difference systems. Int. J. Non-linear Mech. 17(2), 69-84.

Man, F. T. (1970). On the almost sure stability of linear stochastic systems. ASME J. Appl. Mech. 37, 541-543.

McKenna, J. and Morrison, J. A. (1971). Moments of solutions of a class of stochastic differential equations. J. Math. Phys. 12(10), 2126-2136.

Mikhailichenko, A. M., Pustovoitov, N. A., and Sukhorebryi, V. G. (1974); in Dynamic and Stability of Multi-variable Systems. (in Russian) 59-65, Izdatel'stvo Instituta Matematika, AN USSR, Kiev. Combined technique for estimating the probability of the stability of a dynamic system with random parametric perturbations.

Mitchell, R. R. (1972). Sample stability of second order stochastic differential equations with non-singular phase diffusion. IEEE Trans. Automat. Contr. AC 17, 706-707.

Morozan, T. (1966). La stabilite des solutions des systems d'equations differentielles aux parametres aleatoires. Rev. Roumaine Math. Pures Appl. 11, 211-238.

Morrison, J. A. (1970). Calculation of correlation functions of solutions of a stochastic ordinary differential equation. J. Math. Phys. 11(11), 3200-3209.

Moskvin, V. G. (1972). Stability of trivial solutions to Mathieu-Hill stochastic equation. Tr. Mosk. Eng. in-ta. Dinamika i Proch. Mashin, No. 101.

Moskvin, V. G. and Okopnyi, Yu. A. (1972). Investigation of the stability of systems with two degrees-of-freedom in the presence of a random parametric effect. Izv. Vuzov. Mashino-Stroenie, No. 3, 31-35.

Narayanan, S. (1983); in Proc. IUTAM Symp. Random Vibrations and Reliability. (Edited by K. Hennig). 273-283, Akademie-Verlag, Berlin. Stochastic stability of fluid conveying tubes.

Osetinskii, Iu. V. (1970). Random parametric vibrations of a gyro pendulum mounted on a moving platform. (in Russian) Priborostroenie 13(6), 77-79. (Translated into English in Gyroscopic and Navigational Instruments, 20 Jan. 1971, Joint Publications Research Service, Washington, D. C., NTIS).

Payne, H. J. (1968). An approximate method for nearly linear first order stochastic differential equations. Int. J. Contr. 7(5), 451-463.

Phain-Thien, N. and Atkinson, J. D. (1982). On the stability of linear systems: a turbulent flow classification scheme for dilute polymer solutions. Trans. ASME J. Appl. Mech. 49, 247-248.

Potapov, V. D. (1982). Stability of structure elements subjected to stationary loads. J. Appl. Mech. & Tech. Phys. (Zh. Prikl. Mekh. Tekh. Fiz.) 22, 574-578.

Prelewicz, D. A. (1972). Response of linear periodically time-varying systems to random excitation. AIAA J. 10, 1124-1125.

Rabotnikov, Iu. L. (1964). Boundedness of solutions of differential equations with random coefficients whose averages are constant. (in Russian) Zap. Mekh.-Mt. Fak Har'kov. Gos. Univ. i Har'kov. Mat. OVSC 4(30), 75-84.

Roitenberg, L. Ia. (1971); in Non-linear and Optimal Systems. (in Russian) 344-354, Izdatel'stvo, Nauk, Moscow. Effect of random forces on a gyroscopic system with random parametric excitation.

Romanovskii, Yu. M. (1960). Parametric random perturbations in some aeroelasticity problems. Izv. Akad. Nauk USSR Ser. Mekh. Mash. 4, 133-135.

Rosenbloom, A. (1954a); in Proc. Symp. Information Networks, III. 145-153, Polytechnic Institute, Brooklyn. Analysis of linear systems with randomly time-varying parameters.

Rosenbloom, A. (1954b). Ph.D. Thesis, University of California, Los Angeles. Analysis of a randomly time-varying linear system.

Sagirow, P. S. (1970). Stochastic Methods in the Dynamics of Satellites. Lecture Notes, CISM.

Sagirow, P. S. (1972); in Proc. IUTAM Symp. Stochastic Stability of Dynamical Systems. Lecture Notes in Mathematics 294. (Edited by R. F. Curtain). 311-316, Springer-Verlag, Berlin. The stability of a satellite with parametric excitation by the fluctuations of the geometric fluid.

Samuels, J. C. (1960). The buckling of circular cylindrical shells under purely random external pressures. JAS 27, 943-950.

Samuels, J. C. (1963). The dynamics of impulsively and randomly varying systems. ASME J. Appl. Mech. 30, 25-30.

Sawaragi, Y., Sunahara, Y., and Soeda, T. (1962). Statistical studies on the response of non-linear time variant control systems subjected to a suddenly applied stationary Gaussian random input. Mem. Fac. Engr., Kyoto University 24, 465-481.

Schiehlen, W. (1975). Parametric random vibration. Z. Angew. Math. Mech. (ZAMM) 55(4), T67-T68.

Schweiger, W. (1977). On the stability of the random parametric excited Laval shaft. (in German) Mech. Res. Com. 4(1), 29-34.

Schweizer, G. (1962). On the influence of the statistical parameters on the dynamic behavior of vibrating systems. (in German) Regelungstechnik 8 & 10.

Semenov, V. A. and Smirnov, A. I. (1983). On the stability of linear stochastic systems with periodically non-stationary parametric excitation. Mech. Solids (Mekh. Tver. Tela) 18(1), 14-20.

Shapiro, V. E. and Loginov, V. M. (1978). Formulas of differentiation and their use for solving stochastic equations. Physica, Ser. A 91(3/4), 563-574.

Shih, T. Y. (1980). Ph.D. Thesis, University of Illinois at Urbana, Champaign. Random vibration of structures under parametric and non-parametric earthquake loads.

Shih, T. Y. and Lin, Y. K. (1982). Vertical seismic load effect on hysteretic columns. ASCE J. Eng. Mech. Div. 108 (EM2), 242-254.

Shyu, T. P. and Somerset, J. H. (1972); in the 6th Southeastern Conference on Theoretical and Applied Mechanics. (Edited by S. C. Kranc). 453-487, Tampa, Florida. Parametric random vibration of structural systems.

Sobczyk, K. (1983); in Proc. IUTAM Symp. Random Vibrations and Reliability. (Edited by K. Hennig). 317-326, Akademie-Verlag, Berlin. On the normal approximation in stochastic dynamics.

Solodov, A. V. and Petrov, F. S. (1971). Linear Automatic Systems with Variable Parameters. (in Russian) Nauka, Moscow.

Stratonovich, R. L. and Romanovskii, Yu. A. (1965); in Non-linear Transformations of Stochastic Processes. (Edited by P. I. Kuznetsov, R. L. Stratonovich, and V. I. Tikhonov). Paper 27, 327-338. The simultaneous parametric effect of a harmonic and random force on oscillatory systems.

Sun, T. (1979). A finite element method for random differential equations with random coefficients. SIAM J. Num. Analysis 16(6), 1019-1035.

Sur, M. G. (1968). Linear differential equations with randomly perturbed parameters. Amer. Math. Soci. 72(2), 251-276.

Tanaka, K., Onishi, H., and Kaga, M. (1982). Estimation of the variance of steady variation response of structures with random parameters and method to compute the allowable variance of the parameters. Computers & Structures 15(3), 329-334.

Tsarkov, E. F. (1966); in Latviiskii Matematicheskii Ezhegodnik, 327-335 (in Russian). Parametric random effect on linear oscillatory time-lag systems.

Vanmarcke, E. (1979). Some recent developments in random vibration. Appl. Mech. Rev. 32(10), 1197-1202.

Vol'mir, A. S. and Kul'terbaev, Kh. P. (1974). Stochastic stability of forced non-linear shell vibrations. Appl. Math. Mekh. (Prikl. Mat. Mekh., PMM) 38, 840-846.

Vrkoc, I. (1966). On homogeneous linear differential equations with random perturbations. Czech. Math. J. 16(91), 199-230.

Vrkoc, I. (1968). The weak exponential stability and periodic solutions of Itô stochastic equations with small stochastic terms. Czech. Math. J. 18(93), 722-752.

Wedig, W. (1972). Stability conditions of oscillations under random-harmonic parametric excitations. (in German) Ing. Arch. 41, 157-167.

Wedig, W. (1975a). Stability of a stochastic system. (in German) Z. Angew. Math. Mech. 55, T185-T187.

Wedig, W. (1976); in Proc. VII Int. Conf. Non-linear Oscillations. 469-492, Berlin. Moments and probability densities of dynamical systems under stochastic parametric excitation.

Weidenhammer, F. (1966). Conditions for almost sure stability of oscillating systems under random parametric excitations. (in German) Z. Angew. Math. Mech. (ZAMM) 46, 551-553.

Willems, J. L. (1973). Mean square stability criteria for stochastic feedback systems. Int. J. System Sci. 4, 545-564.

Yavin, Y. (1974). On the stochastic stability of a parabolic type system. Int. J. System Sci. 5, 623-637.

Yavin, Y. (1975). On the modeling and stability of a stochastic distributed parameter system. Int. J. System Sci. 6, 301-311.

Yusupov, A. K. (1976). Random parametric oscillations of beams and plates. Izv. Sev. Karkaz Nauch Tsentra Vyssh Shholy. Tekhn. Nauki 1, 69-71.

Zaslavskii, G. M. (1967). Stochastic instability of a non-linear oscillator. J. Appl. Mech. Tech. Phys. (Zh. Prikl Mekh. Tekh. Fiz., PMTF) 8(2), 8-11.

Zeman, J. L. (1972). On the solution of non-linear stochastic mechanical problems. (in German) Acta Mech. 14, 157-169.

Principal Notations

$a_i(\underset{\sim}{x},t)$	first order incremental moment (drift coefficient)
$a_i(t)$	slowly varying response amplitude of coordinate i
B	tip loss factor of a helicopter rotor blade
B(t)	Brownian motion process
$b_{ij}(\underset{\sim}{x},t)$	second order incremental moment (diffusion coefficient)
b_j	quasi-moment of order j
$\underline{C}$	damping matrix coefficient
C_j	coefficient j in the Gram-Charlier expansion (2.95)
$C_x(t,\tau)$	covariance function (autocovariance kernel)
$2D = S_o$	spectral density of the white noise process
E[ ]	expectation
$\underset{\sim}{F}(t)$	forcing excitation vector
F(x)	probability distribution function of a discrete random variable
$F_x(\theta)$	characteristic function of the random variable x with dummy parameter θ
$G_K(y)$	the adjoint Hermite polynomial of order K
$\underline{G}(\underset{\sim}{x},t)$	matrix function usually multiplied by the parametric excitation vector
GLB	Greatest Lower Bound
H(x)	the entropy of the random process x
$H_n(\ )$	Chebychev-Hermite polynomial of order n
$\underline{I}$	unit matrix
inf	infimum

J	the Jacobian (2.94)
$[j\ell,i]$	Christoffel symbol of the first kind (equation 1.13)
$\underline{K}$	stiffness matrix
K,k	integer numbers
LUB	Least Upper Bound
l.i.m.	limit-in-the-mean
$\underline{M}(\underset{\sim}{x},t)$	envelope function matrix
$\underline{M}$	mass matrix
m_K	statistical moment of order K
$\underline{P}$	Liapunov matrix
P	regular pencil
$P(x)$	probability distribution function of the continuous random variable (or process) x
$p(x)$	probability density function
$p(x_i \mid x_{i-1})$	conditional probability density of x_i given x_{i-1}.
$\underline{Q}dt$	a matrix appears in the correction expression of the Itô formula of stochastic differential (4.62)
$\underset{\sim}{q}$	generalized coordinate vector
$R_x(t_1,t_2)$	autocorrelation function of the random process x
$R_{x,y}(t_1,t_2)$	cross-correlation function of the random processes x and y
$\underline{R}$	modal matrix
$r(\tau)$	normalized correlation coefficient (2.43)
$S_x(\omega)$	spectral density function of the random process x(t)
$S_{xy}(\omega)$	cross-spectral density of a pair of random processes x(t) and y(t)
Sup	supremum

t	real time scale
U_t, U_r, U_p	tangential, radial, and perpendicular velocity components of the flow field relative to a helicopter rotor blade
$u(x-x_i)$	the Heaviside unit step function
$\underset{\sim}{u}(x,t)$	spatial response vector
V	energy envelope
$V(\underset{\sim}{x},t)$	Liapunov function
$W(t)$	white noise process
$\underset{\sim}{X}$	state vector of the response (displacement and velocity vectors of the response coordinates)
$\underset{\sim}{Y}$	normal coordinate vector
Z	standardized variable (2.95)
β	flap angle of a helicopter rotor blade
$\delta(x)$	Dirac delta function of x
δ	torsional blade deflection
ε	small parameter
ζ	lag angle of rotor blade
ζ	damping factor
θ	dummy parameter
θ	pitch angle
θ	phase angle
λ	inflow ratio
λ_i	induced inflow ratio
λ_i	characteristic multiplier equation (1.7)
$\lambda[\ ]$	semi-invariant, or cumulant
μ	advance ratio
μ_n	central moment of order n

$\xi(t)$	random (or deterministic) parametric excitation
ρ_i	characteristic exponent
σ^2	the variance
σ	standard deviation
τ	time shift = $t_2 - t_1$
τ	non-dimensional time scale
τ_c	correlation time
$\Phi(t)$	monodromy matrix
$\phi(t)$	slowly varying phase
ψ	azimuth angle
$\chi(\theta)$	natural logarithm of the characteristic function
Ω	excitation frequency
Ω	ensemble space
ω	natural frequency
ω	sample record
$\Vert \; \Vert$	norm
$\vert \; \vert$	absolute value
$[a], \underline{a}$	matrix
$\{b\}, \underset{\sim}{b}$	vector
$\sum$	summation
$\otimes$	Kronecker product

Index

www.ingramcontent.com/pod-product-compliance
Lightning Source LLC
Chambersburg PA
CBHW060321200326
41519CB00011BA/1792

* 9 7 8 0 4 8 6 4 6 2 6 2 2 *